Linear Estimation and Design of Experiments

Linear Estimation and Design of Experiments

D.D. Joshi
Professor of Statistics & Director
Institute of Social Sciences
Agra, India

JOHN WILEY & SONS
New York Chichester Brisbane Toronto Singapore

First published in 1987 by
WILEY EASTERN LIMITED
4835/24 Ansari Road, Daryaganj
New Delhi 110002, India

Distributors:

Australia and New Zealand:
Jacaranda-Wiley Ltd., Jacaranda Press
JOHN WILEY & SONS, INC.
GPO Box 859, Brisbane, Queensland 4001, Australia

Canada:
JOHN WILEY & SONS CANADA LIMITED
22 Worcester Road, Rexdale, Ontario, Canada

Europe and Africa:
JOHN WILEY & SONS LIMITED
Baffins Lane, Chichester, West Sussex, England

South East Asia:
JOHN WILEY & SONS, INC.
05–05 Block B, Union Industrial Building
37 Jalan Pemimpin, Singapore 2057

Africa and South Asia:
WILEY EASTERN LIMITED
4835/24 Ansari Road, Daryaganj
New Delhi 110002, India

North and South America and rest of the world:
JOHN WILEY & SONS, INC.
605 Third Avenue, New York, N.Y. 10158, USA

Library of Congress Cataloging in Publication Data

ISBN 0–470–20740–X John Wiley & Sons, Inc.
ISBN 0–85226–517–4 Wiley Eastern Limited

Printed in India at Prabhat Press, Meerut.

Preface

This text is intended to serve as an introduction to the theory of linear estimation (or the theory of least squares) and its application to the analysis of experimental designs based on fixed-effects linear models. Random-effects or mixed-effects models are not considered but a discussion of recovery of inter-block information in balanced incomplete block designs has been included. The text may be found useful for a first course in experimental designs.

There is no claim to originality except for a systematic use of R.C. Bose's method of linear estimation based on the concept of estimable functions. Following Bose, the minimisation problem in the theory of least squares is tackled through the use of orthogonal projections and not via the differential calculus. As a result, the normal equations of least squares theory are seen to arise naturally from the determination of best linear unbiased estimates of estimable functions, and not from the usual minimisation of the sums of squares of deviations of random variables from their expectations. At the same time, no side conditions need be imposed for obtaining a solution of the normal equations since obtaining a unique solution, for which alone side conditions are needed, is not necessary, as any solution of the normal equations serves the purpose. In my opinion, this method of dealing with the normal equations is to be preferred, since the selection of appropriate side conditions (which must be based on non-estimable functions) is not always an easy matter, specially for beginners.

The distribution of a sum of squares and its decomposition into orthogonal components is obtained directly using Bose's definition of a sum of squares in terms of orthogonal projections, and there is no need to appeal to the theory of quadrate forms with idempotent matrices, nor to Cochran's theorem.

Main effects and interactions in factorial experiments are defined as linear functions of treatment effects and not as linear functions of observations. This approach, also due to R.C. Bose, brings out clearly not only the role of interactions in examining additivity of factor effects but also the fact, which is not explicitly mentioned in most textbooks, that factorial experiments should, as far as possible, be carried out in such designs in which orthogonal treatment contrasts have orthogonal estimates.

A knowledge of matrix and vector algebra, and of basic results in distribution theory and statistical inference, is a pre-requisite for studying this text. A brief summary of these topics is provided in the first three chapters. Orthogonal projections, and the methods for obtaining them, are discussed in more detail since the theoretical results are mostly derived through orthogonal projections.

The theory of linear estimation in fixed-effects models is presented in Chapters 4, 5 and 6. Chapter 7 introduces the basic concepts and terminology of design of experiments. Chapters 8, 9 and 10 deal with the application of linear estimation to the analysis of completely randomised, randomised block, Latin square and Graeco-Latin square designs. The analysis of the general incomplete block design through the well known C-matrix, and its application to balanced incomplete block and partially balanced incomplete block designs is covered in Chapters 11, 12 and 13. This is followed (Chapter 14 to 17) by a discussion of the general asymmetrical, 2^n and 3^n factorial experiments, and confounding in 2^n and 3^n factorials. The last chapter deals with analysis of covariance with m concomitant variables.

At the end of some of the chapters there is a brief discussion, under the heading "Complements and Problems", of some additional topics and some further results. The additional topics may be included in the course by the teacher and the further results may be given as exercises to the students.

I have not included the problem of construction of designs since I consider that to form a part of combinatorial analysis. Besides, even an elementary treatment of the construction of mutually orthogonal Latin Squares and of balanced incomplete block designs would have required a prior discussion of some algebraic structures making the book rather unwieldy. For the same reason, a group-theoretic treatment of confounding in 2^n and 3^n factorials has not been included. However, the basic results are presented without proof as working rules for obtaining confounding arrangements.

Certain other topics, like split-plot and strip-plot designs, fractional replication, etc., have been excluded as they do not fit in with the general scheme of the book.

A study of experimental designs would be incomplete unless accompanied by numerical exercises with a discussion of the interpretation of the results given by the analysis of variance. A judicious selection of such exercises, which should use data derived from actual applications, may be made from many well-known textbooks and from research journals. A number of such exercises are given as illustrative examples in this text. I have, however, not included such exercises as problems for solution as their selection is a matter of individual preference and is best left to the teachers.

This book has grown out of the lectures given by me to my students for a number of years in a one-semester course on design of experiments. The first time I gave lectures on this subject was during my tenure as Senior Visiting Fellow at the University of Leicester many years ago. I was prevailed upon to offer a course on this topic at that time by N.A. Rahman. He not only made me aware of R.C. Bose's method but also advised me at every step to develop an integrated course combining linear estimation theory and its application to experimental designs. He also helped me in selecting suitable numerical examples to be used as illustrations. I am beholden to him for his help but for which I would not have ventured in this field. Needless to add, I alone am responsible for the shortcomings of this book.

D.D. JOSHI

April 1987

Acknowledgements

In some of the illustrative examples in the book numerical data reported by various authors has been used.

Permission to use such data was accorded by the following, and is gratefully acknowledged.

1. Cambridge University Press, Cambridge (U.K.), for the data from p. 85, Vol. 26 (1936) of the *Journal of Agricultural Science*. (Author: R.H. Common).

2. Macmillan Publishing Company, New York (U.S.A.), for a portion of Table V-3 on p. 120 of the book: Experimental Designs by Walter T. Federer, Copyright 1955, renewed 1983 by Macmillan Publishing Company.

3. Editors, *Annals of Applied Biology*, Warwick (U.K.), for the data from p. 341, Vol. 25 (1938). (Author: W.R.S. Ladell).

4. John Wiley & Sons, New York, for the data from p. 456 of the book: Experimental Designs (2nd ed.) by W.G. Cochran & G.M. Cox, Copyright © 1950, 1957 by John Wiley & Sons, Inc.

5. Methuen & Co., London, for the data from p. 104 of the book: Statistical Analysis in Biology by K. Mather.

6. Royal Statistical Society, London, for the data from p. 137 of *Supplement*, *J. Roy. Statist. Soc.*, Vol. 4 (1937) (Author: M.S. Bartlett).

Contents

Preface *v*
Acknowledgements *ix*

Chapter 1. **Vectors and Matrices** 1
1.1 Vectors *1*
1.2 Linear Independence *3*
1.3 Subspaces *3*
1.4 Basis and Dimension *4*
1.5 Linear Forms *6*
1.6 Contrasts *7*
1.7 Matrices *8*
1.8 Partitioning of Matrices *10*
1.9 Determinant of a Matrix *11*
1.10 Rank of a Matrix *12*
1.11 Inverse of a Matrix *13*
1.12 Quadratic Forms *13*
1.13 Simultaneous Linear Equations *14*

Chapter 2. **Linear Transformations and Projections** 18
2.1 Linear Transformation *18*
2.2 Matrices as Linear Transformations *19*
2.3 Orthogonal Matrices and Transformations *20*
2.4 Projections *21*

Chapter 3. **Multivariate Normal and Associated Distributions** 25
3.1 Random Vectors *25*
3.2 Linear Function of Random Variables *26*
3.3 Multivariate Normal Distribution *27*

3.4 The Chi-square Distribution *31*
3.5 The *t*-distribution *32*
3.6 The *F*-distribution *32*

Chapter 4. **Linear Estimation** 34
4.1 The Linear Model *34*
4.2 Estimable Functions *35*
4.3 Estimation and Error Spaces *37*
4.4 Best Estimates *38*
4.5 Normal Equations *40*
4.6 Variance and Covariance of Estimates *43*
4.7 Theory of Least Squares *45*
4.8 Linear Model with Correlated Observations *47*

Chapter 5. **Sums of Squares** 54
5.1 Introduction *54*
5.2 Degrees of Freedom *55*
5.3 Sums of Squares *55*
5.4 Calculations of Sums of Squares *57*
5.5 Distribution of Sums of Squares *59*
5.6 Estimate and Error Sums of Squares *62*

Chapter 6. **Tests of Linear Hypotheses** 65
6.1 Estimable Linear Hypotheses *65*
6.2 The Generalised *F*-test *66*
6.3 The Generalised *t*-test *68*
6.4 Examples *69*
6.5 Tests of Hypotheses and Least Squares Theory *71*

Chapter 7. **Planning of Experiments** 77
7.1 Introduction *77*
7.2 Nomenclature *79*
7.3 Design *80*
7.4 Analysis *82*
7.5 Randomisation *83*
7.6 Missing Plot Technique *84*
7.7 The Dot Notation *86*

Chapter 8. **The Completely Randomised Design** 88
8.1 Introduction *88*
8.2 Normal Equations and Estimates *89*

8.3 Analysis of Variance *89*
8.4 Tests of Hypotheses *90*
8.5 Notes on Computation *92*
8.6 Example *92*

Chapter 9. **Randomised Block Design** **99**
9.1 Introduction *98*
9.2 Normal Equations and Estimates *100*
9.3 Analysis of Variance *102*
9.4 Tests of Hypotheses *103*
9.5 Notes on Computation *104*
9.6 Example *104*
9.7 Efficiency of a Randomised Block Design *109*

Chapter 10. **Latin and Graeco-Latin Square Designs** **114**
10.1 Introduction *114*
10.2 Latin Square Design *115*
10.3 Graeco-Latin Square Design *116*
10.4 Randomisation *117*
10.5 Normal Equations and Estimates *118*
10.6 Analysis of Variance *120*
10.7 Tests of Hypotheses *121*
10.8 Notes on Computation *122*
10.9 Example *122*
10.10 Efficiency of the Latin Square Design *125*

Chapter 11. **Incomplete Block Designs: General Theory** **128**
11.1 Introduction *128*
11.2 Normal Equations *129*
11.3 Adjusted Treatment Totals *130*
11.4 Estimates *132*
11.5 Analysis of Variance *134*
11.6 Example *137*
11.7 Connectedness and Balance *139*

Chapter 12. **Balanced Incomplete Block Design** **142**
12.1 Introduction *142*
12.2 Intrablock Analysis *143*
12.3 Interblock Analysis *147*
12.4 Distributions in Interblock Analysis *153*
12.5 Interblock Estimation of Treatment Contrasts *156*

12.6 Estimating the Weights w and w' *158*
12.7 Notes on Computation *159*
12.8 Recovery of Interblock Information *160*
12.9 Example *161*

Chapter 13. **Partially Balanced Incomplete Block Design** **172**
13.1 Introduction *172*
13.2 Association Schemes *172*
13.3 PBIB Design *174*
13.4 Intrablock Analysis *175*
13.5 Variance of Estimates *178*
13.6 Notes on Computation *178*
13.7 Example *179*

Chapter 14. **The General Factorial Experiment** **183**
14.1 Introduction *183*
14.2 Terminology *186*
14.3 Main Effects and Interactions *186*
14.4 Analysis of Variance in Factorial Experiments *191*
14.5 Notes on Computation *196*
14.6 Example *196*
14.7 Tukey's Test for Non-additivity in a Randomised Block *201*

Chapter 15. **The 2^n Factorial Experiment** **207**
15.1 Notation *207*
15.2 Standard Order for Treatment Combinations *208*
15.3 Main Effects and Interactions *208*
15.4 Calculation of Sums of Squares *210*
15.5 Yates' Method of Computation *211*
15.6 Example *212*

Chapter 16. **The 3^n Factorial Experiment** **217**
16.1 Notation *217*
16.2 Standard Order for Treatment Combinations *217*
16.3 Main Effects and Interactions *218*
16.4 Calculation of Sums of Squares *218*
16.5 Example *220*
16.6 Calculation of Sums of Squares (Alternative Method) *224*
16.7 Example *229*

Chapter 17. **Confounding in Factorial Experiments** **234**
17.1 Introduction *234*
17.2 Confounding Arrangement *235*
17.3 Defining Contrasts *236*
17.4 Confounding in a 2^n Factorial Experiment *237*
17.5 Analysis of Variance for a 2^n Confounded Factorial *240*
17.6 Example *241*
17.7 Confounding in a 3^n Factorial *244*
17.8 Confounding Arrangement for 3^n Factorial Experiments *246*

Chapter 18. **Analysis of Covariance** **252**
18.1 Introduction *252*
18.2 The Analysis of Covariance Model *254*
18.3 Normal Equations and Estimates *254*
18.4 Sums of Squares for Estimates and Error *257*
18.5 Tests of Hypotheses Regarding $\boldsymbol{\beta}$ *258*
18.6 Analysis of Covariance *260*
18.7 Example *262*
18.8 Example *266*

Bibliography **273**
Index **283**

1

Vectors and Matrices

1.1 Vectors

A *vector* **x** is an ordered *n*-tuple of real numbers. Vectors will be written either as columns, called *column-vectors,* or as rows, called *row vectors.* The following notation will be used; **x** standing for a column vector, and **x**′ for the same vector written as a row vector:

$$\mathbf{x} = \begin{bmatrix} x_1 \\ x_2 \\ \vdots \\ x_n \end{bmatrix} \qquad \mathbf{x}' = (x_1, x_2, \ldots, x_n)$$

(small bold face letters **x, y, α, β** will always stand for vectors). The set of all vectors **x** for a given *n*, will be denoted by R_n.

The *null-vector* (or *zero-vector*) is the vector

$$\mathbf{0}' = (0, 0, \ldots, 0)$$

The *sum* of two vectors $\mathbf{x}, \mathbf{y} \in R_n$ is defined as the vector

$$\mathbf{x} + \mathbf{y} = \begin{bmatrix} x_1 + y_1 \\ \cdots\cdots \\ \cdots\cdots \\ \cdots\cdots \\ x_n + y_n \end{bmatrix} \tag{1.1}$$

The *product* $c \cdot \mathbf{x}$ of a vector **x** and a scalar (i.e. a real number) *c* is the vector

$$c\mathbf{x} = \begin{bmatrix} cx_1 \\ \cdots \\ \cdots \\ cx_n \end{bmatrix} \tag{1.2}$$

The following relations are easily verified.

$$
\begin{aligned}
&\text{(i)} \quad \mathbf{x} + \mathbf{y} = \mathbf{y} + \mathbf{x} \\
&\text{(ii)} \quad \mathbf{x} + (\mathbf{y} + \mathbf{z}) = (\mathbf{x} + \mathbf{y}) + \mathbf{z} \\
&\text{(iii)} \quad \mathbf{x} + \mathbf{0} = \mathbf{0} + \mathbf{x} = \mathbf{x} \\
&\text{(iv)} \quad \mathbf{x} + (-1)\,\mathbf{x} = \mathbf{0} \\
&\text{(v)} \quad 1 \cdot \mathbf{x} = \mathbf{x} \\
&\text{(vi)} \quad c(\mathbf{x} + \mathbf{y}) = c\mathbf{x} + c\mathbf{y} \\
&\text{(vii)} \quad (c + d)\,\mathbf{x} = c\mathbf{x} + d\mathbf{x} \\
&\text{(viii)} \quad (cd)\,\mathbf{x} = c(d\mathbf{x}) = d(c\mathbf{x}) = (dc)\mathbf{x}
\end{aligned}
\tag{1.3}
$$

These relations show that R_n, with the addition of vectors and the product of a scalar and a vector as defined by (1.3), forms a vector space over the field of real numbers in the sense of the abstract definition of a vector space.

The *scalar product* (or *inner product*) of two vectors $\mathbf{x}$, $\mathbf{y} \in R_n$ is the scalar quantity

$$x_1 y_1 + \ldots + x_n y_n \tag{1.4}$$

and will be written as $\mathbf{x}'\mathbf{y}$ or $\mathbf{y}'\mathbf{x}$.

We have

$$
\begin{aligned}
&\text{(i)} \quad \mathbf{x}'(\mathbf{y} + \mathbf{z}) = \mathbf{x}'\mathbf{y} + \mathbf{x}'\mathbf{z} \\
&\text{(ii)} \quad \mathbf{x}'(c\mathbf{y}) = (c\mathbf{x})'\mathbf{y} = c(\mathbf{x}'\mathbf{y}) \\
&\text{(iii)} \quad \mathbf{x}'\mathbf{x} \geqslant 0, \text{ with equality, if and only if, } \mathbf{x} = \mathbf{0}
\end{aligned}
\tag{1.5}
$$

The *norm* (or *length*) of the vector $\mathbf{x} \in R_n$, written $\| \mathbf{x} \|$, is defined as

$$\| \mathbf{x} \| = \sqrt{(\mathbf{x}'\mathbf{x})} = \sqrt{(x_1^2 + \ldots + x_n^2)} \tag{1.6}$$

It follows that

$$
\begin{aligned}
&\text{(i)} \quad \| c\mathbf{x} \| = | c | \cdot \| \mathbf{x} \| \\
&\text{(ii)} \quad \| \mathbf{x} + \mathbf{y} \| \leqslant \| \mathbf{x} \| + \| \mathbf{y} \| \\
&\text{(iii)} \quad \| \mathbf{x} \| = 0 \text{ if, and only if, } \mathbf{x} = \mathbf{0}
\end{aligned}
\tag{1.7}
$$

A vector $\mathbf{x}$ may be regarded as a point $(x_1, \ldots, x_n)$ in n-dimensional Euclidian space. The norm $\| \mathbf{x} \|$ is then the distance of the point $\mathbf{x}$ from the origin and $\| \mathbf{x} - \mathbf{y} \|$ is the distance between the points $\mathbf{x}$ and $\mathbf{y}$.

Two vectors $\mathbf{x}, \mathbf{y} \in R_n$ are said to be *orthogonal*, written $\mathbf{x} \perp \mathbf{y}$, if $\mathbf{x}'\mathbf{y} = \mathbf{y}'\mathbf{x} = 0$. The null-vector $\mathbf{0}$ is orthogonal to every vector $\mathbf{x}$ and is the only such vector. It then follows that if $\mathbf{x}, \mathbf{y} \in R_n$ and $\mathbf{x}'\mathbf{z} = \mathbf{y}'\mathbf{z}$ for all $\mathbf{z} \in R_n$, then $\mathbf{x} = \mathbf{y}$.

1.2 Linear Independence

Given vectors $\mathbf{x}_1, \mathbf{x}_2, \ldots, \mathbf{x}_r \in R_n$ and scalars $c_1, c_2, \ldots, c_r$ the vector

$$\mathbf{z} = c_1\mathbf{x}_1 + c_2\mathbf{x}_2 + \ldots + c_r\mathbf{x}_r$$

is called a *linear combination* of the vectors $\mathbf{x}_1, \ldots, \mathbf{x}_r$. The set of all such linear combinations is called the *vector-space spanned* by the vectors $\mathbf{x}_1, \ldots, \mathbf{x}_r$ and denoted by $Sp\,(\mathbf{x}_1, \ldots, \mathbf{x}_r)$.

The vectors $\mathbf{x}_1, \ldots, \mathbf{x}_r \in R_n$ are said to be *linearly dependent* (or to form a *linearly dependent set*) if there exist scalars $c_1, c_2 \ldots c_r$, with at least one $c_i \neq 0$, such that

$$c_1\mathbf{x}_1 + \ldots + c_r\mathbf{x}_r = \mathbf{0}$$

If, on the contrary,

$$c_1\mathbf{x}_1 + \ldots + c_r\mathbf{x}_r = \mathbf{0}$$

implies that $c_1 = c_2 = \ldots = c_r = 0$, the vectors $\mathbf{x}_1, \ldots, \mathbf{x}_r$ are said to be *linearly independent*.

Any set of vectors containing the null-vector is linearly dependent. If $\mathbf{x}_1, \ldots, \mathbf{x}_r \in R_n$ is a set of linearly dependent vectors then at least one of these vectors can be expressed as a linear combination of the remaining. If none of the vectors $\mathbf{x}_1, \ldots, \mathbf{x}_r \in R_n$ can be expressed as a linear combination of the others, then they form a linearly independent set. Consequently, if to a linearly independent set $\mathbf{x}_1, \ldots, \mathbf{x}_r \in R_n$ we add a non-null vector $\mathbf{y} \in R_n$ which is not a linear combination of $\mathbf{x}_1, \ldots, \mathbf{x}_r$, then the set $\mathbf{x}_1, \ldots, \mathbf{x}_r, \mathbf{y}$ is also a linearly independent set of vectors.

Any set of non-null pairwise orthogonal vectors is linearly independent.

1.3 Subspaces

A subset V of R_n is called a *subspace* (also called *subvector-space* or *linear subspace*) of R_n if for all scalars a and b the vector $a\mathbf{x} + b\mathbf{y} \in V$, whenever $\mathbf{x}, \mathbf{y} \in V$. Examples of subspaces are the whole space R_n, the set $\{\mathbf{0}\}$ consisting of the null vector alone, the space spanned by a given set $\{\mathbf{x}_1, \ldots, \mathbf{x}_r\}$ of vectors of R_n.

If V_1 and V_2 are subspaces of R_n, the set $V_1 \cap V_2$ consisting of all vectors belonging to both V_1 and V_2 is also a subspace. Two subspaces will be called *disjoint* if $V_1 \cap V_2$ consists of the null vector $\mathbf{0}$ only. In R_4, the sets consisting of all vectors of the form $(x_1, x_2, 0, 0)$, and $(0, 0, x_3, x_4)$ are disjoint subspaces.

If V_1 and V_2 are subspaces of R_n, the set V of all vectors $\mathbf{x} \in R_n$ which can be written as

$$\mathbf{x} = \mathbf{x}_1 + \mathbf{x}_2, \quad \mathbf{x}_1 \in V_1, \quad \mathbf{x}_2 \in V_2 \tag{1.8}$$

is also a subspace of R_n. The representation of $\mathbf{x} \in V$ as $\mathbf{x}_1 + \mathbf{x}_2$ is unique, if and only if, V_1 and V_2 are disjoint, in which case we shall write V as $V_1 + V_2$. Let V_1 be the space spanned by the vectors $\mathbf{x}_1, \ldots, \mathbf{x}_r$ and V_2 the space spanned by $\mathbf{y}_1, \ldots, \mathbf{y}_s$. If $\mathbf{y}_1, \mathbf{y}_2, \ldots, \mathbf{y}_s$ do not belong to V_1, the subspaces V_1 and V_2 are disjoint and $V_1 + V_2$ is the subspace spanned by the vectors $\mathbf{x}_1, \ldots, \mathbf{x}_r, \mathbf{y}_1 \ldots, \mathbf{y}_s$.

A vector $\mathbf{x} \in R_n$ is said to be *orthogonal* to a subset E of R_n, written $\mathbf{x} \perp E$, if it is orthogonal to every vector of E. If $\mathbf{x}_1, \ldots, \mathbf{x}_r$ span a subspace V of R_n, then $\mathbf{x} \perp V$, if and only if, $\mathbf{x} \perp \mathbf{x}_i$ $(i = 1, \ldots, r)$. Two subsets E and F of R_n will be called *mutually orthogonal*, written $E \perp F$, if every vector of E is orthogonal to every vector of F. If V_1 and V_2 are subspaces spanned by $\mathbf{x}_1, \ldots, \mathbf{x}_r$ and $\mathbf{y}_1, \ldots, \mathbf{y}_s$ respectively, then $V_1 \perp V_2$, if and only if, $\mathbf{x}_i \perp \mathbf{y}_j$ $(i = 1, \ldots, r; j = 1, \ldots, s)$. Since the null vector is the only vector which is orthogonal to itself, it follows that if $V_1 \perp V_2$, then V_1 and V_2 are disjoint subspaces.

Let V be a subspace of R_n. The set W of all vectors $\mathbf{x} \in R_n$, such that $\mathbf{x} \perp V$, is a subspace of R_n. We call W the *orthogonal complement* of V in R_n and denote it by $V^\perp$. If V is a subspace of R_3 consisting of all vectors of the form $(a, b, 0)$, then $V^\perp$ is the subspace consisting of all vectors of the form $(0, 0, c)$. Similarly, if V is the subspace of all vectors (x_1, x_2, x_3) of R_3 for which $x_1 + x_2 + x_3 = 0$, $V^\perp$ is the subspace of all vectors of the form (a, a, a) of R_3.

1.4 Basis and Dimension

Let V be a subspace of R_n. Any set of linearly independent vectors spanning the subspace V is called a *basis* for V. A basis for R_n is provided by the set of n *unit-vectors* $\mathbf{e}_1' = (1, 0, \ldots, 0)$, $\mathbf{e}_2' = (0, 1, 0, \ldots, 0), \ldots, \mathbf{e}_n' = (0, 0, \ldots, 1)$.

Any two bases for a given subspace V contain the same number of vectors. The number of vectors in any basis for V is called the *dimension* of V, written dim (V). Thus dim $(R_n) = n$.

Let $\mathbf{x}$ be a vector belonging to the vector space spanned by the set of vectors $\mathbf{x}_1, \ldots, \mathbf{x}_r$. Then we can write

$$\mathbf{x} = c_1\mathbf{x}_1 + \ldots + c_r\mathbf{x}_r$$

The coefficients $c_1, \ldots, c_r$ in this representation of $\mathbf{x}$ are unique, if and only if, the vectors $\mathbf{x}_1, \ldots, \mathbf{x}_r$ are linearly independent.

Let $\mathbf{x}_1, \ldots, \mathbf{x}_r$ be a basis for a subspace V of R_n. The coefficients $c_1, \ldots, c_r$ in the unique representation of a vector $\mathbf{x} \in V$,

$$\mathbf{x} = c_1\mathbf{x}_1 + \ldots + c_r\mathbf{x}_r$$

are called the *coordinates* of $\mathbf{x}$ relative to the basis $\mathbf{x}_1, \ldots, \mathbf{x}_r$. In particular, if

$$\mathbf{x}' = (x_1, \ldots, x_n)$$

the scalars $x_1, \ldots, x_n$ are the coordinates of the vector $\mathbf{x}$ relative to the basis for R_n provided by the unit-vectors.

A basis $\mathbf{x}_1, \ldots, \mathbf{x}_r$ of a subspace V of R_n is called *orthonormal* if

(i) $\mathbf{x}_i \perp \mathbf{x}_j, \quad i \neq j$ (1.9)

(ii) $\| \mathbf{x}_i \| = 1, \; i = 1, \ldots, r$

The unit-vectors form an orthonormal basis for R_n. An orthonormal basis can be found for any subspace of R_n.

Let V be an r-dimensional subspace of R_n. Then any set of r linearly independent vectors of V forms a basis for V and any set consisting of more than r vectors is linearly dependent.

Let $\mathbf{x}_1, \ldots, \mathbf{x}_r$ be a basis (an orthonormal basis) for a subspace V of R_n. This set can then be extended to provide a basis (an orthonormal basis) $\mathbf{x}_1, \ldots, \mathbf{x}_r, \mathbf{x}_{r+1}, \ldots, \mathbf{x}_n$ for R_n.

Theorem 1.1: If V is a subspace of R_n and dim $(V) = r$, then dim $(V^\perp) = n - r$.

Proof: Let $\mathbf{x}_1, \ldots, \mathbf{x}_r$ be an orthogonal basis for V. We may extend it to get an orthogonal basis $\mathbf{x}_1, \ldots, \mathbf{x}_r, \mathbf{x}_{r+1}, \ldots, \mathbf{x}_n$ for R_n. It is clear that $\mathbf{x}_{r+1}, \ldots, \mathbf{x}_n$ is an orthogonal basis for $V^\perp$. Hence dim $(V^\perp) = n - r$.

Corollary: $R_n = V + V^\perp$ and $(V^\perp)^\perp = V$.

Theorem 1.2: Let V be an r-dimensional subspace of R_n. Then V is the set of all vectors $\mathbf{x} \in R_n$ for which

$$\alpha_i'\mathbf{x} = 0, \; i = 1, \ldots, n - r \qquad (1.10)$$

where the α_i are $n - r$ linearly independent vectors in R_n.

Proof: Choose an orthogonal basis $\alpha_1, \ldots, \alpha_{n-r}$ for $V^\perp$. If $\mathbf{x} \in V$, then $\alpha_i' \mathbf{x} = 0$, $(i = 1, \ldots, n - r)$. Conversely, if $\alpha_i' \mathbf{x} = 0$ for $i = 1, \ldots, n - r$, it follows that $\mathbf{x} \perp V^\perp$, and hence that $\mathbf{x} \in V$.

Theorem 1.3: Let $\alpha_1, \ldots, \alpha_{n-r}$ be linearly independent vectors in R_n. Then the set of all vectors $\mathbf{x} \in R_n$ such that

$$\alpha_i' \mathbf{x} = 0, \quad i = 1, \ldots, n - r \tag{1.11}$$

is an r-dimensional subspace of R_n.

Proof: The set of all such vectors is the orthogonal complement of the subspace spanned by the vectors α_i, and hence is an r-dimensional subspace of R_n.

1.5 Linear Forms

Given real numbers $a_1, \ldots, a_n$ and variables $\mathbf{y}_1, \ldots, \mathbf{y}_n$ the function

$$a_1 y_1 + \ldots + a_n y_n \tag{1.12}$$

of the n variables $y_1, \ldots, y_n$ will be called a *linear form* (or *linear function*) in the variables $y_1, \ldots, y_n$. In vector notation the linear form $a_1 y_1 + \ldots + a_n y_n$ can be written as $\mathbf{a}'\mathbf{y}$ where

$$\mathbf{a}' = (a_1, \ldots, a_n), \quad \mathbf{y}' = (y_1, \ldots, y_n) \tag{1.13}$$

The vector $\mathbf{a}$ will be called the *coefficient vector* of the linear form $\mathbf{a}'\mathbf{y}$. Each of the variables $y_1, \ldots, y_n$ can also be regarded as a linear form, the coefficient vectors being the unit vectors of R_n. Other examples of linear forms are $\bar{y} = (y_1 + \ldots + y_n)/n$, $y_1 - \bar{y}$, $y_1 + y_2 - y_3$, etc. The linear form $a_1 y_1 + a_2 y_2$ in the variables y_1, y_2, can also be regarded as a linear form in $y_1, y_2, \ldots y_n$ with coefficient vector $(a_1, a_2, 0, \ldots, 0)$.

The *sum* of two linear forms $\mathbf{a}'\mathbf{y}$, $\mathbf{b}'\mathbf{y}$ is defined as the linear form $(\mathbf{a} + \mathbf{b})'\mathbf{y}$. Similarly, the result of multiplying the linear form $\mathbf{a}'\mathbf{y}$ by a scalar c is the linear form $(c\mathbf{a})'\mathbf{y}$. There is a one-to-one correspondence between the totality of all linear forms $\mathbf{a}'\mathbf{y}$ in the variables $y_1, \ldots, y_n$ and the totality of all vectors $\mathbf{a} \in R_n$. The addition of two linear forms, or the multiplication of a linear form by a scalar corresponds to the same operation on the coefficient vectors.

It is therefore convenient to speak of *subspaces of linear forms, linear independence of linear forms, orthogonality of linear forms and of subspaces of linear forms*, etc., the terms being defined by use of the corresponding properties of the coefficient vectors. The linear forms $\mathbf{a}_1'\mathbf{y}, \ldots, \mathbf{a}_r'\mathbf{y}$ are linearly independent if the vectors $\mathbf{a}_1, \ldots, \mathbf{a}_r$

are linearly independent. Linear forms are orthogonal if their coefficient vectors are orthogonal. A set of linear forms is a subspace of the space of all linear forms if the coefficient vectors form a subspace of R_n.

If $\mathbf{a}_1'\mathbf{y}, \ldots, \mathbf{a}_r'\mathbf{y}$ are linear forms, then the linear form

$$c_1(\mathbf{a}_1'\mathbf{y}) + \ldots + c_r(\mathbf{a}_r'\mathbf{y}) = (c_1\mathbf{a}_1 + \ldots + c_r\mathbf{a}_r)'\mathbf{y} \qquad (1.14)$$

will be called a *linear combination* of the linear forms $\mathbf{a}_1'\mathbf{y}, \ldots, \mathbf{a}_r'\mathbf{y}$. Since $y_1, \ldots, y_n$ may also be regarded as linear forms, it follows that every linear form $\mathbf{a}'\mathbf{y}$ is a linear combination of the n linear forms $y_1, \ldots, y_n$. The linear forms $y_1, \ldots, y_n$ may therefore be called a *basis* (in fact an *orthogonal basis*) for the space of all linear forms in the variables $y_1, \ldots, y_n$. The space of all linear forms may be said to have *dimension n*. Similarly, the set of all linear forms in the variables $y_1, \ldots, y_r$ is an r-dimensional subspace of the space of all linear forms in $y_1, \ldots, y_n$. If V is the set of all linear forms in $y_1, \ldots, y_r$ and W the set of all linear forms in $y_{r+1}, \ldots, y_n$, then V and W are orthogonal subspaces of linear forms. In fact W is the *orthogonal complement* of V in the space of all linear forms in $y_1, \ldots, y_n$.

1.6 Contrasts

The linear form

$$\mathbf{a}'\mathbf{y} = a_1y_1 + \ldots + a_ny_n$$

will be called a *contrast* in $y_1, \ldots, y_n$ if $a_1 + a_2 + \ldots + a_n = 0$. Examples of contrasts are

$$y_1 - y_2, \quad y_1 - 2y_2 + y_3, \quad 3y_1 + 4y_2 - 7y_3, \quad \text{etc.}$$

Theorem 1.4: The linear form $\mathbf{a}'\mathbf{y}$ is a contrast, if and only if, it is orthogonal to the linear form $y_1 + \ldots + y_n$, or to the linear form $\bar{y} = (y_1 + \ldots + y_n)/n$.

Proof: If $\mathbf{a}'\mathbf{y} = a_1y_1 + \ldots + a_ny_n$ is a contrast then it is orthogonal to $y_1 + y_2 + \ldots + y_n$, since the scalar product of the coefficient vectors $(a_1, \ldots, a_n)$ and $(1, \ldots, 1)$ is

$$a_1 + \ldots + a_n = 0$$

Conversely, if $\mathbf{a}'\mathbf{y}$ is orthogonal to $y_1 + \ldots + y_n$ we have

$$a_1 + \ldots + a_n = 0,$$

so that $\mathbf{a}'\mathbf{y}$ is a contrast.

Theorem 1.5: The contrasts $y_1 - y_2, y_1 - y_3, \ldots, y_1 - y_k$ are linearly independent for all $k = 2, \ldots, n$.

Proof: The coefficient vectors

$$(1, -1, 0, \ldots, 0), (1, 0, -1, \ldots, 0), \ldots, (1, 0, 0, \ldots, -1, 0, \ldots, 0)$$

are linearly independent.

Theorem 1.6: Every contrast $\mathbf{a}'\mathbf{y}$ in $y_1, \ldots, y_n$ can be written as a linear combination of the $n-1$ constrasts $y_1 - y_2$, $y_1 - y_3, \ldots, y_1 - y_n$.

Proof: We have, since $a_1 + \ldots + a_n = 0$,

$$\begin{aligned}\mathbf{a}'\mathbf{y} &= a_1y_1 + \ldots + a_ny_n \\ &= -(a_2 + \ldots + a_n)\,y_1 + a_2y_2 + \ldots + a_ny_n \\ &= -a_2(y_1 - y_2) - \ldots - a_n(y_1 - y_n)\end{aligned}$$

From Theorems 1.5 and 1.6 it follows that the set of all contrasts in $y_1, \ldots, y_n$ is an $(n-1)$-dimensional subspace of the space of all linear forms in $y_1, \ldots, y_n$, and the contrasts $y_1 - y_2, \ldots, y_1 - y_n$ form a basis for this subspace. An orthogonal basis is provided by the contrasts

$$\begin{aligned}&y_1 - y_2 \\ &y_1 + y_2 - 2y_3 \\ &y_1 + y_2 + y_3 - 3y_4 \\ &\cdots\cdots\cdots\cdots\cdots\cdots \\ &y_1 + y_2 + \ldots + y_{n-1} - (n-1)\,y_n\end{aligned} \tag{1.15}$$

Similarly, if $\bar{y} = (y_1 + y_2 + \ldots + y_n)/n$, the n linear forms

$$y_1 - \bar{y}, y_2 - \bar{y}, \ldots, y_n - \bar{y}$$

are all contrasts, and any $n-1$ of them, being linearly independent, provide another basis for the subspace of all contrast in $y_1 \ldots, y_n$.

1.7 Matrices

An $m \times n$ *matrix* $\mathbf{A}$ is a rectangular array of real numbers

$$A = \begin{bmatrix} a_{11} & a_{12} \ldots a_{1n} \\ a_{21} & a_{22} \ldots a_{2n} \\ \cdots & \cdots\cdots \\ a_{m1} & a_{m2} \ldots a_{mn} \end{bmatrix}$$

with m rows and n columns. An $m \times m$ matrix is called a *square matrix* of order m. A *diagonal matrix* is a square matrix **A** with $a_{ij} = 0$ for $i \neq j$. An *upper triangular* (*lower triangular*) *matrix* is a square matrix **A** for which $a_{ij} = 0$ for $i > j\,(i < j)$. If **A** is a square matrix, the elements a_{ii} are called the *diagonal elements*. Matrices will be denoted by bold face capital letters.

The *transpose* **A'** of an $m \times n$ matrix **A** is the $n \times m$ matrix obtained from **A** by writing the rows as columns and the columns as rows.

A column vector **x** may be regarded as an $n \times 1$ matrix. Its transpose is then the vector **x'** written as a row vector. This is the reason for writing the row vector as **x'**.

The matrix **A** is called *symmetric* if $\mathbf{A} = \mathbf{A}'$, *skew-symmetric* if $\mathbf{A} = -\mathbf{A}'$.

An $m \times n$ matrix **A** with elements a_{ij} will occasionally be written briefly as $\mathbf{A} = [a_{ij}]_{m \times n}$, or simply as $[a_{ij}]$, if no confusion regarding m and n is possible.

The *sum* of two $m \times n$ matrices $\mathbf{A} = [a_{ij}]$, $\mathbf{B} = [b_{ij}]$ is the $m \times n$ matrix $\mathbf{C} = [c_{ij}]$ with

$$c_{ij} = a_{ij} + b_{ij},\ i = 1, \ldots, m,\ j = 1, 2, \ldots, n \tag{1.16}$$

We shall write $\mathbf{C} = \mathbf{A} + \mathbf{B}$, and we have evidently

$$\mathbf{C}' = [\mathbf{A} + \mathbf{B}]' = \mathbf{A}' + \mathbf{B}' \tag{1.17}$$

If $\mathbf{A} = [a_{ij}]$ is an $m \times n$ matrix, and c a scalar, the *product* $c\mathbf{A}$ is the $m \times n$ matrix $[ca_{ij}]$. This agrees with the earlier definition of the product $c \cdot \mathbf{x}$ of a scalar c and a vector **x**, **x** being regarded as an $n \times 1$ matrix. We have

$$(c\mathbf{A})' = c \cdot \mathbf{A}' \tag{1.18}$$

Given an $m \times n$ matrix **A** and an $n \times p$ matrix **B**, $\mathbf{A} = [a_{ij}]$, $\mathbf{B} = [b_{ij}]$, the *product* **AB** is the $m \times p$ matrix $\mathbf{C} = [c_{ij}]$ with

$$c_{ij} = \sum_{k=1}^{n} a_{ik} b_{kj} \tag{1.19}$$

It should be noted that the product **AB** is not always defined and that the multiplication is not commutative. The reason for writing the scalar product of two vectors $\mathbf{x}, \mathbf{y} \in R_n$ as $\mathbf{x}'\mathbf{y}$ (or $\mathbf{y}'\mathbf{x}$) is now clear.

The *null-matrix*, denoted by **0**, is a matrix with all its elements equal to zero. The *identity matrix* $\mathbf{I}_r$ is a square matrix of order r with all the diagonal elements equal to unity and the other elements equal to zero. The identity matrix $\mathbf{I}_r$, will usually be written simply as $\mathbf{I}$, the order r being clear from the context.

The following relations are easily verified:

$$
\begin{aligned}
&\text{(i)} \quad \mathbf{A(B+C)} = \mathbf{AB} + \mathbf{AC} \\
&\text{(ii)} \quad \mathbf{(B+C)\,D} = \mathbf{BD} + \mathbf{CD} \\
&\text{(iii)} \quad (c\mathbf{A})\,\mathbf{B} = \mathbf{A}(c\mathbf{B}) = c(\mathbf{AB}) = c\mathbf{AB} \\
&\text{(iv)} \quad \mathbf{(AB)'} = \mathbf{B'A'} \\
&\text{(v)} \quad \mathbf{(AB)\,C} = \mathbf{A(BC)} \\
&\text{(vi)} \quad \mathbf{I}_m \mathbf{A}_{m\times n} = \mathbf{A}_{m\times n} \mathbf{I}_n = \mathbf{A}_{m\times n}
\end{aligned}
\tag{1.20}
$$

1.8 Partitioning of Matrices

Sometimes it is convenient to represent a matrix by the juxtaposition of two or more matrices in a partitioned form. A partitioned matrix is represented usually as

$$
\mathbf{A} = \begin{bmatrix} \mathbf{P} & \vdots & \mathbf{Q} \\ \cdots & & \cdots \\ \mathbf{R} & \vdots & \mathbf{S} \end{bmatrix} = \begin{bmatrix} \mathbf{P} & \mathbf{Q} \\ \mathbf{R} & \mathbf{S} \end{bmatrix} \tag{1.21}
$$

with or without dotted lines. In the above representation the *submatrices* $\mathbf{P}$ and $\mathbf{Q}$ have the same number of rows, $\mathbf{P}$ and $\mathbf{R}$ have the same number of columns, etc. The matrix $\mathbf{A}$ may be partitioned into any number of submatrices.

The product of two partitioned matrices is obtained by the usual rule of multiplying matrices treating the submatrices as elements. Thus, if

$$
\mathbf{A} = \begin{bmatrix} \mathbf{P} & \mathbf{Q} \\ \mathbf{R} & \mathbf{S} \end{bmatrix}, \quad \mathbf{B} = \begin{bmatrix} \mathbf{E} \\ \mathbf{G} \end{bmatrix}, \quad \mathbf{C} = \begin{bmatrix} \mathbf{E} & \mathbf{F} \\ \mathbf{G} & \mathbf{H} \end{bmatrix}
$$

then,

$$
\mathbf{AB} = \begin{bmatrix} \mathbf{PE} + \mathbf{QG} \\ \mathbf{RE} + \mathbf{SG} \end{bmatrix}, \quad \mathbf{AC} = \begin{bmatrix} \mathbf{PE} + \mathbf{QG} & \mathbf{PF} + \mathbf{QH} \\ \mathbf{RE} + \mathbf{SG} & \mathbf{RF} + \mathbf{SH} \end{bmatrix} \tag{1.22}
$$

it being assumed that the partitions are so chosen, that the various products $\mathbf{PE}$, $\mathbf{RF}$ etc. of the submatrices are defined.

A matrix $\mathbf{A} = [a_{ij}]_{m \times n}$ may be partitioned in terms of its row and column vectors as

$$\mathbf{A} = [\mathbf{c}_1 \ldots \mathbf{c}_n] \tag{1.23}$$

and

$$\mathbf{A} = \begin{bmatrix} \mathbf{r}_1' \\ \vdots \\ \mathbf{r}_m' \end{bmatrix} \tag{1.24}$$

where $\mathbf{c}_j$ is the j-th column vector of $\mathbf{A}$, and $\mathbf{r}_i'$ is the i-th row vector of $\mathbf{A}$.

With this notation we have

$$\mathbf{Ax} = x_1\mathbf{c}_1 + \ldots + x_n\mathbf{c}_n \tag{1.25}$$

and

$$\mathbf{x}'\mathbf{A} = x_1\mathbf{r}_1' + \ldots + x_m\mathbf{r}_m' \tag{1.26}$$

showing that the vectors $\mathbf{Ax}$ and $\mathbf{x}'\mathbf{A}$ are linear combinations of the column and row vectors of $\mathbf{A}$ respectively.

1.9 Determinant of a Matrix

A *determinant* of a square matrix $\mathbf{A} = [a_{ij}]_{m \times m}$, written $|\mathbf{A}|$, is a real number defined by

$$|\mathbf{A}| = \Sigma \pm (a_{1i}, a_{2j}, \ldots, a_{mp}) \tag{1.27}$$

where the summation is taken over all permutations $(i, j, \ldots, p)$ of $(1, 2, \ldots, m)$ with a plus sign if $(i, j, \ldots, p)$ is an even permutation and a minus sign if it is an odd permutation.

The *cofactor* of the element a_{ij}, denoted by A_{ij}, is defined to be $(-1)^{i+j}$ times the determinant of the matrix obtained from $\mathbf{A}$ by omitting the i-th row and the j-th column.

The following properties are well known (all matrices are square matrices):

(i) $|\mathbf{A}'| = |\mathbf{A}|$

(ii) $|\mathbf{AB}| = |\mathbf{A}| \cdot |\mathbf{B}|$

(iii) $|\mathbf{A}|$ changes sign if any two rows (or columns) are interchanged.

(iv) $|\mathbf{A}| = \sum_i a_{ri} A_{ri}$ for any $r = \sum_r a_{ri} A_{ri}$ for any i (1.28)

(v) $\sum_i a_{ri} A_{si} = 0$ if $r \neq s$

(vi) If $\mathbf{A}$ is a diagonal or a triangular matrix, $|\mathbf{A}|$ is the product of the diagonal elements

1.10 Rank of a Matrix

The *rank* of an $m \times n$ matrix **A** is the number of linearly independent rows, each row being regarded as a vector belonging to R_n.

The columns of **A** can be regarded as vectors in R_m. The number of linearly independent rows of **A** equals the number of linearly independent columns, and the rank of **A** can also be defined as the number of linearly independent columns of **A**.

The vector space spanned by the row vectors of an $m \times n$ matrix **A** is called the *row space* of **A** and shall be denoted by $R(\mathbf{A})$. Similarly, the space spanned by the column vectors of **A** is called the *column space* of **A** and denoted by $C(\mathbf{A})$. Obviously, the dimension of both $R(\mathbf{A})$ and $C(\mathbf{A})$ equals the rank of **A**. We also have $C(\mathbf{A}'\mathbf{A}) = C(\mathbf{A}') = R(\mathbf{A})$, and $R(\mathbf{A}'\mathbf{A}) = R(\mathbf{A}) = C(\mathbf{A}')$.

The maximum rank of an $m \times n$ matrix equals min (m, n) and a matrix with the maximum possible rank is called a matrix of *full rank*.

A square matrix of order m is called *non-singular* if it has full rank, otherwise it is called *singular*.

The following results are true:

(i) rank $(\mathbf{AB}) \leqslant$ min (rank $(\mathbf{A})$, rank $(\mathbf{B})$)

(ii) rank $(\mathbf{A} + \mathbf{B}) \leqslant$ rank $(\mathbf{A})$ + rank $(\mathbf{B})$

(iii) the rank of a matrix is unaltered by multiplication by a non-singular matrix; that is, if **A**, **B**, **C** are matrices such that **AB** and **BC** exist, and **A** and **C** are nonsingular, then

$$\text{rank}\,(\mathbf{AB}) = \text{rank}\,(\mathbf{BC}) = \text{rank}\,(\mathbf{B})$$

(iv) rank $(\mathbf{A})$ is equal to the maximum order of all nonsingular square sub-matrices of **A**

(v) a square matrix **A** is non-singular, if and only if, $|\mathbf{A}| \neq 0$ (1.29)

(vi) rank $(\mathbf{AA}')$ = rank $(\mathbf{A}'\mathbf{A})$ = rank $(\mathbf{A})$ = rank $(\mathbf{A}')$

(vii) if **A** and **B** are square matrices of order n, then

$$\text{rank}\,(\mathbf{AB}) \geqslant \text{rank}\,(\mathbf{A}) + \text{rank}\,(\mathbf{B}) - n$$

(viii) if **A** and **B** are square matrices and $\mathbf{AB} = 0$, then either $\mathbf{A} = \mathbf{0}$, or $\mathbf{B} = \mathbf{0}$, or **A** and **B** are both singular

1.11 Inverse of a Matrix

If **A** is a square matrix of order m, its *inverse*, written $\mathbf{A}^{-1}$, is a square matrix of order m, such that

$$\mathbf{A}^{-1}\mathbf{A} = \mathbf{A}\mathbf{A}^{-1} = \mathbf{I}_m \tag{1.30}$$

The inverse exists, if and only if, **A** is non-singular and then

$$\mathbf{A}^{-1} = [\mathbf{A}_{ij}/|\mathbf{A}|]' \tag{1.31}$$

i.e. the (i, j)-th element of $\mathbf{A}^{-1}$ is $\mathbf{A}_{ji}/|\mathbf{A}|$.

We have the following relations:

(i) $(\mathbf{A}^{-1})^{-1} = \mathbf{A}$

(ii) if **A** is non-singular, then $(\mathbf{A}')^{-1} = (\mathbf{A}^{-1})'$

(iii) if **A** and **B** are non-singular, then so is their product (if defined) and $(\mathbf{AB})^{-1} = \mathbf{B}^{-1}\mathbf{A}^{-1}$

1.12 Quadratic Forms

Given a matrix $A_{m\times n}$ and two vectors $\mathbf{x} \in R_m$, $\mathbf{y} \in R_n$, we have

$$\mathbf{x}'\mathbf{A}\mathbf{y} = \sum_{i=1}^{m}\sum_{j=1}^{n} a_{ij}\, x_i\, y_j a_{12} \tag{1.32}$$

If **A** is square of order m and $\mathbf{x} = \mathbf{y}$, we get

$$\mathbf{x}'\mathbf{A}\mathbf{x} = a_{11}x_1^2 + \ldots + a_{mm}x_m^2 + (a_{12} + a_{21})\, x_1x_2 + \ldots + (a_{m-1,\,m} + a_{m,\,m-1})\, x_{m-1}\, x_m \tag{1.33}$$

If **A** is symmetric, the above expression becomes

$$\mathbf{x}'\mathbf{A}\mathbf{x} = a_{11}x_1^2 + \ldots + a_{mm}x_m^2 + 2a_{12}\, x_1x_2 + \ldots + 2a_{m-1,\,m}\, x_{m-1}\, x_m$$

$$= \sum_{i=1}^{m}\sum_{j=1}^{m} a_{ij}\, x_i\, x_j \tag{1.34}$$

and is then called a *quadratic form* in the m variables $x_1, x_2 \ldots, x_m$, or simply a quadratic form in **x**. To every quadratic form corresponds a symmetric matrix and vice versa. The matrix is called the *matrix of the quadratic form*.

The quadratic form $\mathbf{x}'\mathbf{A}\mathbf{x}$ and the matrix **A** of the form are called *positive definite* if $\mathbf{x}'\mathbf{A}\mathbf{x} > 0$ for all $\mathbf{x} \neq \mathbf{0}$, *positive semi-definite* if $\mathbf{x}'\mathbf{A}\mathbf{x} \geqslant 0$ for all **x**. We shall use the abbreviations p.d. and p.s.d.

for positive definite and positive semi-definite respectively. The following results are well known.

(i) If **A** is a p.s.d. matrix then $a_{ii} \geqslant 0$ and if $a_{ii} = 0$ then $a_{ij} = 0$ for all j, and $a_{ji} = 0$ for all j.

(ii) If **P** is a non-singular matrix and **A** is a p.d. (p.s.d.) matrix, then **P′AP** is also a p.d. (p.s.d) matrix.

(iii) A matrix **A** is a p.d. matrix if, and only if, there exists a non-singular matrix **P** such that **A** = **P′P**. It follows that a p.d. matrix is non-singular.

(iv) **A** = $[a_{ij}]$ is a p.d. matrix, if and only if,

$$a_{11} > 0, \quad \begin{vmatrix} a_{11} & a_{12} \\ a_{21} & a_{22} \end{vmatrix} > 0, \quad \begin{vmatrix} a_{11} & a_{12} & a_{13} \\ a_{21} & a_{22} & a_{23} \\ a_{31} & a_{32} & a_{33} \end{vmatrix} > 0 \quad (1.35)$$

(v) If **A** is an $m \times n$ matrix and rank (**A**) $= m < n$, then **AA′** is p.d. and **A′A** is p.s.d.

(vi) If **A** is an $m \times n$ matrix and rank (**A**) $= k < m < n$, then both **AA′** and **A′A** are p.s.d.

1.13 Simultaneous Linear Equations

The set of m linear equations in n unknowns

$$\sum_{j=1}^{n} a_{ij} x_j = b_i, \; i = 1, \ldots, m$$

can be formulated in compact form as

$$\mathbf{Ax} = \mathbf{b} \quad (13.6)$$

where **A** is the $m \times n$ matrix $[a_{ij}]$, $\mathbf{x}' = (x_1, \ldots, x_n)$ and $\mathbf{b}' = (b_1 \ldots, b_m)$. The matrix **A** is called the *coefficient matrix* and the matrix [**A** ⋮ **b**] is called the *augmented matrix* of the given set of equations.

Equation (1.36) are called *consistent* if there exists a vector $\mathbf{u} \in R_n$ such that $\mathbf{Au} = \mathbf{b}$, and the vector **u** is called a solution of the equations. If no such vector can be found, the equations have no solution and are called *inconsistent*. A necessary and sufficient condition for equations (1.36) to be consistent is that $\mathbf{b} \in C(\mathbf{A})$, or equivalently, that the coefficient matrix and the augmented matrix have the same rank.

Consistent equations may have only one solution, or more than one in which case the number of solutions is infinite. If the number of equations is less than the number of unknowns there cannot be a unique solution. If the number of unknowns does not exceed the number of equations the solution may be unique. If the coefficient matrix is a square matrix, the equations have a unique solution, if and only if, the coefficient matrix is non-singular.

The set of equations (1.36) is called homogeneous if $\mathbf{b} = \mathbf{0}$, and non-homogeneous if $\mathbf{b} \neq \mathbf{0}$. Homogeneous equations are always consistent as $\mathbf{0}$ is a solution. If equations (1.36) are consistent, the general solution is given by $\mathbf{x}_0 + \mathbf{v}$, where $\mathbf{x}_0$ is any given particular solution of (1.36), and $\mathbf{v}$ any solution of the homogeneous system $\mathbf{Ax} = \mathbf{0}$. As $\mathbf{v}$ ranges over all solutions of the homogeneous system $\mathbf{x}_0 + \mathbf{v}$ gives all solutions of (1.36). We get the same set of solutions of (1.36) if $\mathbf{x}_0$ is replaced by any other particular solution of (1.36).

COMPLEMENTS AND PROBLEMS

1. If $\mathbf{x}$ and $\mathbf{y}$ are non-null vectors such that $\mathbf{u}'\mathbf{x} = 0$ implies $\mathbf{u}'\mathbf{y} = 0$, then $\mathbf{x} = a\mathbf{y}$ for some real number a.

2. The subset

$$V = \{\mathbf{x} \mid \mathbf{x} \in R_n,\ x_n = a_1x_1 + \ldots + a_{n-1}\, x_{n-1}\}$$

where $a_1, a_2, \ldots, a_{n-1}$ are given constants, is a subspace of R_n of dimension $n - 1$.

More generally,

$$V = \{\mathbf{x} \mid \mathbf{x} \in R_n,\ \mathbf{Ax} = 0\}$$

where $\mathbf{A}$ is a given $m \times n$ matrix of rank r, is a subspace of R_n of dimension of $n - r$.

3. Consider a set of linear forms $\sum_{j=1}^{n} l_{ij} y_{ij}$ $(i = 1, \ldots., m)$. If

$$\sum_{i=1}^{m} l_{ij} = 0,\ j = 1, \ldots, n$$

and

$$\sum_{j=1}^{n} l_{ij} = 0,\ i = 1, \ldots, m$$

then the subspace spanned by the given linear forms has dimension $(m - 1)(n - 1)$.

4. The rows (columns) of an orthogonal matrix $\mathbf{A}$ form an orthonormal basis for the row space (column space) of $\mathbf{A}$.

5. A square matrix $\mathbf{A}$ of order n is an orthogonal matrix, if and only if,

$$\| \mathbf{Ax} \| = \| \mathbf{x} \|$$

for all $\mathbf{x} \in R_n$.

6. Suppose $\mathbf{A}$ is an $m \times n$ matrix. Then $\mathbf{ABA} = \mathbf{A}$ implies $\mathbf{AB} = \mathbf{I}$ if rank of $\mathbf{A}$ is m, and $\mathbf{BA} = \mathbf{I}$ if rank of $\mathbf{A}$ is n.

7. Given N variables y_{ij} ($j = 1, 2, \ldots, n_i$; $i = 1, \ldots, k$; $N = \Sigma n_i$), let

$$\bar{y}_{i.} = \frac{1}{n_i} \sum_{j=1}^{n_i} y_{ij},\ \bar{y} = \frac{1}{N} \sum_i \sum_j y_{ij}.$$

Then,

(a) the linear forms $\bar{y}_{i.}$ ($i = 1, \ldots, k$) are linearly independent and the space spanned by them contains the linear form $\bar{y}$

(b) the linear form $\bar{y}$ is orthogonal to the linear form $\bar{y}_{r.} - \bar{y}_{s.}$

(c) the linear forms $\bar{y}_{r.} - \bar{y}_{s.}$ and $\bar{y}_{t.} - \bar{y}_{u.}$ are linearly independent if any three of r, s, t, u are unequal

(d) the linear forms $\bar{y}_{i.} - \bar{y}_{p.}$ ($i, p = 1, 2, \ldots, k$) span a $(k - 1)$-dimensional space of linear forms

(e) the iinear forms $\bar{y}_{1.} - \bar{y}_{2.}, \bar{y}_{1.} - \bar{y}_{3.}, \ldots, \bar{y}_{1.} - \bar{y}_{k.}$ constitute a basis for the space of (d)

8. Given N variables y_{ijk} ($i = 1, \ldots, a$; $j = 1, \ldots, b$; $k = 1, \ldots, c$; $N = abc$), let

$$\bar{y} = \frac{1}{N} \sum_{ijk} y_{ijk},\ \bar{y}_{ij.} = \frac{1}{c} \sum_k y_{ijk},\ \bar{y}_{\cdot jk} = \frac{1}{a} \sum_i y_{ijk}$$

$$\bar{y}_{i\cdot k} = \frac{1}{b} \sum_j y_{ijk},\ \bar{y}_{i..} = \frac{1}{bc} \sum_{j,k} y_{ijk},\ \bar{y}_{\cdot j.} = \frac{1}{ca} \sum_{i,k} y_{ijk}$$

$$\bar{y}_{\cdot\cdot k} = \frac{1}{ab} \sum_{i,j} y_{ijk}$$

Then,

(a) the linear form $\bar{y}$ belongs to the space spanned by $\bar{y}_{ij.}$ ($i = 1, \ldots, a$, $j = 1, \ldots, b$) and also to the space spanned by $\bar{y}_{i..}$ ($i = 1, \ldots, a$)

(b) the linear form $\bar{y}_{i..}$ belongs to the space spanned by $\bar{y}_{ij.}$ ($j = 1, \ldots, b$) and also to the space spanned by $\bar{y}_{i\cdot k}$ ($k = 1, \ldots, c$)

(c) the linear forms $\bar{y}_{ij.}$ ($i = 1, \ldots, a$; $j = 1, \ldots, b$) are mutually orthogonal

(d) the linear forms $\bar{y}_{i..}$ ($i = 1, \ldots, a$) are mutually orthogonal

(e) $\bar{y}$ is orthogonal to $\bar{y}_{ij.} - \bar{y}_{pq.}$ and also to $\bar{y}_{i..} - \bar{y}_{p..}$

(f) $\bar{y}_{i..}$ is orthogonal to $\bar{y}_{ij.} - \bar{y}_{ip.}$ and also to $\bar{y}_{i\cdot k} - \bar{y}_{i\cdot q}$

(g) the linear forms $\bar{y}_{i..} - \bar{y}_{p..}$ ($i,p = 1, \ldots, a$) span an $(a - 1)$ dimensional space of linear forms

(h) the linear forms $\bar{y}_{1..} - \bar{y}_{2..}, \bar{y}_{1..} - \bar{y}_{3..}, \ldots, \bar{y}_{1..} - \bar{y}_{a..}$ constitute a basis for the space of (g)

(i) the linear forms $\bar{y}_{ij\cdot} - \bar{y}_{pq\cdot}$ $(i, p = 1, \ldots, a; j, q = 1, \ldots, b)$ span an $(ab - 1)$-dimensional space of linear forms.

9. If V_1, V_2 are two orthogonal subspaces of R_n, we have

$$\text{div}\,(V_1 + V_2) = \text{div}\,(V_1) + \text{div}\,(V_2)$$

2

Linear Transformations and Projections

2.1 Linear Transformation

A *transformation* L from R_n to R_m, written $L: R_n \to R_m$, is a mapping which associates to each vector $\mathbf{x} \in R_n$ a unique vector $\mathbf{y} \in R_m$. The vector $\mathbf{y}$ is called the *image of* $\mathbf{x}$ under the transformation L and denoted by $L(\mathbf{x})$.

The transformation $L: R_n \to R_m$ is called a *linear transformation* if for all $\mathbf{x}_1, \mathbf{x}_2 \in R_n$ and all real numbers a_1, a_2,

$$L(a_1 \mathbf{x}_1 + a_2 \mathbf{x}_2) = a_1 L(\mathbf{x}_1) + a_2 L(\mathbf{x}_2) \tag{2.1}$$

For example, the transformation that associates to the vector $(x_1, x_2, x_3) \in R_3$ the vector $(x_1 + x_2, x_3) \in R_2$ is a linear transformation from R_3 to R_2. Similarly, the transformation which carries $(x_1, x_2) \in R_2$ to $(x_1, x_2, x_1 + x_2) \in R_3$ is a linear transformation from R_2 to R_3.

Let $L: R_n \to R_m$ be given. The set of all $\mathbf{x} \in R_n$ for which $L(\mathbf{x}) = \mathbf{0}$ is a subspace of R_n. We call it the *null space* of the transformation L. Similarly, the set of all images $L(\mathbf{x})$, i.e. the set of all vectors $\mathbf{y} \in R_m$ such that $\mathbf{y} = L(\mathbf{x})$ for some $\mathbf{x} \in R_n$, is a subspace of R_m. It is called the *image space* of the transformation L.

The transformation $L: R_n \to R_m$ is called a *one-to-one* transformation if $\mathbf{x}_1, \mathbf{x}_2 \in R_n$ and $\mathbf{x}_1 \neq \mathbf{x}_2$ implies $L(\mathbf{x}_1) \neq L(\mathbf{x}_2)$. In other words, different vectors have different images under the transformation. Obviously, a transformation L is one-to-one, if and only if, $L(\mathbf{x}) = \mathbf{0}$ implies $\mathbf{x} = \mathbf{0}$, i.e., if and only if, the null space has dimension zero. The transformation $L: R_n \to R_m$ is called a transformation from R_n onto R_m, or simply an *onto* transformation, if every $\mathbf{y} \in R_m$ is the image of some $\mathbf{x} \in R_n$. A transformation $L: R_n \to R_m$ is onto, if and only if, the image space has dimension m.

A transformation which is both one-to-one and onto is called a *non-singular* transformation. If $L: R_n \to R_m$ is non-singular, then for every $\mathbf{y} \in R_m$ there exists a unique $\mathbf{x} \in R_n$ such that $\mathbf{y} = L(\mathbf{x})$. Hence, we can define a transformation $M: R_m \to R_n$ associating to each $\mathbf{y} \in R_m$ the unique $\mathbf{x} \in R_n$ for which $L(\mathbf{x}) = \mathbf{y}$. The transformation M is called the inverse of L and denoted by L^{-1}. If L is a linear transformation so is L^{-1}.

Given a linear transformation $L: R_n \to R_m$ we have

$$\dim(\text{Image Space}) + \dim(\text{Null Space}) = n$$

If L is one-to-one, the dimension of the image space must be equal to n so that $m \geqslant n$. It then follows that if L is non-singular we have $m = n$. The dimension of the image space is called the *rank* of the transformation and that of the null space its *nullity*.

2.2 Matrices as Linear Transformations

Let $\mathbf{A} = [a_{ij}]$ be an $m \times n$ matrix, and consider the multiplication of $\mathbf{A}$ by a vector $\mathbf{x} \in R_n$ on the right. The result is a vector $\mathbf{y} \in R_m$ given by

$$\mathbf{y} = \mathbf{A}\mathbf{x} = \begin{bmatrix} a_{11}x_1 + \ldots + a_{1n}x_n \\ \cdots\cdots\cdots\cdots \\ \cdots\cdots\cdots\cdots \\ a_{m1}x_1 + \ldots + a_{mn}x_n \end{bmatrix}$$

The properties of matrix multiplication yield

$$\mathbf{A}(a_1\mathbf{x}_1 + a_2\mathbf{x}_2) = a_1\mathbf{A}\mathbf{x}_1 + a_2\mathbf{A}\mathbf{x}_2 \tag{2.2}$$

which shows that the matrix $\mathbf{A}$ can be regarded as a linear transformation from R_n to R_m associating to each vector $\mathbf{x} \in R_n$ the vector $\mathbf{y} = \mathbf{A}\mathbf{x} \in R_m$.

If instead we consider the product $\mathbf{x}'\mathbf{A}$, where now $\mathbf{x} \in R_m$ and $\mathbf{x}'\mathbf{A} \in R_n$, we may regard the matrix $\mathbf{A}$ as a linear transformation from R_m to R_n, associating to each vector $\mathbf{x} \in R_m$ the vector $\mathbf{y}' = \mathbf{x}'\mathbf{A} \in R_n$. Since $\mathbf{x}'\mathbf{A} = \mathbf{y}'$ can be written as $\mathbf{A}'\mathbf{x} = \mathbf{y}$, the two ways to regard a matrix $\mathbf{A}$ as a linear transformation are essentially the same. We shall therefore consider the $m \times n$ matrix $\mathbf{A}$ as representing a linear transformation from R_n to R_m which associates to each $\mathbf{x} \in R_n$ the vector $\mathbf{A}\mathbf{x} \in R_m$.

Theorem 2.1: Let $L: R_n \to R_m$ be a linear transformation. Then, there exists an $m \times n$ matrix **A** such that

$$L(\mathbf{x}) = \mathbf{A}\mathbf{x} \tag{2.3}$$

for all $\mathbf{x} \in R_n$.

Proof: Let $\mathbf{e}_1, \ldots, \mathbf{e}_n$ be the unit vectors in R_n, and let $\mathbf{y}_1, \ldots, \mathbf{y}_n$ be their images under L. Given $\mathbf{x}' = (x_1, \ldots, x_n) \in R_n$ we can write

$$\mathbf{x} = x_1\mathbf{e}_1 + \ldots + x_n\mathbf{e}_n$$

Then

$$L(\mathbf{x}) = x_1 L(\mathbf{e}_1) + \ldots + x_n L(\mathbf{e}_n) = x_1\mathbf{y}_1 + \ldots + x_n\mathbf{y}_n$$

If we put

$$\mathbf{A} = [\mathbf{y}_1 \cdots \mathbf{y}_n] \tag{2.4}$$

then **A** is an $m \times n$ matrix and

$$\mathbf{A}\mathbf{x} = x_1\mathbf{y}_1 + \ldots + x_n\mathbf{y}_n = L(\mathbf{x}) \tag{2.5}$$

Thus every linear transformation $L: R_n \to R_m$ can be represented by means of an $m \times n$ matrix **A**. The matrix **A** may be called the *matrix of the linear transformation L*; the columns of **A** are the images under L of the unit vectors of R_n.

If the $m \times n$ matrix **A** is used to define a linear transformation from $R_n \to R_m$, the image space of the transformation is the column space $C(\mathbf{A})$ of the matrix **A**, and the null space is the orthogonal complement (in R_n) $R(\mathbf{A})^{\perp}$ of the row space $R(\mathbf{A})$. It is then clear that the rank of the transformation defined by **A** is equal to the rank of the matrix **A**.

2.3 Orthogonal Matrices and Transformations

A square matrix **A** is called an *orthogonal matrix* if $\mathbf{A}'\mathbf{A} = \mathbf{A}\mathbf{A}' = \mathbf{I}$, or equivalently, if $\mathbf{A}^{-1} = \mathbf{A}'$. An orthogonal matrix is non-singular. If **A** is orthogonal, $\mathbf{A}'$ is also orthogonal.

If **A** is an $n \times n$ orthogonal matrix, the transformation which carries the vector $\mathbf{x} \in R_n$ into the vector $\mathbf{A}\mathbf{x} \in R_n$ is called an *orthogonal transformation*.

A matrix $\mathbf{A} = [a_{ij}]$ is orthogonal, if and only if, the row vectors of **A** form an orthonormal set. It then follows that the column vectors also form an orthonormal set.

If $\mathbf{x}$ and $\mathbf{y}$ are two given vectors and $\mathbf{A}$ an orthogonal matrix, then

$$(\mathbf{Ax})'(\mathbf{Ay}) = \mathbf{x}'\mathbf{A}'\mathbf{Ay} = \mathbf{x}'\mathbf{Iy} = \mathbf{x}'\mathbf{y} \tag{2.6}$$

It follows that the length (or norm) of any vector $\mathbf{x}$ is invariant under orthogonal tranformations. If the vectors $\mathbf{x}$ are regarded as points in a Euclidean space then the distance between any two points is invariant under orthogonal transformations.

Conversely, if a square matrix $\mathbf{A}$ is such that $(\mathbf{Ax})'(\mathbf{Ax}) = \mathbf{x}'\mathbf{x}$ for all $\mathbf{x}$, then $\mathbf{A}$ is an orthogonal matrix.

Suppose $\mathbf{x}_1, \ldots, \mathbf{x}_m$ are m orthonormal vectors in R_n $(m < n)$. Then it is possible to find $n-m$ vectors $\mathbf{x}_{m+1}, \ldots, \mathbf{x}_n$ such that the set of n vectors $\mathbf{x}_1, \ldots, \mathbf{x}_n$ is an orthonormal set. Thus if the row vectors of an $m \times n$ matrix $\mathbf{A}$ form an orthonormal set it is possible to add $n—m$ rows to the matrix $\mathbf{A}$ to get a new matrix which is orthogonal.

2.4 Projections

Let V_1 be a subspace of R_n and $V_2 = V_1^{\perp}$ the orthogonal complement of V_1 in R_n. Then, every $\mathbf{x} \in R_n$ can be uniquely written as

$$\mathbf{x} = \mathbf{x}_1 + \mathbf{x}_2 \tag{2.7}$$

where $\mathbf{x}_1 \in V_1$, $\mathbf{x}_2 \in V_2$ and consequently $\mathbf{x}_1 \perp \mathbf{x}_2$. The transformation $P_1 : R_n \to R_n$ which associates to each $\mathbf{x} \in R_n$ the unique vector of $\mathbf{x}_1 \in V_1$ is called the *orthogonal projection,* or simply *projection,* of R_n on V_1. The vector $\mathbf{x}_1$ is called the projection of $\mathbf{x}$ on V_1. Similarly, the transformation $P_2 : R_n \to R_n$ for which $P_2(\mathbf{x}) = \mathbf{x}_2$ is called the projection on V_2. If V_1 is spanned by a single vector $\boldsymbol{\alpha}$, then the projection of $\mathbf{x}$ on V_1 will also be called the projection of $\mathbf{x}$ on $\boldsymbol{\alpha}$.

A vector $\mathbf{x}_1$ is the projection of $\mathbf{x}$ on V_1, if and only if, $\mathbf{x} \in V_1$ and $(\mathbf{x} - \mathbf{x}_1) \perp V_1$.

Consider the projection of $\mathbf{x}$ on $\boldsymbol{\alpha}$; it belongs to the subspace spanned by $\boldsymbol{\alpha}$ and hence may be written as $a\boldsymbol{\alpha}$. Now $(\mathbf{x} - a\boldsymbol{\alpha}) \perp \boldsymbol{\alpha}$, so that $\mathbf{x}'\boldsymbol{\alpha} - a\boldsymbol{\alpha}'\boldsymbol{\alpha} = 0$, i.e. $a = (\mathbf{x}'\boldsymbol{\alpha})/(\boldsymbol{\alpha}'\boldsymbol{\alpha})$.

Hence, the projection of $\mathbf{x}$ on $\boldsymbol{\alpha}$ is the vector

$$\frac{(\mathbf{x}'\boldsymbol{\alpha})}{(\boldsymbol{\alpha}'\boldsymbol{\alpha})} \cdot \boldsymbol{\alpha} \tag{2.8}$$

Theorem 2.2: Let $P : R_n \to R_n$ be the projection on V. Then P is a linear transformation.

Proof: Take $\mathbf{x}, \mathbf{y} \in R_n$, and let $\mathbf{x}_1, \mathbf{y}_1$ be their projections. Then

$$P(\mathbf{x}) = \mathbf{x}_1 \in V$$

$$P(\mathbf{y}) = \mathbf{y}_1 \in V$$

so that

$$a\mathbf{x}_1 + b\mathbf{y}_1 \in V$$

Further,

$$(\mathbf{x} - \mathbf{x}_1) \perp V, (\mathbf{y} - \mathbf{y}_1) \perp V$$

and hence $(a\mathbf{x} + b\mathbf{y}) - (a\mathbf{x}_1 + b\mathbf{y}_1) = a(\mathbf{x} - \mathbf{x}_1) + b(\mathbf{y} - \mathbf{y}_1) \perp V$

Thus $a\mathbf{x}_1 + b\mathbf{y}_1$ is the projection of $a\mathbf{x} + b\mathbf{y}$ on V, and hence P is linear.

Theorem 2.3: Let $P: R_n \to R_n$ be the projection on V. Then V is the image space of the transformation P and $V^\perp$ is the null space.

Proof: We have $P(\mathbf{x}) = \mathbf{x}$, if and only if, $\mathbf{x} \in V$, and $P(\mathbf{x}) = \mathbf{0}$, if and only if, $\mathbf{x} \in V^\perp$.

Theorem 2.4: A square matrix $\mathbf{A}$ of order n represents a projection if, and only if,

(i) $\mathbf{A}^2 = \mathbf{A}$

(ii) $\mathbf{A} = \mathbf{A}'$

Proof: Suppose the transformation taking $\mathbf{x} \in R_n$ to $\mathbf{Ax} \in R_n$ is a projection on some subspace V. Then $\mathbf{Ax} \in V$ for all $\mathbf{x} \in R_n$ and hence $\mathbf{A}(\mathbf{Ax}) = \mathbf{Ax}$, so that we have $\mathbf{A}^2 = \mathbf{A}$. Since $(\mathbf{x} - \mathbf{Ax}) \perp V$, $(x - \mathbf{Ax})'\mathbf{Ax} = 0$, i.e. $\mathbf{x}'(\mathbf{A} - \mathbf{A}'\mathbf{A})\,\mathbf{x} = 0$, for all $\mathbf{x} \in R_n$. It follows that

$$\mathbf{A} - \mathbf{A}'\mathbf{A} = \mathbf{0}$$

or that $\mathbf{A}$ is symmetric.

Conversely, suppose $\mathbf{A}^2 = \mathbf{A}$, and $\mathbf{A} = \mathbf{A}'$. For every $\mathbf{x} \in R_n$ we can write $\mathbf{x} = \mathbf{Ax} + (\mathbf{x} - \mathbf{Ax})$.

But

$$\mathbf{Ax} \in C(\mathbf{A}),$$

and $$(\mathbf{x} - \mathbf{Ax})'\mathbf{A} = \mathbf{x}'\mathbf{A} - \mathbf{x}'\mathbf{A}'\mathbf{A} = \mathbf{x}'\mathbf{A} - \mathbf{x}'\mathbf{A} = \mathbf{0}$$

Thus, we see that $(\mathbf{x} - \mathbf{Ax}) \perp C(\mathbf{A})$.

This shows that the vector $\mathbf{Ax}$ is the projection of $\mathbf{x}$ on $C(\mathbf{A})$, which completes the proof.

Theorem 2.5: Let V_1 and V_2 be subspaces of R_n with $V_2 \subset V_1$. If $\mathbf{x}_1$ and $\mathbf{x}_2$ are projections of $\mathbf{x}$ on V_1 and V_2 respectively, then $\mathbf{x}_2$ is the projection of $\mathbf{x}_1$ on V_2.

Proof: It is enough to prove that $(\mathbf{x}_1 - \mathbf{x}_2) \perp V_2$. Now $(\mathbf{x} - \mathbf{x}_1) \perp V_1$ and therefore also orthogonal to V_2, and $(\mathbf{x} - \mathbf{x}_2) \perp V_2$. Thus,

$$(\mathbf{x}_1 - \mathbf{x}_2) = (\mathbf{x} - \mathbf{x}_2) - (\mathbf{x} - \mathbf{x}_1) \perp V_2$$

Theorem 2.5 says that if $V_2 \subset V_1$, then the projection of $\mathbf{x}$ on V_2 may be obtained by first getting the projection $\mathbf{x}_1$ of $\mathbf{x}$ on V_1, and then the projection of $\mathbf{x}_1$ on V_2.

Theorem 2.6: Let V_1 and V_2 be mutually orthogonal subspaces of R_n, and $\mathbf{x}_1$ and $\mathbf{x}_2$ the projections of $\mathbf{x} \in R_n$ on V_1 and V_2 respectively. Then $\mathbf{x}_1 + \mathbf{x}_2$ is the projection of $\mathbf{x}$ on $V_1 + V_2$.

Proof: $\mathbf{x}_1 + \mathbf{x}_2 \in V_1 + V_2$, and thus it is enough to prove that $(\mathbf{x} - \mathbf{x}_1 - \mathbf{x}_2) \perp (V_1 + V_2)$. Now, $(\mathbf{x} - \mathbf{x}_1) \perp V_1, \mathbf{x}_2 \in V_2$ and therefore $\mathbf{x}_2 \perp V_1$, so that $(\mathbf{x} - \mathbf{x}_1 - \mathbf{x}_2) \perp V_1$.

Similarly, by considering $\mathbf{x} - \mathbf{x}_2$ and $\mathbf{x}_1$ we see that

$$(\mathbf{x} - \mathbf{x}_1 - \mathbf{x}_2) \perp V_2$$

from which it follows that

$$(\mathbf{x} - \mathbf{x}_1 - \mathbf{x}_2) \perp (V_1 + V_2)$$

Theorem 2.7: Let V be a subspace of R_n spanned by the vectors $\mathbf{a}_1, \ldots, \mathbf{a}_m$ and A the $n \times m$ matrix $[\mathbf{a}_1 \ldots \mathbf{a}_m]$ whose columns are the vectors $\mathbf{a}_1, \ldots, \mathbf{a}_m$. Then, the projection of $\mathbf{x} \in R_n$ on V, is the vector $\mathbf{A}\hat{\boldsymbol{\alpha}}$, where $\hat{\boldsymbol{\alpha}}$ is any solution of the equations

$$\mathbf{A}'\mathbf{A}\boldsymbol{\alpha} = \mathbf{A}'\mathbf{x} \tag{2.9}$$

Proof: The projection belongs to V and can be written as $\mathbf{A}\boldsymbol{\alpha}$ for some vector $\boldsymbol{\alpha}$. If $\mathbf{A}\boldsymbol{\alpha}$ is the projection of $\mathbf{x}$ on V, we must have $(\mathbf{x} - \mathbf{A}\boldsymbol{\alpha}) \perp V$, i.e. $(\mathbf{x} - \mathbf{A}\boldsymbol{\alpha}) \perp C(\mathbf{A})$.

Thus

$$\mathbf{A}'(\mathbf{x} - \mathbf{A}\boldsymbol{\alpha}) = \mathbf{0},$$

or

$$\mathbf{A}'\mathbf{A}\boldsymbol{\alpha} = \mathbf{A}'\mathbf{x}$$

which completes the proof.

The equations $\mathbf{A}'\mathbf{A}\boldsymbol{\alpha} = \mathbf{A}'\mathbf{x}$ are consistent. If $\hat{\boldsymbol{\alpha}}$ is any solution, then $\mathbf{A}\hat{\boldsymbol{\alpha}}$ is the projection of $\mathbf{x}$ on $V = C(\mathbf{A})$. If $\boldsymbol{\alpha}^*$ is any other solution we have $\mathbf{A}\hat{\boldsymbol{\alpha}} = \mathbf{A}\boldsymbol{\alpha}^*$ so that the projection is uniquely determined by any solution.

Theorems 2.5, 2.6 and 2.7 are useful in obtaining the projections of a given vector in specific cases

Theorem 2.8: Let V be any subspace of R_n and $\mathbf{x}$ a given vector in R_n. Then

$$\inf_{\mathbf{y} \in V} \| \mathbf{x} - \mathbf{y} \| = \| \mathbf{x} - \mathbf{x}_0 \|$$

where $\mathbf{x}_0 \in V$ is the projection of $\mathbf{x}$ on V.

Proof: $\mathbf{x} - \mathbf{x}_0$ is orthogonal to V and hence orthogonal to $\mathbf{x}_0 - \mathbf{y}$. Therefore

$$\| \mathbf{x} - \mathbf{y} \|^2 = \| \mathbf{x} - \mathbf{x}_0 + \mathbf{x}_0 - \mathbf{y} \|^2$$
$$= \| \mathbf{x} - \mathbf{x}_0 \|^2 + \| \mathbf{x}_0 - \mathbf{y} \|^2$$

Thus

$$\| \mathbf{x} - \mathbf{y} \|^2 \geqslant \| \mathbf{x} - \mathbf{x}_0 \|^2$$

with equality, if and only if, $\mathbf{y} = \mathbf{x}_0$.

Corollary: For given y and A, the expression

$$(\mathbf{y} - \mathbf{A}\boldsymbol{\beta})' (\mathbf{y} - \mathbf{A}\boldsymbol{\beta}) = \| \mathbf{y} - \mathbf{A}\boldsymbol{\beta} \|^2$$

is minimised by that value of $\boldsymbol{\beta}$ for which $\mathbf{A}\boldsymbol{\beta}$ is the projection of $\mathbf{y}$ on $C(\mathbf{A})$.

3

Multivariate Normal and Associated Distributions

3.1 Random Vectors

Given n random variables $y_1, y_2, \ldots, y_n$, we may write them together as a vector $\mathbf{y}$, where

$$\mathbf{y}' = (y_1, y_2, \ldots, y_n) \tag{3.1}$$

We then call $\mathbf{y}$ a *random vector* with values in R_n or an *n-dimensional random vector.* Associated with each random vector $\mathbf{y}$ we have

(i) the *expectation vector*

$$\eta' = E[\mathbf{y}]' = (E(y_1), E(y_2), \ldots, E(y_n)) \tag{3.2}$$

(ii) the *dispersion matrix* $\mathbf{D}(\mathbf{y})$,

$$\mathbf{D}(\mathbf{y}) = \begin{bmatrix} \text{Var}(y_1) & \text{Cov}(y_1, y_2) \ldots \text{Cov}(y_1, y_n) \\ \text{Cov}(y_2, y_1) & \text{Var}(y_2) \ldots\ldots \text{Cov}(y_2, y_n) \\ \cdots & \cdots \\ \text{Cov}(y_n, y_1) & \text{Cov}(y_n, y_2) \ldots \text{Var}(y_n) \end{bmatrix} \tag{3.3}$$

The dispersion matrix (also called the *variance-covariance matrix*) is symmetric. If the variables $y_1, y_2, \ldots, y_n$ are independently distributed the dispersion matrix is a diagonal matrix, the diagonal elements being the variances of $y_1, y_2, \ldots, y_n$. If the variables are independent and have common variance σ^2, then $\mathbf{D}(\mathbf{y}) = \sigma^2\mathbf{I}$.

Given two random vectors $\mathbf{x}$, $\mathbf{y}$ with

$$\mathbf{x}' = (x_1, x_2, \ldots, x_m)$$

$$\mathbf{y}' = (y_1, y_2, \ldots, y_n)$$

we define the $m \times n$ matrix **Cov** (**x**, **y**) as

$$\mathbf{Cov}\,(\mathbf{x}, \mathbf{y}) = \begin{bmatrix} \text{Cov}\,(x_1, y_1) & \text{Cov}\,(x_1, y_2) \ldots & \text{Cov}\,(x_1, y_n) \\ \text{Cov}\,(x_2, y_1) & \text{Cov}\,(x_2, y_2) \ldots & \text{Cov}\,(x_2, y_n) \\ \cdots & \cdots & \cdots \\ \text{Cov}\,(x_m, y_1) & \text{Cov}\,(x_m, y_2) \ldots & \text{Cov}\,(x_m, y_n) \end{bmatrix} \quad (3.4)$$

And we have $\mathbf{Cov}\,(\mathbf{y}, \mathbf{y}) = \mathbf{D}\,(\mathbf{y})$.

3.2 Linear Functions of Random Variables

If we have n random variables $y_1, y_2, \ldots, y_n$, then a linear function

$$\lambda_1 y_1 + \lambda_2 y_2 + \ldots + \lambda_n y_n$$

of these random variables, the λ_i being constants, is also a random variable. Introducing the random vector **y** and the vector $\boldsymbol{\lambda}' = (\lambda_1, \ldots, \lambda_n)$, we can write the linear combination as $\boldsymbol{\lambda}'\mathbf{y}$.

For the expectation and variance of the random variable $\boldsymbol{\lambda}'\mathbf{y}$ we have

$$E(\boldsymbol{\lambda}'\mathbf{y}) = \lambda_1 E(y_1) + \ldots + \lambda_n E(y_n) = \boldsymbol{\lambda}' E(\mathbf{y}) \quad (3.5)$$

and

$$\begin{aligned} \text{Var}\,(\boldsymbol{\lambda}'\mathbf{y}) &= \text{Var}\,(\lambda_1 y_1 + \ldots + \lambda_n y_n) \\ &= \sum_i \lambda_i^2 \,\text{Var}\,(y_i) + \sum_{i<j} 2\lambda_i y_j \,\text{Cov}\,(y_i, y_j) \\ &= \boldsymbol{\lambda}'\mathbf{D}(\mathbf{y})\boldsymbol{\lambda} \end{aligned} \quad (3.6)$$

If the random variables $y_1, y_2, \ldots, y_n$ are independent and have common variance σ^2, $\mathbf{D}\,(\mathbf{y}) = \sigma^2\mathbf{I}$ and hence

$$\text{Var}\,(\boldsymbol{\lambda}'\,\mathbf{y}) = (\boldsymbol{\lambda}'\boldsymbol{\lambda})\sigma^2$$

Theorem 3.1: The dispersion matrix **D**(**y**) is positive semi definite.

Proof: The quadratic form

$$\boldsymbol{\lambda}'\mathbf{D}(\mathbf{y})\boldsymbol{\lambda} = \text{Var}\,(\boldsymbol{\lambda}'\mathbf{y}) \geqslant 0 \quad (3.7)$$

for all $\boldsymbol{\lambda}$.

Corollary: If the random variables $y_1, \ldots, y_n$ do not satisfy any linear relation

$$\boldsymbol{\lambda}'\mathbf{y} = \text{constant} \quad (\text{with } \boldsymbol{\lambda} \neq \mathbf{0}) \quad (3.8)$$

then **D**(**y**) is positive definite, and conversely.

If $\boldsymbol{\lambda}'\mathbf{y}$ and $\boldsymbol{\mu}'\mathbf{y}$ are two linear functions of the random variables $y_1, y_2, \ldots, y_n$, then

$$E(a\boldsymbol{\lambda}'\mathbf{y} + b\boldsymbol{\mu}'\mathbf{y}) = E[(a\boldsymbol{\lambda} + b\boldsymbol{\mu})'\mathbf{y}]$$
$$= (a\boldsymbol{\lambda} + b\boldsymbol{\mu})'E(\mathbf{y}) = a\boldsymbol{\lambda}'E(\mathbf{y}) + b\boldsymbol{\mu}'E(\mathbf{y}) \qquad (3.9)$$

a and b being real numbers. Also

$$\text{Cov}(\boldsymbol{\lambda}'\mathbf{y}, \boldsymbol{\mu}'\mathbf{y}) = \text{Cov}(\lambda_1 y_1 + \ldots + \lambda_n y_n, \mu_1 y_1 + \ldots + \mu_n y_n)$$
$$= \boldsymbol{\lambda}'\mathbf{D}(\mathbf{y})\boldsymbol{\mu} = \boldsymbol{\mu}'\mathbf{D}(\mathbf{y})\boldsymbol{\lambda} \qquad (3.10)$$

If $\mathbf{D}(\mathbf{y}) = \sigma^2\mathbf{I}$, we have

$$\text{Cov}(\boldsymbol{\lambda}'\mathbf{y}, \boldsymbol{\mu}'\mathbf{y}) = (\boldsymbol{\lambda}'\boldsymbol{\mu})\sigma^2 = (\boldsymbol{\mu}'\boldsymbol{\lambda})\sigma^2 \qquad (3.11)$$

Consider now m linear functions

$$\begin{aligned} \mathbf{l}_1'\mathbf{y} &= l_{11}y_1 + \ldots + l_{1n}y_n \\ \mathbf{l}_2'\mathbf{y} &= l_{21}y_1 + \ldots + l_{2n}y_n \\ &\cdots\cdots\cdots\cdots \\ \mathbf{l}_m'\mathbf{y} &= l_{m1}y_1 + \ldots + l_{mn}y_n \end{aligned} \qquad (3.12)$$

of the random variables $y_1, y_2, \ldots, y_n$. Introducing the $n \times m$ matrix $\boldsymbol{\Lambda} = [\mathbf{l}_1 \ldots \mathbf{l}_m]$, we can write the set of m linear combinations as $\boldsymbol{\Lambda}'\mathbf{y}$.

Theorem 3.2: $E(\boldsymbol{\Lambda}'\mathbf{y}) = \boldsymbol{\Lambda}'E(\mathbf{y})$

$$\mathbf{D}(\boldsymbol{\Lambda}'\mathbf{y}) = \boldsymbol{\Lambda}'\mathbf{D}(\mathbf{y})\boldsymbol{\Lambda} \qquad (3.13)$$

Proof: $\boldsymbol{\Lambda}'\mathbf{y}$ can be considered as a random vector with coordinates,

$$u_i = l_{i1}y_1 + \ldots + l_{in}y_n, \; i = 1, \ldots, m$$

Now,

$$E(u_i) = l_{i1}E(y_1) + \ldots + l_{in}E(y_n) = \mathbf{l}_i'E(\mathbf{y})$$

i.e.

$$E(\boldsymbol{\Lambda}'\mathbf{y}) = \boldsymbol{\Lambda}'E(\mathbf{y})$$

Further let $\mathbf{D}(\boldsymbol{\Lambda}'\mathbf{y}) = [\sigma_{ij}]_{m\times m}$. Then

$$\sigma_{ii} = \text{Var}(u_i) = \mathbf{l}_i'\mathbf{D}(\mathbf{y})\mathbf{l}_i,$$
$$\sigma_{ij} = \text{Cov}(u_i, u_j) = \mathbf{l}_i'\mathbf{D}(\mathbf{y})\mathbf{l}_j$$

i.e. $$\mathbf{D}(\boldsymbol{\Lambda}'\mathbf{y}) = \boldsymbol{\Lambda}'\mathbf{D}(\mathbf{y})\boldsymbol{\Lambda}$$

3.3 Multivariate Normal Distribution

Suppose $x_1, x_2, \ldots, x_n$ are n independent normal variates with mean zero and variance unity. Writing these as a random vector $\mathbf{x}$, we have $E(\mathbf{x}) = \mathbf{0}$, $\mathbf{D}(\mathbf{x}) = \mathbf{I}$.

Consider now the random variables $y_1, y_2, \ldots, y_m$ where each y_i is a non-homogeneous linear combination of the x_j's,

$$
\begin{aligned}
y_1 &= a_{11}x_1 + \ldots + a_{1n}x_n + \eta_1 \\
&\cdots\cdots\cdots\cdots\cdots\cdots\cdots\cdots \\
y_m &= a_{m1}x_1 + \ldots + a_{mn}x_n + \eta_m
\end{aligned} \tag{3.14}
$$

Writing $\mathbf{A} = [a_{ij}]_{m\times n}$, $\boldsymbol{\eta}' = (\eta_1, \ldots, \eta_m)$, $\mathbf{y}' = (y_1, \ldots, y_m)$ and $\mathbf{x}' = (x_1, \ldots, x_n)$, we have

$$\mathbf{y} = \mathbf{A}\mathbf{x} + \boldsymbol{\eta} \tag{3.15}$$

Now it is well known that each y_i is a normal variate with mean η_i and variance $a_{i1}^2 + \ldots + a_{in}^2$. Also Cov $(y_i, y_j) = a_{i1}a_{j1} + \ldots + a_{in}\,a_{jn}$. Thus $E(\mathbf{y}) = \boldsymbol{\eta}$, $\mathbf{D}(\mathbf{y}) = \mathbf{A}\mathbf{A}'$.

We say that the random vector $\mathbf{y}$ has the *m-variate multivariate normal distribution* with mean $\boldsymbol{\eta}$ and dispersion matrix $\mathbf{A}\mathbf{A}'$. As in the case of a single normal variable the probability distribution of $\mathbf{y}$ is completely determined by the mean vector $\boldsymbol{\eta}$ and the dispersion matrix $\mathbf{A}\mathbf{A}'$.

Conversely, given any vector $\boldsymbol{\eta} \in R_m$ and a positive semi-definite $m \times m$ matrix $\boldsymbol{\Sigma} = [\sigma_{ij}]$, we can define a random vector $\mathbf{y}$ having a multivariate normal distribution (i.e. each y_i is a non-homogeneous linear combination of independent standard normal variates) with mean $\boldsymbol{\eta}$ and dispersion matrix $\boldsymbol{\Sigma}$.

If $\boldsymbol{\Sigma}$ is singular the distribution is called a *singular multivariate normal distribution*. In this case, if rank of $\boldsymbol{\Sigma}$ is k, the entire probability mass of the distribution lies in a k-dimensional linear manifold of R_m but not in any manifold of dimension less than k.

If $\boldsymbol{\Sigma}$ is non-singular, i.e. has an inverse $\boldsymbol{\Sigma}^{-1}$, the distribution is called a *non-singular distribution*. In this case the random variables $y_1, y_2, \ldots, y_m$ have a joint density function (in m variables)

$$\frac{\sqrt{|\boldsymbol{\Sigma}^{-1}|}}{(2\pi)^{m/2}}\, e^{-\frac{1}{2}(\mathbf{y}-\boldsymbol{\eta})'\boldsymbol{\Sigma}^{-1}(\mathbf{y}-\boldsymbol{\eta})} \tag{3.16}$$

This density function is called the multivariate normal density function.

For example, consider the bivariate case with $\mathbf{y}' = (y_1, y_2)$. If $E(y_1) = \eta_1$, $E(y_2) = \eta_2$, Var $(y_1) = \sigma_1^2$, Var $(y_2) = \sigma_2^2$, Cov $(y_1, y_2) = \rho\sigma_1\sigma_2$, ρ being the correlation coefficient of y_1 and y_2, then

$$\boldsymbol{\Sigma} = \begin{bmatrix} \sigma_1^2 & \rho\sigma_1\sigma_2 \\ \rho\sigma_1\sigma_2 & \sigma_2^2 \end{bmatrix}, \quad |\boldsymbol{\Sigma}| = \sigma_1^2\sigma_2^2(1-\rho^2) \tag{3.17}$$

The distribution is non-singular if $\rho^2 < 1$, and then

$$\Sigma^{-1} = \begin{bmatrix} 1/\sigma_1^2(1-\rho^2) & -\rho/\sigma_1\sigma_2(1-\rho^2) \\ -\rho/\sigma_1\sigma_2(1-\rho^2) & 1/\sigma_2^2(1-\rho^2) \end{bmatrix}, \; |\Sigma^{-1}| = \frac{1}{\sigma_1^2\sigma_2^2(1-\rho^2)} \tag{3.18}$$

and
$$(\mathbf{y}-\boldsymbol{\eta})'\Sigma^{-1}(\mathbf{y}-\boldsymbol{\eta}) = \left\{\frac{(y_1-\eta_1)^2}{\sigma_1^2} - \frac{2\rho(y_1-\eta_1)(y_2-\eta_2)}{\sigma_1\sigma_2} + \frac{(y_2-\eta_2)^2}{\sigma_2^2}\right\}\frac{1}{1-\rho^2} \tag{3.19}$$

Thus the density function is the familiar bivariate normal density function. If $\rho^2 = 1$, Σ is singular and the density function does not exist. The entire probability mass is then concentrated in a straight line passing through (η_1, η_2) if rank $(\Sigma) = 1$, and at the point (η_1, η_2) if rank $(\Sigma) = 0$.

If $\mathbf{y}' = (y_1, y_2)$ has the bivariate normal distribution, then each y_i is a normal variable and the conditional distribution of y_2 given y_1, or that of y_1 given y_2, is again a normal distribution.

In the same way, if $\mathbf{y}' = (y_1, \ldots, y_m)$ has the multivariate normal distribution with mean $\boldsymbol{\eta}$ and dispersion matrix $\Sigma = [\sigma_{ij}]$, each y_i is a normal variable with mean η_i and variance σ_{ii}. Further, any subset $(y_{i_1}, \ldots, y_{i_k})$, $k \leqslant m$, has the multivariate normal distribution with means $\eta_{i_1}, \ldots, \eta_{i_k}$ and variance-covariance matrix $[\sigma_{i_r i_s}]$ obtained from Σ by retaining only the rows and columns corresponding to the variables $y_{i_1}, \ldots, y_{i_k}$.

If $\mathbf{y}$ is partitioned as $\mathbf{y} = \begin{bmatrix} \mathbf{y}_1 \\ \cdots \\ \mathbf{y}_2 \end{bmatrix}$ and $\begin{bmatrix} \eta_1 \\ \cdots \\ \eta_2 \end{bmatrix}$, $\begin{bmatrix} \Sigma_{11} & \vdots & \Sigma_{12} \\ \cdots & & \cdots \\ \Sigma_{21} & \vdots & \Sigma_{22} \end{bmatrix}$

are the corresponding partitions of $\boldsymbol{\eta}$ and Σ, then the conditional distribution of $\mathbf{y}_1$ given $\mathbf{y}_2$, or of $\mathbf{y}_2$ given $\mathbf{y}_1$, is again a multivariate normal distribution. As in the bivariate case, y_i, y_j are independent random variables, if and only if, Cov $(y_i, y_j) = \sigma_{ij} = 0$. More generally, the necessary and sufficient condition, that each member of $\mathbf{y}_1$ be independent of each member of $\mathbf{y}_2$ in the partition above, is $\Sigma_{12} = 0$. In particular the variables $y_1, y_2, \ldots, y_m$ are mutually independent, if and only if, Σ is a diagonal matrix.

Suppose $\mathbf{y}' = (y_1, \ldots, y_m)$ has the multivariate normal distribution with mean $\boldsymbol{\eta}$ and dispersion matrix Σ. Consider the random variable

$$\mathbf{l}'\mathbf{y} = l_1 y_1 + \ldots + l_m y_m \tag{3.20}$$

where $\mathbf{l} \in R_m$.

Then $\mathbf{l}'\mathbf{y}$ is a normal variable with mean

$$\mathbf{l}'\boldsymbol{\eta} = l_1\eta_1 + \ldots + l_m \eta_m \tag{3.21}$$

and variance

$$\mathbf{l}'\boldsymbol{\Sigma}\mathbf{l} = l_1^2\sigma_{11} + \ldots + l_m^2\sigma_{mm} + 2l_1 l_2\sigma_{12} + \ldots + 2l_{m-1}\, l_m\sigma_{m-1,\,m} \tag{3.22}$$

Further, given two vectors $\mathbf{l}, \mathbf{m} \in R_m$, the random variables $u_1 = \mathbf{l}'\mathbf{y}$, $u_2 = \mathbf{m}'\mathbf{y}$ have the bivariate normal distribution with means $\mathbf{l}'\boldsymbol{\eta}$, $\mathbf{m}'\boldsymbol{\eta}$, variances $\mathbf{l}'\boldsymbol{\Sigma}\mathbf{l}$, $\mathbf{m}'\boldsymbol{\Sigma}\mathbf{m}$, and covariance $\mathbf{l}'\boldsymbol{\Sigma}\mathbf{m} = \mathbf{m}'\boldsymbol{\Sigma}\mathbf{l}$. More generally, given k linear functions

$$\begin{aligned} u_1 &= a_{11}y_1 + \ldots + a_{1m}y_m = \mathbf{a}_1'\mathbf{y} \\ &\cdots\cdots\cdots\cdots\cdots\cdots \\ u_k &= a_{k1}y_1 + \ldots + a_{km}y_m = \mathbf{a}_k'\mathbf{y} \end{aligned} \tag{3.23}$$

which we may write as

$$\mathbf{u} = \mathbf{A}'\mathbf{y} \qquad \mathbf{A} = [\mathbf{a}_1 \ldots \mathbf{a}_k] \tag{3.24}$$

it can be shown that $\mathbf{u}$ has the multivariate normal distribution with mean $\mathbf{A}'\boldsymbol{\eta}$ and dispersion matrix $\mathbf{A}'\boldsymbol{\Sigma}\mathbf{A}$.

The following theorem will be of use later.

Theorem 3.3: Let $y_1, y_2, \ldots, y_m$ be independent normal variables with means $\eta_1, \ldots, \eta_m$ and common variance σ^2, and $\mathbf{l}_1, \ldots, \mathbf{l}_k$ an orthonormal set of vectors in R_m. Then the random variables

$$\begin{aligned} u_1 &= \mathbf{l}_1'\mathbf{y} \\ &\cdots\cdots \\ u_k &= \mathbf{l}_k'\mathbf{y} \end{aligned} \tag{3.25}$$

are independent normal variables with means $\mathbf{l}_i'\boldsymbol{\eta}$ and common variance σ^2.

Proof: $\mathbf{y}' = (y_1, \ldots, y_m)$ has the multivariate normal distribution with mean $\boldsymbol{\eta}$ and dispersion matrix $\sigma^2\mathbf{I}$. Hence each u_i is a normal variable with mean $\mathbf{l}_i'\boldsymbol{\eta}$ and variance $(\mathbf{l}_i'\mathbf{l}_i)\,\sigma^2 = \sigma^2$. Further, for $i \neq j$,

$$\text{Cov}\,(u_i, u_j) = \mathbf{l}_i'\mathbf{D}(\mathbf{y})\,\mathbf{l}_j = (\mathbf{l}_i'\mathbf{l}_j)\,\sigma^2 = 0$$

so that the variables $u_1, u_2, \ldots, u_k$ are independent.

3.4 The Chi-square Distribution

Let $x_1, x_2, \ldots, x_k$ be independent normal variables with means zero and variance unity. Then the distribution of the random variable $x_1^2 + x_2^2 + \ldots + x_k^2$ is called the *χ^2-distribution* with *k degrees of freedom* (written d.f.).

If the random variables $x_1, x_2, \ldots, x_k$ have means $\eta_1, \ldots, \eta_k$ and variance unity, the distribution of the random variable

$$x_1^2 + x_2^2 + \ldots + x_k^2 \tag{3.26}$$

is called the *non-central χ^2-distribution* with k d.f. and *non-centrality parameter*

$$\lambda = \eta_1^2 + \eta_2^2 + \ldots + \eta_k^2 \tag{3.27}$$

It then follows that if the independent normal variables $x_1, x_2 \ldots, x_k$ have means $\eta_1, \ldots, \eta_k$ and common variance σ^2, the random variable

$$(1/\sigma^2) \sum_{i=1}^{k} x_i^2 \tag{3.28}$$

has the non-central χ^2-distribution with k d.f. and non-centrality parameter $\lambda = (1/\sigma^2) \sum_{i=1}^{k} \eta_i^2$. If $\eta_1 = \ldots = \eta_k = 0$, the distribution of $(1/\sigma^2)(x_1^2 + \ldots + x_k^2)$ is the χ^2-distribution with k d.f.

The non-central χ^2-distribution has the additive property similar to that of the χ^2-distribution, viz. if $u_1, \ldots, u_m$ are independent random variables and u_i has the non-central χ^2-distribution with k_i d.f. and non-centrality parameter λ_i ($i = 1, 2, \ldots, m$), then the random variable

$$u = u_1 + u_2 + \ldots + u_m \tag{3.29}$$

has the non-central χ^2-distribution with $k = k_1 + \ldots + k_m$ d.f. and non-centrality parameter $\lambda = \lambda_1 + \ldots + \lambda_m$.

If u has the χ^2-distribution with k d.f., we have

$$E(u) = k, \qquad \mathrm{Var}(u) = 2k \tag{3.30}$$

If the distribution of u is the non-central χ^2-distribution with k d.f. and non-centrality parameter λ, then

$$E(u) = k + \lambda, \qquad \mathrm{Var}(u) = 2k + 4\lambda \tag{3.31}$$

The non-central χ^2-distribution can be approximated by means of a χ^2-distribution (Patnaik's approximation). This is done by replacing the random variable u, which has the non-central χ^2-distribution with k d.f. and non-centrality parameter λ by $c \cdot v$, where v has the χ^2-distribution with p d.f. The constants c and p are obtained by equating the moments of the two distributions. Thus

$$c \cdot p = k + \lambda$$

$$c^2 \cdot 2p = 2k + 4\lambda \tag{3.32}$$

giving

$$c = \frac{k + 2\lambda}{k + \lambda}$$

$$p = \frac{(k + \lambda)^2}{k + 2\lambda} \tag{3.33}$$

Note that p, the d.f. for v, need no longer be an integer.

3.5 The *t*-distribution

If x has the normal distribution with mean zero and variance unity, u has the χ^2-distribution with n d.f., and u and x are independent random variables, then the distribution of the random variable

$$t = \frac{x}{\sqrt{(u/n)}} \tag{3.34}$$

is called the *t-distribution* with n d.f.

If $E(x) = \delta \neq 0$, then the distribution of $\frac{x}{\sqrt{(u/n)}}$ is called the *non-central t-distribution* with n d.f. and *non-centrality parameter* δ.

3.6 The *F*-distribution

If u_1, u_2 are independent random variables with the χ^2-distribution with n_1, n_2 d.f. respectively, the distribution of the random variable

$$F = \frac{u_1/n_1}{u_2/n_2} \tag{3.35}$$

is called the *F-distribution* with (n_1, n_2) d.f.

If the distribution of u_1 is the non-central χ^2-distribution with n_1 d.f. and non-centrality paramter λ, the distribution of F is called the *non-central F-distribution* with (n_1, n_2) d.f. and *non-centrality parameter* λ.

COMPLEMENTS AND PROBLEMS

1. If **x** has the multivariate normal distribution with mean vecror $\boldsymbol{\mu}$ and non-singular dispersion matrix $\boldsymbol{\Sigma}$, then

$$(\mathbf{x}-\boldsymbol{\mu})' \, \boldsymbol{\Sigma}^{-1} \, (\mathbf{x}-\boldsymbol{\mu})$$

has the chi-square distribution with degrees of freedom equal to the rank of $\boldsymbol{\Sigma}$.

2. If **x** has the multivariate normal distribution with mean vector **0** and dispersion matrix I, the quadratic form $\mathbf{x}'A\mathbf{x}$ has the chi-square distribution with degrees of freedom equal to the rank of A, if and only if, $A^2 = A$.

3. If **x** has the normal distribution with mean vector $\boldsymbol{\mu}$ and dispersion matrix $\sigma^2 I$, then the quadratic form $\frac{1}{\sigma^2}(\mathbf{x}'A\mathbf{x})$, with $A^2 = A$, has the non-central chi-square distribution with degrees of freedom equal to the rank of A, and non-centrality parameter equal to $\frac{1}{\sigma^2}(\boldsymbol{\mu}'A\boldsymbol{\mu})$.

REFERENCES

Patnaik (1949) gave the approximation of non-central chi-square in terms of the chi-square distribution. The non-central chi-square distribution tables are given in Fix (1949), Johnson and Pearson (1965).

Applications of the non-central t-distribution are discussed in Johnson and Welch (1940); for tables see Resnikoff and Lieberman (1957).

Press (1966) discusses the approximation of linear combination of non-central chi-square variables.

For a slightly more detailed discussion of the ideas of this chapter see Scheffe (1959).

4

Linear Estimation

4.1 The Linear Model

Suppose we have n observations or measurements. In the *linear model* we assume that these observations are values taken by n random variables $y_1, \ldots, y_n$, satisfying the following conditions.

(i) The expectation $E(y_i)$ of y_i is a linear combination of p unknown *parameters* $\beta_1, \ldots, \beta_p$,

$$E(y_i) = a_{i1}\beta_1 + \cdots + a_{ip}\beta_p, \; i = 1, \ldots, n \tag{4.1}$$

where the a_{ij} are known constants.

(ii) The variables y_i have all the same variance σ^2, and are uncorrelated. The value of σ^2 is not known.

(iii) The variables y_i are normally distributed.

Introducing the independent normal random variables $e_1, \ldots, e_n$ with zero expectations and common variance σ^2, we may write

$$y_i = a_{i1}\beta_1 + \ldots + a_{ip}\beta_p + e_i, \; i = 1, \ldots, n \tag{4.2}$$

It is clear that the variables y_i of Eq. (4.2) satisfy the conditions (i), (ii) and (iii) above. Equations (4.2) will be called the *model equations* of the linear model.

The model equations can be compactly written in matrix notation as

$$\mathbf{y} = \mathbf{A}\boldsymbol{\beta} + \mathbf{e} \tag{4.3}$$

where $\mathbf{y}' = (y_1, \ldots, y_n)$, $\mathbf{A}$ is the $n \times p$ matrix $[a_{ij}]$, $\boldsymbol{\beta}' = (\beta_1, \ldots, \beta_p)$, and $\mathbf{e}' = (e_1, \ldots, e_n)$. The vector $\mathbf{y}$ is called the *observation vector*, the matrix $\mathbf{A}$ the *coefficient matrix*, and $\boldsymbol{\beta}$ the *parameter vector*. The random vector $\mathbf{e}$, usually called the *error vector*, is $\mathbf{N}(\mathbf{0}, \sigma^2\mathbf{I})$, and consequently the random vector **y** is N $(\mathbf{A}\boldsymbol{\beta}, \sigma^2\mathbf{I})$. Alternatively, we may describe the linear model as a normal random vector **y** with

$$E(\mathbf{y}) = \mathbf{A}\boldsymbol{\beta},\ \mathbf{D}(\mathbf{y}) = \sigma^2\mathbf{I} \tag{4.4}$$

It should be remembered that whereas $\boldsymbol{\beta}$ and σ^2 are unknown, the matrix **A** is known.

The statistical analysis of linear models is mainly concerned with the estimation of the unknown parameters $\boldsymbol{\beta}$ and σ^2 in terms of the given observations on the random variables $y_1, \ldots, y_n$ and with the formulation of tests of hypotheses regarding these parameters.

4.2 Estimable Functions

A function $f(y_1, \ldots, y_n)$ of the random variables $y_1, y_2, \ldots y_n$ is called an *unbiased estimate* of the function $g(\beta_1, \ldots, \beta_p)$ of the parameters $\beta_1, \ldots, \beta_p$ if

$$E[f(y_1, \ldots, y_n)] = g(\beta_1, \ldots, \beta_p) \tag{4.5}$$

for all $\boldsymbol{\beta} \in R_p$.

The estimate $f(y_1, \ldots, y_n)$ is called a *linear estimate* if

$$f(y_1, \ldots, y_n) = l_1 y_1 + \ldots + l_n y_n = \mathbf{l}'\mathbf{y} \tag{4.6}$$

for some $\mathbf{l} \in R_n$. Similarly, the function $g(\beta_1, \ldots, \beta_p)$ is called a linear parametric function if

$$g(\beta_1, \ldots, \beta_p) = \lambda_1 \beta_1 + \ldots + \lambda_p \beta_p = \boldsymbol{\lambda}'\boldsymbol{\beta} \tag{4.7}$$

We shall only be concerned with the problem of finding linear unbiased estimates $\mathbf{l}'\mathbf{y}$ of linear parametric functions $\boldsymbol{\lambda}'\boldsymbol{\beta}$. In addition, an unbiased estimate of σ^2 will also be obtained, but it will not be a linear estimate.

If $\boldsymbol{\lambda}' = (0, 0, \ldots, 1, \ldots, 0)$ is the ith unit vector of R_p, $\boldsymbol{\lambda}'\boldsymbol{\beta} = \beta_i$, and the estimate $\mathbf{l}'\mathbf{y}$ of $\boldsymbol{\lambda}'\boldsymbol{\beta}$ will be called an *estimate* of β_i.

Given a linear model $\mathbf{y} = \mathbf{A}\boldsymbol{\beta} + \mathbf{e}$ it may not be possible to find unbiased linear estimates for all the parameters $\beta_1, \ldots, \beta_p$, or equivalently, for all linear parametric functions $\boldsymbol{\lambda}'\boldsymbol{\beta}$. Thus in the model

$$y_i = \mu + \alpha_i + e_i,\ i = 1, 2, 3$$

there is no linear estimate of μ. For, if $\mathbf{l}'\mathbf{y}$ is an estimate of μ,

$$E(\mathbf{l}'\mathbf{y}) = (l_1 + l_2 + l_3)\,\mu + l_1\alpha_1 + l_2\alpha_2 + l_3\alpha_3$$

For this to be equal to μ for all values of $\mu, \alpha_1, \alpha_2, \alpha_3$, we must have $l_1 = l_2 = l_3 = 0$. Similarly, there are no linear estimates of the parameters $\alpha_1, \alpha_2, \alpha_3$. But the parametric functions $\mu + \alpha_1$, $\alpha_1 - \alpha_2$, $\alpha_1 - 2\alpha_2 + \alpha_3$ have linear estimates y_1, $y_1 - y_2$, $y_1 - 2y_2 + y_3$ respectively.

Substituting $\mu + \alpha_i$ by β_i, the model becomes

$$y_i = \beta_i + e_i \ (i = 1, 2, 3)$$

and now the parameters β_1, β_2 and β_3 have linear estimates y_1, y_2 and y_3 respectively. Thus all the parameters in the second representation can be estimated separately, but that is not the case for the first representation of the model.

We say that the second model is a *reparametrisation* of the first model. It can be shown that every linear model in which all the parameters cannot be separately estimated can be so reparametrised that all the parameters of the new model are estimable. In the reparametrised form of the model linear unbiased estimates will exist for all linear parametric functions. A statistical analysis of the original model can be carried out in terms of the analysis of the reparametrised model.

We shall, however, follow a different approach (due to R.C. Bose) using the concept of estimable parametric functions.

A linear function $\boldsymbol{\lambda}'\boldsymbol{\beta}$ of the parameters is said to be an *estimable parametric function* (or simply *estimable*) if there exists a linear function $\mathbf{l}'\mathbf{y}$ of the observations $\mathbf{y}$ such that

$$E(\mathbf{l}'\mathbf{y}) = \boldsymbol{\lambda}'\boldsymbol{\beta} \qquad (4.8)$$

for all $\boldsymbol{\beta} \in R_p$.

Theorem 4.1: $\boldsymbol{\lambda}'\boldsymbol{\beta}$ is estimable, if and only if, $\boldsymbol{\lambda} \in R(\mathbf{A})$.

Proof: $\boldsymbol{\lambda}'\boldsymbol{\beta}$ is estimable, if and only if, there exists a vector $\mathbf{l}$ such that $E(\mathbf{l}'\mathbf{y}) = \boldsymbol{\lambda}'\boldsymbol{\beta}$, i.e. $\mathbf{l}'\mathbf{A}\boldsymbol{\beta} = \boldsymbol{\lambda}'\boldsymbol{\beta}$ for all $\boldsymbol{\beta} \in R_p$. Hence $\boldsymbol{\lambda}'\boldsymbol{\beta}$ is estimable, if and only if, $\boldsymbol{\lambda}' = \mathbf{l}'\mathbf{A}$ for some $\mathbf{l}$, i.e. $\boldsymbol{\lambda} \in R(\mathbf{A})$.

Theorem 4.2: If $\mathbf{l}_1'\mathbf{y}$, $\mathbf{l}_2'\mathbf{y}$ are estimates of $\boldsymbol{\lambda}_1'\boldsymbol{\beta}$, $\boldsymbol{\lambda}_2'\boldsymbol{\beta}$ then $a_1(\mathbf{l}_1'\mathbf{y}) + a_2(\mathbf{l}_2'\mathbf{y})$ is an estimate of $a_1(\boldsymbol{\lambda}_1'\boldsymbol{\beta}) + a_2(\boldsymbol{\lambda}_2'\boldsymbol{\beta})$.

Proof:
$$E[a_1(\mathbf{l}_1'\mathbf{y}) + a_2(\mathbf{l}_2'\mathbf{y})] = a_1 E(\mathbf{l}_1'\mathbf{y}) + a_2 E(\mathbf{l}_2'\mathbf{y})$$
$$= a_1(\boldsymbol{\lambda}_1'\boldsymbol{\beta}) + a_2(\boldsymbol{\lambda}_2'\boldsymbol{\beta})$$

Corollary: The set of all estimable functions $\boldsymbol{\lambda}'\boldsymbol{\beta}$ is a subspace of dimension r (= rank $\mathbf{A}$) of the vector space of linear forms in the variables $\beta_1, \ldots, \beta_p$.

From the corollary to Theorem 4.2 we see immediately that the parameters $\beta_1, \ldots, \beta_p$ are estimable, if and only if, rank $\mathbf{A} = p$.

Theorem 4.3: $\boldsymbol{\lambda}'\boldsymbol{\beta}$ is estimable, if and only if,

$$\text{rank } \mathbf{A} \quad = \text{rank } (\mathbf{A}' : \boldsymbol{\lambda}) \tag{4.9}$$

$$\text{rank } \mathbf{A}'\mathbf{A} = \text{rank } (\mathbf{A}'\mathbf{A} : \boldsymbol{\lambda}) \tag{4.10}$$

Condition (4.) says that $\boldsymbol{\lambda} \in C(\mathbf{A}')$ and (4.10) that $\boldsymbol{\lambda} \in C(\mathbf{A}'\mathbf{A})$. Since $C(\mathbf{A}'\mathbf{A}) = C(\mathbf{A}') = R(\mathbf{A})$, thosc conditions are equivalent to the condition $\boldsymbol{\lambda} = R(\mathbf{A})$ of Theorem 4.1.

4.3 Estimation and Error Spaces

The column space $C(\mathbf{A})$ of the coefficient matrix is called the *estimation space* of the linear model. We shall denote it by V_s. The estimation space V_s is a subspace of R_n; its orthogonal complement in R_n is called the *error space*. We denote the error space by V_e. Reasons for adopting this terminology will become clear as we proceed.

Theorem 4.4: $E(\mathbf{l}'\mathbf{y}) = 0$ for all $\boldsymbol{\beta} \in R_p$, if and only if, $\mathbf{l} \in V_e$.

Proof: $E(\mathbf{l}'\mathbf{y}) = \mathbf{l}'\mathbf{A}\boldsymbol{\beta}$.

Hence $E(\mathbf{l}'\mathbf{y}) = 0$ for all $\boldsymbol{\beta} \in R_p$, if and only if, $\mathbf{l}'\mathbf{A} = \mathbf{0}'$, i.e. $\mathbf{l}$ is orthogonal to the columns of $\mathbf{A}$, i.e. $\mathbf{l} \in V_e$.

Linear functions $\mathbf{l}'\mathbf{y}$ of the observations for which $E(\mathbf{l}'\mathbf{y}) = 0$ for all values of the parameters $\boldsymbol{\beta}$ are called *error functions*.

Theorem 4.5: Let $\mathbf{l}'\mathbf{y}$ be an estimate of the estimable function $\boldsymbol{\lambda}'\boldsymbol{\beta}$, and $\mathbf{l}_s$ be the projection of $\mathbf{l}$ on V_s. Then $\mathbf{l}_s'\mathbf{y}$ is also an estimate of $\boldsymbol{\lambda}'\boldsymbol{\beta}$ and

$$\text{Var } (\mathbf{l}_s'\mathbf{y}) \leqslant \text{Var } (\mathbf{l}'\mathbf{y}) \tag{4.11}$$

Proof: If $\mathbf{l}_e$ denotes the projection of $\mathbf{l}$ on the error space V_e, we have $E(\mathbf{l}_e'\mathbf{y}) = 0$. Then,

$$E(\mathbf{l}'\mathbf{y}) = E[(\mathbf{l}_s + \mathbf{l}_e)'\mathbf{y}] = E(\mathbf{l}_s'\mathbf{y}) = \boldsymbol{\lambda}'\boldsymbol{\beta}$$

Also,

$$\begin{aligned} \text{Var } (\mathbf{l}'\mathbf{y}) &= \sigma^2 (\mathbf{l}'\mathbf{l}) = \sigma^2 (\mathbf{l}_s'\mathbf{l}_s + \mathbf{l}_e'\mathbf{l}_e) \\ &\geqslant \sigma^2 (\mathbf{l}_s'\mathbf{l}_s) = \text{Var } (\mathbf{l}_s'\mathbf{y}) \end{aligned}$$

with equality, if and only if, $\mathbf{l}_e = 0$ i.e. $\mathbf{l} \in V_s$.

Theorem 4.6: If $\mathbf{l}'\mathbf{y}$, $\mathbf{m}'\mathbf{y}$ are two estimates of the estimable function $\boldsymbol{\lambda}'\boldsymbol{\beta}$ and $\mathbf{l}_s$ and $\mathbf{m}_s$ the projections of $\mathbf{l}$ and $\mathbf{m}$ respectively on V_s, then $\mathbf{l}_s = \mathbf{m}_s$.

Proof: $E(\mathbf{l}_s'\mathbf{y}) = E(\mathbf{m}_s'\mathbf{y}) = \boldsymbol{\lambda}'\boldsymbol{\beta}$.

Hence

$$E(\mathbf{l}_s'\mathbf{y} - \mathbf{m}_s'\mathbf{y}) = 0$$

for all $\boldsymbol{\beta}$, i.e.

$$\mathbf{l}_s - \mathbf{m}_s \in V_e$$

But $\mathbf{l}_s$ and $\mathbf{m}_s$ both belong to the estimation space so that

$$\mathbf{l}_s - \mathbf{m}_s \in V_s$$

Thus we must have

$$\mathbf{l}_s - \mathbf{m}_s = \mathbf{0}.$$

4.4 Best Estimates

Theorems 4.5 and 4.6 show that once we have found any unbiased estimate $\mathbf{l}'\mathbf{y}$ of $\boldsymbol{\lambda}'\boldsymbol{\beta}$ we can obtain another estimate $\mathbf{l}_s'\mathbf{y}$ which is better in the sense that it has a smaller variance than $\mathbf{l}'\mathbf{y}$. The estimate $\mathbf{l}_s'\mathbf{y}$ so obtained is unique and has the smallest variance among all unbiased linear estimates of $\boldsymbol{\lambda}'\boldsymbol{\beta}$.

The unbiased minimum variance linear estimate $\mathbf{l}'\mathbf{y}$ of an estimable function $\boldsymbol{\lambda}'\boldsymbol{\beta}$ is called the *best linear unbiased estimate* (BLUE) of $\boldsymbol{\lambda}'\boldsymbol{\beta}$. We shall call it the *best estimate* or if there is no room for confusion simply the *estimate* of $\boldsymbol{\lambda}'\boldsymbol{\beta}$.

From the results proved above it is clear that if $\mathbf{l} \in V_s$ and $E[\mathbf{l}'\mathbf{y}] = \boldsymbol{\lambda}'\boldsymbol{\beta}$, then $\mathbf{l}'\mathbf{y}$ is the best estimate of $\boldsymbol{\lambda}'\boldsymbol{\beta}$. Conversely, if $\mathbf{l}'\mathbf{y}$ is the best estimate of $\boldsymbol{\lambda}'\boldsymbol{\beta}$, then $\mathbf{l} \in V_s$.

Theorem 4.7: Let $\mathbf{l}_1'\mathbf{y}$, and $\mathbf{l}_2'\mathbf{y}$ be the best estimates of $\boldsymbol{\lambda}_1'\boldsymbol{\beta}$ and $\boldsymbol{\lambda}_2'\boldsymbol{\beta}$ respectively. Then, $(a_1\mathbf{l}_1 + a_2\mathbf{l}_2)'\mathbf{y}$ is the best estimate of $(a_1\boldsymbol{\lambda}_1 + a_2\boldsymbol{\lambda}_2)'\boldsymbol{\beta}$.

Proof: We have

$$E[(a_1\mathbf{l}_1 + a_2\mathbf{l}_2)'\mathbf{y}] = a_1\boldsymbol{\lambda}_1'\boldsymbol{\beta} + a_2\boldsymbol{\lambda}_2'\boldsymbol{\beta}$$

Since $\mathbf{l}_1, \mathbf{l}_2 \in V_s$, $a_1\mathbf{l}_1 + a_2\mathbf{l}_2 \in V_s$, which completes the proof of the theorem.

Theorem 4.8: Let $\mathbf{l}_i'\mathbf{y}$ be the best estimate of $\boldsymbol{\lambda}_i'\boldsymbol{\beta}$, $i = 1, \ldots, k$. If $\boldsymbol{\lambda}_1, \ldots, \boldsymbol{\lambda}_k$ are linearly independent, so are $\mathbf{l}_1, \ldots, \mathbf{l}_k$ and conversely.

Proof: Suppose $\boldsymbol{\lambda}_1, \ldots, \boldsymbol{\lambda}_k$ are linearly independent, and let $b_1\mathbf{l}_1 + \ldots + b_k\mathbf{l}_k = \mathbf{0}$. Then,

$$E[(b_1\mathbf{l}_1 + \ldots + b_k\mathbf{l}_k)'\mathbf{y}] = (b_1\boldsymbol{\lambda}_1 + \ldots + b_k\boldsymbol{\lambda}_k)'\boldsymbol{\beta} = 0$$

for all $\boldsymbol{\beta} \in R_p$.

Hence,

$$b_1\boldsymbol{\lambda}_1 + \ldots + b_k\boldsymbol{\lambda}_k = \mathbf{0}$$

Since the $\boldsymbol{\lambda}_i$'s are linearly independent, we must have $b_1 = \ldots = b_k = 0$, which shows that $\mathbf{l}_1, \ldots, \mathbf{l}_k$ are linearly independent. (Note that this part of the theorem holds good even if the $\mathbf{l}_i'\mathbf{y}$'s are unbiased and not necessarily the best, estimates of $\boldsymbol{\lambda}_i'\boldsymbol{\beta}$.)

Conversely, suppose $\mathbf{l}_1, \ldots, \mathbf{l}_k$ are linearly independent, and let $a_1\boldsymbol{\lambda}_1 + \ldots + a_k\boldsymbol{\lambda}_k = \mathbf{0}$. Then,

$$E[a_1\mathbf{l}_1 + \ldots + a_k\mathbf{l}_k)'\mathbf{y}] = (a_1\boldsymbol{\lambda}_1 + \ldots + a_k\boldsymbol{\lambda}_k)'\boldsymbol{\beta} = 0$$

for all $\boldsymbol{\beta} \in R_p$. Hence

$$a_1\mathbf{l}_1 + \ldots + a_k\mathbf{l}_k \in V_e$$

But,

$$a_1\mathbf{l}_1 + \ldots + a_k\mathbf{l}_k \in V_s$$

so that we must have

$$a_1\mathbf{l}_1 + \ldots + a_k\mathbf{l}_k = \mathbf{0}$$

Linear independence of $\mathbf{l}_1, \ldots, \mathbf{l}_k$ then implies that $a_1 = \ldots = a_k = 0$, which shows that $\boldsymbol{\lambda}_1, \ldots, \boldsymbol{\lambda}_k$ are linearly independent.

We see from Theorem 4.7 that the set of all best estimates $\mathbf{l}'\mathbf{y}$ of estimable functions is a subspace of the vector space of all linear forms in the variables $y_1, \ldots, y_n$. There is a one-to-one correspondence between estimable functions $\boldsymbol{\lambda}'\boldsymbol{\beta}$ and their best estimates $\mathbf{l}'\mathbf{y}$; linearly independent estimable functions have linearly independent best estimates and vice versa. Thus, the subspace of all best estimates has the same dimension as the subspace of all estimable functions, i.e. r (= rank $\mathbf{A}$).

We note, however, that the best estimates of orthogonal parametric functions need not be orthogonal; nor does the orthogonality of the best estimates imply that of the parametric functions of which they are the estimates. This is readily established by considering the model

$$y_1 = \beta_1 + \beta_2 + e_1$$
$$y_2 = \beta_2 + e_2$$

The estimation space, i.e. the space spanned by the vectors (1, 0) and (1, 1) is R_2. Thus any unbiased estimate of $\boldsymbol{\lambda}'\boldsymbol{\beta}$ is the best estimate. The best estimates of β_1, β_2 being $y_1 - y_2$ and y_2, are not orthogonal. Similarly, though $y_1 - y_2$ and $y_1 + y_2$ are orthogonal, they are estimates of non-orthogonal functions β_1 and $\beta_1 + 2\beta_2$.

Theorem 4.9: If $\boldsymbol{\lambda}'\boldsymbol{\beta}$ is estimable, its best estimate is $\boldsymbol{\lambda}'\hat{\boldsymbol{\beta}}$, where $\hat{\boldsymbol{\beta}}$ is any solution of the equations

$$\mathbf{A}'\mathbf{A}\,\boldsymbol{\beta} = \mathbf{A}'\mathbf{y} \tag{4.12}$$

Proof: Let $\mathbf{l}'\mathbf{y}$ be an unbiased estimate of $\boldsymbol{\lambda}'\boldsymbol{\beta}$. Then

$$E\,(\mathbf{l}'\mathbf{y}) = \mathbf{l}'\mathbf{A}\boldsymbol{\beta} = \boldsymbol{\lambda}'\boldsymbol{\beta}$$

for all $\boldsymbol{\beta} \in R_p$, so that $\mathbf{l}'\mathbf{A} = \boldsymbol{\lambda}'$. To find the best estimate of $\boldsymbol{\lambda}'\boldsymbol{\beta}$ we have to project $\mathbf{l}$ on V_s. The projection of $\mathbf{l}$ on V_s may be written as $\mathbf{Ab}$ where, from Theorem 2.7, $\mathbf{b}$ is any solution of

$$\mathbf{A}'\mathbf{A}\mathbf{b} = \mathbf{A}'\mathbf{l} = \boldsymbol{\lambda}$$

Then, the best estimate of $\boldsymbol{\lambda}'\boldsymbol{\beta}$ is given by

$$(\mathbf{Ab})'\mathbf{y} = \mathbf{b}'\mathbf{A}'\mathbf{y} = \mathbf{b}'\mathbf{A}'\mathbf{A}\hat{\boldsymbol{\beta}} = (\mathbf{A}'\mathbf{Ab})'\,\hat{\boldsymbol{\beta}} = \boldsymbol{\lambda}'\hat{\boldsymbol{\beta}}$$

It is easy to verify that the equations $\mathbf{A}'\mathbf{A}\boldsymbol{\beta} = \mathbf{A}'\mathbf{y}$ are consistent, so that a solution $\hat{\boldsymbol{\beta}}$ always exists, and that even if the equations have more than one solution, the best estimate $\boldsymbol{\lambda}'\hat{\boldsymbol{\beta}}$ is uniquely determined be any one solution. Consistency follows from the fact that

$$\mathbf{A}'\mathbf{y} \in C\,(\mathbf{A}') = C\,(\mathbf{A}'\mathbf{A})$$

Further, if $\boldsymbol{\beta}_1$, $\boldsymbol{\beta}_2$ are two solutions of the equations $\mathbf{A}'\mathbf{A}\boldsymbol{\beta} = \mathbf{A}'\mathbf{y}$, we have

$$\mathbf{A}'\mathbf{A}\,(\boldsymbol{\beta}_1 - \boldsymbol{\beta}_2) = \mathbf{0}$$

Thus $\boldsymbol{\beta}_1 - \boldsymbol{\beta}_2$ is orthogonal to $R\,(\mathbf{A}'\mathbf{A})$ and therefore to $R\,(\mathbf{A})$. Since $\boldsymbol{\lambda}'\boldsymbol{\beta}$ is estimable, $\boldsymbol{\lambda} \in R\,(\mathbf{A})$, and hence is orthogonal to $\boldsymbol{\beta}_1 - \boldsymbol{\beta}_2$. Thus

$$\boldsymbol{\lambda}'\,(\boldsymbol{\beta}_1 - \boldsymbol{\beta}_2) = 0$$

or,

$$\boldsymbol{\lambda}'\boldsymbol{\beta}_1 = \boldsymbol{\lambda}'\boldsymbol{\beta}_2$$

which shows that all solutions lead to the same estimate.

4.5. Normal Equations

The equations

$$\mathbf{A}'\mathbf{A}\boldsymbol{\beta} = \mathbf{A}'\mathbf{y} \tag{4.13}$$

are called the *normal equations* of the linear model $\mathbf{y} = \mathbf{A\beta} + \mathbf{e}$.

To write down the normal equations we simply multiply both sides of the model equations by $\mathbf{A}'$ ignoring the error-vector $\mathbf{e}$. There wlll be as many normal equations as there are parameters in the model.

The ith column of $\mathbf{A}$ is the column of coefficients of the parameter β_i in the model equations. Thus the ith normal equation, called the normal equation for β_i, is obtained by combining linearly the model equations by means of the coefficients of β_i. If, in particular, the matrix $\mathbf{A}$ is a matrix of 0's and 1's only, the ith normal equation is obtained by simply adding together those model equations which contain β_i.

The importance of the normal equations resides in the fact that they enable us to determine immediately the class of all estimable functions and their best estimates. We have seen (Theorem 4.3) that $\boldsymbol{\lambda}'\boldsymbol{\beta}$ is estimable, if and only if, $\boldsymbol{\lambda} \in C(\mathbf{A}'\mathbf{A}) = R(\mathbf{A}'\mathbf{A})$. Thus the estimable functions are functions of the form $\mathbf{c}'\mathbf{A}'\mathbf{A}\boldsymbol{\beta}$, where $\mathbf{c} \in R_p$ is arbitrary; further, the best estimate of $\mathbf{c}'\mathbf{A}'\mathbf{A}\boldsymbol{\beta}$ is $\mathbf{c}'\mathbf{A}'\mathbf{A}\hat{\boldsymbol{\beta}}$, where $\hat{\boldsymbol{\beta}}$ is any solution of the normal equations $\mathbf{A}'\mathbf{A}\boldsymbol{\beta} = \mathbf{A}'\mathbf{y}$. Thus the best estimate of $\mathbf{c}'\mathbf{A}'\mathbf{A}\boldsymbol{\beta}$ is $\mathbf{c}'\mathbf{A}'\mathbf{y}$. In other words, an estimable function is simply a linear combination of the left hand side $\mathbf{A}'\mathbf{A}\boldsymbol{\beta}$ of the normal equations, and its best estimate is the *same* linear combination of the right hand side $\mathbf{A}'\mathbf{y}$.

We note that each row of $\mathbf{A}$ is an estimable function, i.e. $a_{i1}\beta_1 + \dots + a_{ip}\beta_p$ is estimable for $i = 1, \dots, n$. Further, y_i is an estimate of $a_{i1}\beta_1 + \dots + a_{ip}\beta_p$, though not necessarily the best. The class of all estimable functions can also be represented as $\mathbf{c}'\mathbf{A}\boldsymbol{\beta}$, with $\mathbf{c} \in R_n$ being arbitrary. The advantage of using the normal equations to represent estimable functions as $\mathbf{c}'\mathbf{A}'\mathbf{A}\boldsymbol{\beta}$ is that we get at the same time the best estimates $\mathbf{c}'\mathbf{A}'\mathbf{y}$.

EXAMPLE 1: Let the model equations be

$$y_1 = 2\alpha_1 + 3\alpha_2 + e_1$$

$$y_2 = 3\alpha_1 + 4\alpha_2 + e_2$$

$$y_3 = 4\alpha_1 + 5\alpha_2 + e_3$$

The first normal equation is obtained by multiplying the equations (ignoring the error terms) successively by 2, 3 and 4 (the coefficients of α_1) and adding; the second by using 3, 4 and 5 (the coefficients of α_2). The normal equations are

$$29\alpha_1 + 38\alpha_2 = 2y_1 + 3y_2 + 4y_3$$

$$38\alpha_1 + 50\alpha_2 = 3y_1 + 4y_2 + 5y_3$$

The estimable parameteric functions are a_1 $(29\alpha_1+38\alpha_2)+a_2$ $(38\alpha_1+50\alpha_2)$, their best estimates being a_1 $(2y_1+3y_2+4y_3)+a_2$ $(3y_1+4y_2+5y_3)$, where a_1, a_2 are arbitrary real numbers. We may write an estimable function as $(29a_1+38a_2)\ \alpha_1+(38a_1+50a_2)\ \alpha_2$ and its best estimate as $(2a_1+3a_2)\ y_1+(3a_1+4a_2)\ y_2+(4a_1+5a_2)\ y_3$.

In particular, α_1 is estimable (take $a_1 = 25/3$ and $a_2 = -19/3$) and its best estimate is

$$(-7y_1 - y_2 + 5y_3)/3$$

Similarly, α_2 is estimable.

EXAMPLE 2: Let the model equations be

$$y_{11} = \mu + \tau_1 + \alpha_1 + e_{11}$$

$$y_{12} = \mu + \tau_1 + \alpha_2 + e_{12}$$

$$y_{13} = \mu + \tau_1 + \alpha_3 + e_{13}$$

$$y_{21} = \mu + \tau_2 + \alpha_1 + e_{21}$$

$$y_{22} = \mu + \tau_2 + \alpha_2 + e_{22}$$

$$y_{23} = \mu + \tau_2 + \alpha_3 + e_{23}$$

Here, we have six parameters. The matrix A has 0's and 1's only (presence or absence of a parameter in the model equations), and the six normal equations are

$$6\mu + 3\tau_1 + 3\tau_2 + 2\alpha_1 + 2\alpha_2 + 2\alpha_3 = \sum_{i,j} y_{ij} \text{ (adding for } \mu)$$

$$3\mu + 3\tau_1 \qquad + \alpha_1 + \alpha_2 + \alpha_3 = \sum_{j} y_{1j} \text{ (adding for } \tau_1)$$

$$3\mu \qquad + 3\tau_2 + \alpha_1 + \alpha_2 + \alpha_3 = \sum_{j} y_{2j} \text{ (adding for } \tau_2)$$

$$2\mu + \tau_1 + \tau_2 + 2\alpha_1 \qquad = \sum_{i} y_{i1} \text{ (adding for } \alpha_1)$$

$$2\mu + \tau_1 + \tau_2 \qquad + 2\alpha_2 \qquad = \sum_{i} y_{i2} \text{ (adding for } \alpha_2)$$

$$2\mu + \tau_1 + \tau_2 \qquad + 2\alpha_3 = \sum_{i} y_{i3} \text{ (adding for } \alpha_3)$$

The estimable parametric functions are

$$c_1(6\mu + 3\tau_1 + 3\tau_2 + 2\alpha_1 + 2\alpha_2 + 2\alpha_3)$$
$$+ c_2(3\mu + 3\tau_1 + \alpha_1 + \alpha_2 + \alpha_3) + c_3(3\mu + 3\tau_2 + \alpha_1 + \alpha_2 + \alpha_3)$$
$$+ c_4(2\mu + \tau_1 + \tau_2 + 2\alpha_1) + c_5(2\mu + \tau_1 + \tau_2 + 2\alpha_2)$$
$$+ c_6(2\mu + \tau_1 + \tau_2 + 2\alpha_3)$$

i.e.

$$(6c_1 + 3c_2 + 3c_3 + 2c_4 + 2c_5 + 2c_6)\mu + (3c_1 + 3c_2 + c_4 + c_5 + c_6)\tau_1$$
$$+(3c_1 + 3c_3 + c_4 + c_5 + c_6)\tau_2 + (2c_1 + c_2 + c_3 + 2c_4)\alpha_1$$
$$+ (2c_1 + c_2 + c_3 + 2c_5)\alpha_2 + (2c_1 + c_2 + c_3 + 2c_6)\alpha_3$$

i.e.

$$a_0\mu + a_1\tau_1 + a_2\tau_2 + b_1\alpha_1 + b_2\alpha_2 + b_3\alpha_3$$

with

$$a_0 = a_1 + a_2 = b_1 + b_2 + b_3$$

In particular, the contrasts $l_1\tau_1 + l_2\tau_2$ $(l_1 + l_2 = 0)$ and $m_1\alpha_1 + m_2\alpha_2 + m_3\alpha_3$ $(m_1 + m_2 + m_3 = 0)$ are estimable. Using the second and third equations, the best estimate of the contrast $l_1\tau_1 + l_2\tau_2$ is $\frac{1}{3}[l_1 \sum_j y_{1j} + l_2 \sum_j y_{2j}]$. Similarly, using the last three equations the best estimate of the contrast $m_1\alpha_1 + m_2\alpha_2 + m_3\alpha_3$ is

$$\tfrac{1}{2}[m_1 \sum_i y_{i1} + m_2 \sum_i y_{i2} + m_3 \sum_i y_{i3}]$$

A general contrast

$$a_0\mu + a_1\tau_1 + a_2\tau_2 + b_1\alpha_1 + b_2\alpha_2 + b_3\alpha_3,$$

where,

$$a_0 + a_1 + a_2 + b_1 + b_2 + b_3 = 0$$

is estimable if

$$a_0 = a_1 + a_2 = b_1 + b_2 + b_3$$

that is, if

$$a_0 = 0,\ a_1 + a_2 = 0,\ b_1 + b_2 + b_3 = 0$$

i.e., if it is a contrast composed of contrasts involving τ_1 and τ_2 and contrasts involving α_1, α_2 and α_3 alone.

The functions μ, $\tau_1 + \tau_2$ and $\alpha_1 + \alpha_2 + \alpha_3$ are not estimable.

4.6 Variances and Covariances of Estimates

If the rank of the coefficient matrix **A** is p, the matrix **A′A** is non-singular and the normal equations

$$\mathbf{A'A\beta} = \mathbf{A'y} \tag{4.14}$$

have a unique solution

$$\hat{\boldsymbol{\beta}} = (\mathbf{A'A})^{-1}\,\mathbf{A'y} \tag{4.15}$$

In this case the parameters $\beta_1, \ldots, \beta_p$ are estimable and consequently all linear parametric functions are estimable. The dispersion matrix of the estimates $\hat{\boldsymbol{\beta}}$ of $\boldsymbol{\beta}$ is then given by

$$D(\hat{\boldsymbol{\beta}}) = (\mathbf{A'A})^{-1}\,\mathbf{A'}\,D(\mathbf{y})\,\mathbf{A}(\mathbf{A'A})^{-1} = (\mathbf{A'A})^{-1}\,\sigma^2 \tag{4.16}$$

It follows that if $\boldsymbol{\lambda}'\hat{\boldsymbol{\beta}}$ and $\boldsymbol{\mu}'\hat{\boldsymbol{\beta}}$ are the estimates of $\boldsymbol{\lambda}'\boldsymbol{\beta}$ and $\boldsymbol{\mu}'\boldsymbol{\beta}$ we have

$$\text{Var}\,(\boldsymbol{\lambda}'\hat{\boldsymbol{\beta}}) = \boldsymbol{\lambda}'\,D(\hat{\boldsymbol{\beta}})\,\boldsymbol{\lambda} = \sigma^2\,[\boldsymbol{\lambda}'\,(\mathbf{A'A})^{-1}\,\boldsymbol{\lambda}] \tag{4.17}$$

and,

$$\begin{aligned}\text{Cov}\,(\boldsymbol{\lambda}'\hat{\boldsymbol{\beta}}, \boldsymbol{\mu}'\hat{\boldsymbol{\beta}}) &= \boldsymbol{\lambda}'\,D(\hat{\boldsymbol{\beta}})\,\boldsymbol{\mu} \\ &= \sigma^2\,[\boldsymbol{\lambda}'\,(\mathbf{A'A})^{-1}\,\boldsymbol{\mu}] = \sigma^2\,[\boldsymbol{\mu}'\,(\mathbf{A'A})^{-1}\,\boldsymbol{\lambda}]\end{aligned} \tag{4.18}$$

However, if the rank of the coefficient matrix $\mathbf{A}$ is $r\ (< p)$, the matrix $\mathbf{A'A}$ is singular and the normal equations do not have a unique solution. If $\hat{\boldsymbol{\beta}}$ is any solution of the normal equations we may write

$$\hat{\boldsymbol{\beta}} = \mathbf{BA'y} \tag{4.19}$$

We now show that results (4.17) and (4.18) still hold provided $(\mathbf{A'A})$ is replaced by $\mathbf{B}$. (The matrix $\mathbf{B}$ is a generalised inverse of $\mathbf{A'A}$.) We have from (4.14) and (4.19)

$$\mathbf{A'A\,BA'y} = \mathbf{A'y} \tag{4.20}$$

Taking expectations on both sides we get

$$\mathbf{A'ABA'A\beta} = \mathbf{A'A\beta} \tag{4.21}$$

for all $\boldsymbol{\beta} \in R_p$. Hence, $\mathbf{B}$ satisfies the relation

$$\mathbf{A'ABA'A} = \mathbf{A'A} \tag{4.22}$$

Now let $\boldsymbol{\lambda}'\boldsymbol{\beta}$, $\boldsymbol{\mu}'\boldsymbol{\beta}$ be estimable functions; we can then write $\boldsymbol{\lambda}' = \mathbf{a'A'A}$, $\boldsymbol{\mu}' = \mathbf{b'A'A}$ for some $\mathbf{a}$ and $\mathbf{b}$. The best estimates are $\boldsymbol{\lambda}'\hat{\boldsymbol{\beta}}$ and $\boldsymbol{\mu}'\hat{\boldsymbol{\beta}}$. We have

$$\begin{aligned}\text{Var}(\boldsymbol{\lambda}'\hat{\boldsymbol{\beta}}) &= \boldsymbol{\lambda}'\,\mathbf{D}\,(\hat{\boldsymbol{\beta}})\,\boldsymbol{\lambda} \\ &= \boldsymbol{\lambda}'\,\mathbf{BA}'\,\mathbf{D}(\mathbf{y})\,\mathbf{AB}'\boldsymbol{\lambda} \\ &= \sigma^2\,[\boldsymbol{\lambda}'\mathbf{BA}'\mathbf{AB}'\boldsymbol{\lambda}] \\ &= \sigma^2\,[\mathbf{a}'\mathbf{A}'\mathbf{ABA}'\mathbf{AB}'\boldsymbol{\lambda}] \\ &= \sigma^2\,[\mathbf{a}'\mathbf{A}'\mathbf{AB}'\boldsymbol{\lambda}] \qquad \text{(from 4.22)} \\ &= \sigma^2\,[\boldsymbol{\lambda}'\mathbf{B}'\boldsymbol{\lambda}] = \sigma^2\,[\boldsymbol{\lambda}'\mathbf{B}\boldsymbol{\lambda}] \end{aligned} \tag{4.23}$$

since $\boldsymbol{\lambda}'\mathbf{B}'\boldsymbol{\lambda}$ being a scalar is equal to its transpose.

Similarly, we get

$$\text{Cov}(\boldsymbol{\lambda}'\hat{\boldsymbol{\beta}}, \boldsymbol{\mu}'\hat{\boldsymbol{\beta}}) = \sigma^2\,(\boldsymbol{\lambda}'\mathbf{B}\boldsymbol{\mu}) = \sigma^2\,(\boldsymbol{\mu}'\mathbf{B}\boldsymbol{\lambda}) \tag{4.24}$$

Thus $\sigma^2\mathbf{B}$ may be formally considered as the dispersion matrix of $\hat{\boldsymbol{\beta}}$ so long as Eqs. (4.23) and (4.24) are applied to estimates of estimable functions only.

4.7 Theory of Least Squares

Consider the expression

$$(\mathbf{y} - \mathbf{A}\boldsymbol{\beta})'(\mathbf{y} - \mathbf{A}\boldsymbol{\beta}) = \sum_{i=1}^{n} (y_i - a_{i1}\beta_1 - \ldots - a_{ip}\beta_p)^2 \tag{4.25}$$

This is the sum of the squares of the deviations of the observations y_i from their expected values $a_{i1}\beta_1 + \ldots + a_{ip}\beta_p$. Let us try to obtain the values of the parameters β_i which make this sum a minimum. If $\mathbf{y}_0$ is the projection of $\mathbf{y}$ on $C(\mathbf{A})$, $(\mathbf{y} - \mathbf{y}_0) \perp C(\mathbf{A})$. We can write

$$\mathbf{y} - A\boldsymbol{\beta} = (\mathbf{y} - \mathbf{y}_0) + (\mathbf{y}_0 - A\boldsymbol{\beta})$$

Now, $\mathbf{y}_0$ and $A\boldsymbol{\beta}$ both belong to $C(\mathbf{A})$, so that

$$(\mathbf{y} - \mathbf{y}_0) \perp (\mathbf{y}_0 - A\boldsymbol{\beta})$$

Hence,

$$\begin{aligned}(\mathbf{y} - \mathbf{A}\boldsymbol{\beta})'(\mathbf{y} - \mathbf{A}\boldsymbol{\beta}) &= (\mathbf{y} - \mathbf{y}_0)'(\mathbf{y} - \mathbf{y}_0) + (\mathbf{y}_0 - \mathbf{A}\boldsymbol{\beta})'(\mathbf{y}_0 - \mathbf{A}\boldsymbol{\beta}) \\ &\geqslant (\mathbf{y} - \mathbf{y}_0)'(\mathbf{y} - \mathbf{y}_0)\end{aligned}$$

Thus, the minimum value of (4.25) is obtained if $\boldsymbol{\beta}$ is so chosen that $\mathbf{A}\boldsymbol{\beta}$ is equal to the projection $\mathbf{y}_0$ of $\mathbf{y}$ on $C(\mathbf{A})$, i.e. $\boldsymbol{\beta}$ satisfies

$$\mathbf{A}'\mathbf{A}\boldsymbol{\beta} = \mathbf{A}'\mathbf{y} \tag{4.26}$$

Thus we get the normal equations obtained earlier. If $\hat{\boldsymbol{\beta}}$ is any solution of these equations the minimum value of $(\mathbf{y} - \mathbf{A}\boldsymbol{\beta})' (\mathbf{y} - \mathbf{A}\boldsymbol{\beta})$ is $(\mathbf{y} - \mathbf{A}\hat{\boldsymbol{\beta}})' (\mathbf{y} - \mathbf{A}\hat{\boldsymbol{\beta}})$. The minimum value so obtained is unique, i.e. it does not depend on the particular solution $\hat{\boldsymbol{\beta}}$. This follows from the fact that $\mathbf{A}\boldsymbol{\beta}$ is a set of estimable functions and $\mathbf{A}\hat{\boldsymbol{\beta}}$ being the set of best estimates of $\mathbf{A}\boldsymbol{\beta}$ is the same no matter which solution $\hat{\boldsymbol{\beta}}$ of the normal equations is used.

If the matrix $\mathbf{A}'\mathbf{A}$, a square matrix of order p, is non-singular, the unique solution

$$\hat{\boldsymbol{\beta}} = (\mathbf{A}'\mathbf{A})^{-1} \mathbf{A}'\mathbf{y} \tag{4.27}$$

of the normal equations is called the *least squares estimate* of $\boldsymbol{\beta}$. If $\hat{\boldsymbol{\beta}}' = (\hat{\beta}_1, \ldots, \hat{\beta}_p)$, then $\hat{\beta}_i$ is called the least squares estimate of β_i. Thus we see that the least squares estimate is the same as the best linear unbiased estimate.

If $\mathbf{A}'\mathbf{A}$ is singular, there are an infinity of solutions of the normal equations and we cannot talk of the least squares estimate of $\boldsymbol{\beta}$. However, we have determined earlier that if $\boldsymbol{\lambda}'\boldsymbol{\beta}$ is estimable, $\boldsymbol{\lambda}'\hat{\boldsymbol{\beta}}$ is unique for all solutions $\hat{\boldsymbol{\beta}}$ of the normal equations. We may, therefore, call $\boldsymbol{\lambda}'\hat{\boldsymbol{\beta}}$ the least squares estimate of $\boldsymbol{\lambda}'\boldsymbol{\beta}$ whenever the latter is estimable. (It can also be shown that if $\boldsymbol{\lambda}'\hat{\boldsymbol{\beta}}$ is the same for all solutions $\hat{\boldsymbol{\beta}}$ of the normal equations, $\boldsymbol{\lambda}'\boldsymbol{\beta}$ is estimable.)

The vector $\mathbf{y} - \mathbf{A}\hat{\boldsymbol{\beta}}$ is called the *residual vector* and the quantity $(\mathbf{y} - \mathbf{A}\hat{\boldsymbol{\beta}})' (\mathbf{y} - \mathbf{A}\hat{\boldsymbol{\beta}})$, which is the minimum value of the sum of squares of deviations of the observations from their expected values, is called the *residual sum of squares*. As already discussed the residual vector and the residual sum of squares do not depend on the particular solution $\hat{\boldsymbol{\beta}}$ used.

Theorem 4.10: $E(\mathbf{y} - \mathbf{A}\hat{\boldsymbol{\beta}}) = \mathbf{0}$.

Proof: Consider the ith component

$$y_i - a_{i1}\hat{\beta}_1 - \ldots - a_{ip}\hat{\beta}_p$$

of $y - \mathbf{A}\boldsymbol{\beta}$. Since $a_{i1}\beta_1 + \ldots + a_{ip}\beta_p$ is estimable, $a_{i1}\hat{\beta}_1 + \ldots + a_{ip}\hat{\beta}_p$ is its best estimate. Thus

$$E(a_{i1}\hat{\beta}_1 + \ldots + a_{ip}\hat{\beta}_p) = a_{i1}\beta_1 + \ldots + a_{ip}\beta_p = E(y_i)$$

Hence

$$E(\mathbf{y} - \mathbf{A}\hat{\boldsymbol{\beta}}) = \mathbf{0} \tag{4.28}$$

Theorem 4.10 says that the n components of the residual vector are all error functions. These are in fact all the error functions essentially, in the sense that, every error function is a linear combination of the error functions obtained from the residual vector. This is proved by the following theorem.

Theorem 4.11: The set $\mathbf{y} - \mathbf{A}\hat{\boldsymbol{\beta}}$ contains $n - r$ linearly independent linear functions of $y_1, \ldots, y_r$, and no more, where r is the rank of $\mathbf{A}$.

Proof: Putting $\mathbf{A}\hat{\boldsymbol{\beta}} = \mathbf{B}\mathbf{y}$ we have

$$\mathbf{y} - \mathbf{A}\hat{\boldsymbol{\beta}} = (\mathbf{I} - \mathbf{B})\mathbf{y} \tag{4.29}$$

Thus, it is enough to prove that rank $(\mathbf{I} - \mathbf{B}) = n - r$.

Now $\mathbf{A}\boldsymbol{\beta}$ is a set of estimable functions of which r are linearly independent. Hence, from theorem 4.8, the set $\mathbf{A}\hat{\boldsymbol{\beta}}$ of best estimates of $\mathbf{A}\boldsymbol{\beta}$ has r linearly independent estimates, i e. rank $\mathbf{B} = r$. Now $\mathbf{I} = (\mathbf{I} - \mathbf{B}) + \mathbf{B}$, so that $n \leqslant$ rank $(\mathbf{I} - \mathbf{B})$ + rank $\mathbf{B}$, i.e. rank $(\mathbf{I} - \mathbf{B}) \geqslant n - r$.

The error space has dimension $n - r$ so that the space of error functions also has dimension $n - r$.

Since $\mathbf{y} - \mathbf{A}\hat{\boldsymbol{\beta}} = (\mathbf{I} - \mathbf{B})\mathbf{y}$ are error functions we must have

$$\text{rank } (\mathbf{I} - \mathbf{B}) \leqslant n - r$$

and the theorem is proved.

Thus we find that if $\hat{\boldsymbol{\beta}}$ is any solution of the normal equations, all the error functions can be obtained from $\mathbf{y} - \mathbf{A}\hat{\boldsymbol{\beta}}$, and all the best estimates from $\mathbf{A}\hat{\boldsymbol{\beta}}$. In other words, every error function is of the form $\mathbf{c}'(\mathbf{y} - \mathbf{A}\hat{\boldsymbol{\beta}})$, and every best estimate of the form $\mathbf{d}'A\hat{\boldsymbol{\beta}}$ where $\mathbf{c}$ and $\mathbf{d}$ are arbitrary.

4.8 Linear Model with Correlated Observations

We now consider the model

$$\mathbf{y} = \mathbf{A}\boldsymbol{\beta} + \mathbf{e} \tag{4.30}$$

where the random vector $\mathbf{e}$ is normally distributed with expectation vector $\mathbf{0}$ and dispersion matrix $\boldsymbol{\Sigma}$. It follows that $\mathbf{y}$ is normally distributed with $E(\mathbf{y}) = \mathbf{A}\boldsymbol{\beta}$, $\mathbf{D}(\mathbf{y}) = \boldsymbol{\Sigma}$. So far we had taken $\boldsymbol{\Sigma} = \sigma^2\mathbf{I}$ so that the e_i, and hence the y_i, were independent random variables. The random variables y_i are no longer independent, and the problem of estimating linear parametric functions has to be examined afresh.

As before, we call $\boldsymbol{\lambda}'\boldsymbol{\beta}$ estimable if there exists a linear function $\mathbf{l}'\mathbf{y}$ of the observations such that

$$E(\mathbf{l}'\mathbf{y}) = \mathbf{l}'\mathbf{A}\boldsymbol{\beta} = \boldsymbol{\lambda}'\boldsymbol{\beta} \tag{4.31}$$

It is clear that $\boldsymbol{\lambda}'\boldsymbol{\beta}$ is estimable if, and only if, $\boldsymbol{\lambda} \in R(\mathbf{A})$.

We assume that Σ is non-singular which means that it is positive definite, since it is a variance-covariance matrix. Thus we can write

$$\boldsymbol{\Sigma} = \mathbf{H}'\mathbf{H} \tag{4.32}$$

for some non-singular matrix $\mathbf{H}$. Putting

$$\mathbf{y} = \mathbf{H}'\mathbf{z} \quad \text{or} \quad \mathbf{z} = \mathbf{H}'^{-1}\mathbf{y} \tag{4.34}$$

we get

$$E(\mathbf{z}) = \mathbf{H}'^{-1}\, E(\mathbf{y}) = \mathbf{H}'^{-1}\mathbf{A}\boldsymbol{\beta} \tag{4.34}$$

$$\mathbf{D}(\mathbf{z}) = \mathbf{H}'^{-1}\,\mathbf{D}(\mathbf{y})\mathbf{H}^{-1} = \mathbf{H}'^{-1}\boldsymbol{\Sigma}\mathbf{H}^{-1} = \mathbf{I}$$

Thus the variables $\mathbf{z}$ are uncorrelated and all have variance equal to one, so that the earlier theory can be applied to $\mathbf{z}$.

The normal equations using $\mathbf{z}$ are

$$(\mathbf{H}'^{-1}\mathbf{A})'\,(\mathbf{H}'^{-1}\mathbf{A})\boldsymbol{\beta} = (\mathbf{H}'^{-1}\mathbf{A})'\mathbf{z} \tag{4.36}$$

i.e.

$$\mathbf{A}'\boldsymbol{\Sigma}^{-1}\mathbf{A}\boldsymbol{\beta} = \mathbf{A}'\mathbf{H}^{-1}\mathbf{z} \tag{4.37}$$

Since

$$\mathbf{A}'\mathbf{H}^{-1}\mathbf{z} = \mathbf{A}'\mathbf{H}^{-1}\mathbf{H}'^{-1}\mathbf{y} = \mathbf{A}'\boldsymbol{\Sigma}^{-1}\mathbf{y} \tag{4.38}$$

we get from (4.37)

$$\mathbf{A}'\boldsymbol{\Sigma}^{-1}\mathbf{A}\boldsymbol{\beta} = \mathbf{A}'\boldsymbol{\Sigma}^{-1}\mathbf{y} \tag{4.39}$$

which we call the normal equations for the correlated model. The normal equations (4.39) can also be obtained, using least squares theory, by minimizing

$$(\mathbf{z} - \mathbf{H}'^{-1}\mathbf{A}\boldsymbol{\beta})' (\mathbf{z} - \mathbf{H}'^{-1}\mathbf{A}\boldsymbol{\beta})$$

that is

$$[\mathbf{H}'^{-1}(\mathbf{y} - \mathbf{A}\boldsymbol{\beta})]' [\mathbf{H}'^{-1}(\mathbf{y} - \mathbf{A}\boldsymbol{\beta})]$$

$$(\mathbf{y} - \mathbf{A}\boldsymbol{\beta})' \boldsymbol{\Sigma}^{-1} (\mathbf{y} - \mathbf{A}\boldsymbol{\beta}) \tag{4.40}$$

The values of $\boldsymbol{\beta}$ which minimize (4.40) are the solutions of the normal equations (4.39).

Using the normal equations (4.39) we see that every estimable function is of the form

$$\mathbf{c}'\mathbf{A}'\boldsymbol{\Sigma}^{-1}\mathbf{A}\boldsymbol{\beta} \tag{4.41}$$

and its best estimate is

$$\mathbf{c}'\mathbf{A}'\mathbf{H}^{-1}\mathbf{z} \quad \text{or} \quad \mathbf{c}'\mathbf{A}'\boldsymbol{\Sigma}^{-1}\mathbf{y} \tag{4.42}$$

where $\mathbf{c}$ is arbitrary. We see that in the case of correlated observations the best estimates depend on the variance-covariance matrix $\boldsymbol{\Sigma}$.

If $\boldsymbol{\lambda}'\boldsymbol{\beta}$ is estimable it can be written as $\mathbf{c}'\mathbf{A}'\boldsymbol{\Sigma}^{-1}\mathbf{A}\boldsymbol{\beta}$ for some $\mathbf{c}$. Now if $\hat{\boldsymbol{\beta}}$ is any solution of the normal equations (4.39) we have

$$\boldsymbol{\lambda}'\hat{\boldsymbol{\beta}} = \mathbf{c}'\mathbf{A}'\boldsymbol{\Sigma}^{-1}\mathbf{A}\hat{\boldsymbol{\beta}} = \mathbf{c}'\mathbf{A}'\boldsymbol{\Sigma}^{-1}\mathbf{y} \tag{4.43}$$

which is the best estimate of $\boldsymbol{\lambda}'\boldsymbol{\beta}$.

In this case too we call $\mathbf{l}'\mathbf{y}$ an error function if $E(\mathbf{l}'\mathbf{y}) = 0$ for all $\boldsymbol{\beta}$. Evidently, $\mathbf{l}'\mathbf{y}$ is an error function if, and only if, $\mathbf{l}'\mathbf{A} = \mathbf{0}$. If r is the rank of A, there are n-r linearly independent error functions.

Let $\mathbf{l}'\mathbf{y}$ be an error function and $\mathbf{c}'\mathbf{A}'\boldsymbol{\Sigma}^{-1}\mathbf{y}$ the best estimate of an estimable function. Then

$$\text{Cov}\,(\mathbf{l}'\mathbf{y}, \mathbf{c}'\mathbf{A}'\boldsymbol{\Sigma}^{-1}\mathbf{y}) = \mathbf{l}'D(\mathbf{y})\boldsymbol{\Sigma}^{-1}\mathbf{A}\mathbf{c} = \mathbf{l}'\boldsymbol{\Sigma}\boldsymbol{\Sigma}^{-1}\mathbf{A}\mathbf{c} = \mathbf{l}'\mathbf{A}\mathbf{c} = 0$$

so that, as before, the error functions are uncorrelated with the estimates.

COMPLEMENTS AND PROBLEMS

1. In the linear model let $\mathbf{c}_1, \mathbf{c}_2, \ldots, \mathbf{c}_p$ denote the column vectors of the coefficient matrix A.

   (a) β_1 is estimable if, and only if, $\mathbf{c}_1$ does not belong to the vector space spanned by the remaining columns of A.

   (b) If β_1 is estimable its best estimate is

$$(\mathbf{c}_1 - \mathbf{d}_1)'\mathbf{y}/(\mathbf{c}_1 - \mathbf{d}_1)' (\mathbf{c}_1 - \mathbf{d}_1)$$

where $\mathbf{d}_1$ is the orthogonal projection of $\mathbf{c}_1$ on Sp $\{\mathbf{c}_2, \ldots, \mathbf{c}_p\}$. In particular, if $\mathbf{c}_1$ is orthogonal to $\mathbf{c}_2, \ldots, \mathbf{c}_p$, the best estimate of β_1 is $(\mathbf{c}_1' \mathbf{y})/(\mathbf{c}_1' \mathbf{c}_1)$.

(c) More generally, β_1 and β_2 are estimable if, and only if, $\mathbf{c}_1 - \mathbf{d}_1$ and $\mathbf{c}_2 - \mathbf{d}_2$ are linearly independent, where $\mathbf{d}_1$, $\mathbf{d}_2$ are respectively the orthogonal projections of $\mathbf{c}_1$, $\mathbf{c}_2$ on Sp $\{\mathbf{c}_3, \ldots, \mathbf{c}_p\}$.

(d) $\lambda_1 \beta_1 + \lambda_2 \beta_2$ is estimable if, and only if, $\mathbf{c}_1$, $\mathbf{c}_2$ do not belong to Sp $\{\lambda_2 \mathbf{c}_1 - \lambda_1 \mathbf{c}_2, \mathbf{c}_3, \ldots, \mathbf{c}_p\}$.

2. (a) If $\boldsymbol{\rho}$ denotes the vector whose components are the row totals of A, and $\boldsymbol{\alpha}$ the orthogonal projection of the vector (1, 1, ..., 1) on C(A), a necessary and sufficient condition for contrasts, and contrasts alone, to be estimated by contrasts, is $\boldsymbol{\rho} = a\boldsymbol{\alpha}$ for some $a \neq 0$.

(b) If the row totals of A are all equal, then contrasts, and contrasts alone, are estimated by contrasts. (This is a sufficient but not a necessary condition.) This condition is satisfied by the coefficient matrix of all standard design models.

3. *Effect of additional observations.* Suppose original observations are given by the model (called model A)

$$\mathbf{y}_1 = \mathbf{A}_1 \boldsymbol{\beta} + \mathbf{e}_1,$$

and the additional observations by the model (called model B)

$$\mathbf{y}_2 = \mathbf{A}_2 \boldsymbol{\beta} + \mathbf{e}_2$$

Taking all the variables together we get the model (called model C)

$$\begin{bmatrix} \mathbf{y}_1 \\ \cdots \\ \mathbf{y}_2 \end{bmatrix} = \begin{bmatrix} \mathbf{A}_1 \\ \cdots \\ \mathbf{A}_2 \end{bmatrix} \boldsymbol{\beta} + \begin{bmatrix} \mathbf{e}_1 \\ \cdots \\ \mathbf{e}_2 \end{bmatrix}$$

If $\boldsymbol{\lambda}' \boldsymbol{\beta}$ is estimable in A it is also estimable in C. Suppose $\mathbf{l}_1' \mathbf{y}_1$ is the best estimator of $\boldsymbol{\lambda}'\boldsymbol{\beta}$ in A then $\mathbf{l}_1' \mathbf{y}_1$ is also an estimator of $\boldsymbol{\lambda}'\boldsymbol{\beta}$ in C though it may not be the best.

If $\mathbf{l}_1' \mathbf{y}_1$ and $\mathbf{l}_2' \mathbf{y}_2$ are error functions in A an B respectively, then

$$\mathbf{l}_1' \mathbf{y}_1 + \mathbf{l}_2' \mathbf{y}_2$$

is an error function in C. If all error functions in C are of this form, then the best estimator of $\boldsymbol{\lambda}' \boldsymbol{\beta}$ in A continues to be the best estimator of $\boldsymbol{\lambda}'\boldsymbol{\beta}$ in C. The condition which ensures that the estimators of all the estimable functions in A do not change with the introduction of additional observations is thus seen to be that the error degrees of freedom in C are equal to the sum of the error degrees of freedom in A and B. Or, equivalently, the condition is that the row spaces of $\mathbf{A}_1$ and $\mathbf{A}_2$ be disjoint. For example, if we take model A as

$$y_1 = \beta_1 + \beta_2 + e_1, \; y_2 = \beta_1 + 2\beta_2 + e_2$$

and model B as

$$y_3 = \beta_2 + \beta_3 + e_3,\ y_4 = 2\beta_2 + \beta_3 + e_4$$

then error degrees of freedom in models A, B and C are 0, 0, 1 respectively. The estimate of β_1 is $2y_1 - y_2$ in model A. but changes to $\frac{5}{4}y_1 - \frac{1}{4}y_2 + \frac{3}{4}y_3 - \frac{3}{4}y_4$ in model C. Thus, even though β_1 is not estimable in the additional observations represented by model B, the use of the additional observations has changed the estimate of β_1.

4. *Two sets of parameters.* Suppose the parameters $\boldsymbol{\beta}$ are divided into two sets $\boldsymbol{\beta}_1$ and $\boldsymbol{\beta}_2$ and the linear model $\mathbf{y} = \mathbf{A}\boldsymbol{\beta} + \mathbf{e}$ is formulated accordingly as

$$\mathbf{y} = \mathbf{A}_1\boldsymbol{\beta}_1 + \mathbf{A}_2\boldsymbol{\beta}_2 + \mathbf{e}$$

If the function $\boldsymbol{\lambda}_1'\boldsymbol{\beta}_1$, involving only the parameters from the set $\boldsymbol{\beta}_1$, is estimable in this model, then it is also estimable in the restricted model $\mathbf{y} = \mathbf{A}_1\boldsymbol{\beta}_1 + \mathbf{e}$. However, the converse is not true. For example, α_1 and α_2 are not estimable in the model $y_1 = \mu + \alpha_1 + e_1$, $y_2 = \mu + \alpha_2 + e_2$, but are estimable in the restricted model $y_1 = \alpha_1 + e_1$, $y_2 = \alpha_2 + e_2$.

The function $\mathbf{l}'\mathbf{y}$ is an error function in the model $\mathbf{y} = \mathbf{A}\boldsymbol{\beta} + \mathbf{e}$ if, and only if, it is an error function in both the restricted models $\mathbf{y} = \mathbf{A}_1\boldsymbol{\beta}_1 + \mathbf{e}$, and $\mathbf{y} = \mathbf{A}_2\boldsymbol{\beta}_2 + \mathbf{e}$.

Let the estimation space in $\mathbf{y} = \mathbf{A}\boldsymbol{\beta} + \mathbf{e}$ be denoted by V, and in $\mathbf{y} = \mathbf{A}_1\boldsymbol{\beta}_1 + \mathbf{e}$ by V_1. Let V_0 denote the orthogonal complement of V_1 in V. Then the coefficient vectors of the best estimates of the estimable functions of the form $\boldsymbol{\lambda}_2'\boldsymbol{\beta}_2$ (i.e., not involving parameters from the set $\boldsymbol{\beta}_1$) belong to V_0 and conversely. (By the orthogonal complement of V_1 in V we mean the set of all vectors of V which are orthogonal to V_1)

5. The following are equivalent

(i) $\boldsymbol{\lambda}'\boldsymbol{\beta}$ is estimable

(ii) $\boldsymbol{\lambda}'\hat{\boldsymbol{\beta}}$ is invariant for all solutions $\hat{\boldsymbol{\beta}}$ of the normal equations

$$\mathbf{A}'\mathbf{A}\boldsymbol{\beta} = \mathbf{A}'\mathbf{y}.$$

Since $\boldsymbol{\lambda} \in R(\mathbf{A}) = R(\mathbf{A}'\mathbf{A})$, we have $\boldsymbol{\lambda}' = \mathbf{c}'\mathbf{A}'\mathbf{A}$ for some $\mathbf{c} \in R_p$

Hence, for any solution $\hat{\boldsymbol{\beta}}$ of the normal equations

$$\boldsymbol{\lambda}'\hat{\boldsymbol{\beta}} = \mathbf{c}'\mathbf{A}'\mathbf{A}\hat{\mathbf{B}} = \mathbf{c}'\mathbf{A}'\mathbf{y}$$

which does not depend on the particular solution $\hat{\boldsymbol{\beta}}$.

Conversely, let (ii) hold. If $\hat{\boldsymbol{\beta}}$ is a solution of the normal equations, and $\mathbf{A}'\mathbf{A}\mathbf{c} = 0$, then $\hat{\boldsymbol{\beta}} - \mathbf{c}$ is also a solution. Hence, $\boldsymbol{\lambda}'\hat{\boldsymbol{\beta}} = \boldsymbol{\lambda}'(\hat{\boldsymbol{\beta}} - \mathbf{c})$ Thus $\boldsymbol{\lambda}'\mathbf{c} = 0$ whenever $\mathbf{A}'\mathbf{A}\mathbf{c} = 0$, that is $\mathbf{c} \in R(\mathbf{A}'\mathbf{A})^\perp = R(\mathbf{A})^\perp \Rightarrow \mathbf{c} \perp \boldsymbol{\lambda}$. Therefore $\boldsymbol{\lambda} \in R(\mathbf{A})$.

Since $\mathbf{A}\boldsymbol{\beta}$ is a set of estimable parametric functions, it follows from the above that $(\mathbf{y} - \mathbf{A}\hat{\boldsymbol{\beta}})'(\mathbf{y} - \mathbf{A}\hat{\boldsymbol{\beta}})$ is invariant for all solutions $\hat{\boldsymbol{\beta}}$ of the normal equations.

6. The sum of squares of the deviations of the observations y_i from the expected values is given by

$$S = \sum_{i=1}^{n} (y_i - a_{i1}\beta_1 \ldots a_{ip}\beta_p)^2 = (\mathbf{y} - \mathbf{A}\boldsymbol{\beta})'(\mathbf{y} - \mathbf{A}\boldsymbol{\beta})$$

Equating the partial derivatives of S with respect to the β_j's to zero we get the equations

$$\sum_{i=1}^{n} a_{ij}(y_i - a_{i1}\beta_1 \ldots a_{ip}\beta_p) = 0, \; j = 1, \ldots, p$$

which are the normal equations,

$$\mathbf{A}'\mathbf{A}\boldsymbol{\beta} = \mathbf{A}'\mathbf{y}$$

It can now be shown that the minimum value of S over the β_j's is given by

$$(\mathbf{y} - \mathbf{A}\hat{\boldsymbol{\beta}})'(\mathbf{y} - \mathbf{A}\hat{\boldsymbol{\beta}})$$

7. In the linear model (which arises in the "method of paired comparisons")

$$y_1 = \mu_1 - \mu_2 + e_1$$
$$y_2 = \mu_2 - \mu_3 + e_2$$
$$\ldots\ldots\ldots\ldots\ldots\ldots$$
$$y_n = \mu_n - \mu_1 + e_n,$$

the mean $\bar{y}.$ is the only error function; all elementary contrasts are estimable. Since $y_j - \bar{y}.$ is a contrast, it is orthoganal to $\bar{y}.$, and hence is the best estimate of its expected value $\mu_j - \mu_{j+1}$ ($j = 1, \ldots, n$; $\mu_{n+1} = \mu_1$). The best estimate of a general contrast $\Sigma t_j \mu_j$ may be written as

$$t_1(y_1 - \bar{y}.) + (t_1 + t_2)(y_2 - \bar{y}.) + \ldots + (t_1 + \ldots + t_{n-1})(y_{n-1} - \bar{y}.).$$

A solution of the normal equations is given by

$$\boldsymbol{\mu} = \left(\frac{n+1}{2}\bar{y}.\right)\mathbf{i} - \frac{1}{n}\mathbf{N}\mathbf{y},$$

where $\boldsymbol{\mu}' = (\mu_1, \ldots, \mu_n)$, $\mathbf{i}' = (1, 1, \ldots, 1)$, and $\mathbf{N}$ is the matrix

$$\begin{vmatrix} 1 & 2 & 3 & - & - & - & n \\ n & 1 & 2 & - & - & - & n-1 \\ - & - & - & - & - & - & - \\ 2 & 3 & 4 & - & - & - & 1 \end{vmatrix}$$

REFERENCES

The concepts of error space, estimation space, and estimable functions are due to Bose (1944). See also Bose (1949 a), Scheffe (1959), Chakrabarti (1962), Rao (1965).

For a different approach using generalised inverses of matrics see Rao (1971), Rao and Mitra (1971).

The model with correlated observations is discussed in Aitken (1933, 1935). See also McElroy (1967), Watson (1967), Kruskal (1968), Rao and Mitra (1969). Atiqullah (1969) discusses the situation in which there are restrictions on the parameters.

A special algorithm for solving the normal equations is given in Bradu (1965). Hill (1969) discusses least squares from a Bayesian viewpoint.

5

Sums of Squares

5.1 Introduction

In the linear model

$$\mathbf{y} = \mathbf{A}\boldsymbol{\beta} + \mathbf{e},\ E(\mathbf{e}) = \mathbf{0},\ \mathbf{D}(\mathbf{e}) = \sigma^2\mathbf{I} \tag{5.1}$$

consider a linear function $\mathbf{l}'\mathbf{y}$ of the observations. The function $\mathbf{l}'\mathbf{y}$ is normally distributed with mean $\mathbf{l}'\mathbf{A}\boldsymbol{\beta}$ and variance $(\mathbf{l}'\mathbf{l})\,\sigma^2$. Hence $\mathbf{l}'\mathbf{y}/\sigma\sqrt{\mathbf{l}'\mathbf{l}}$ is normally distributed with mean $\mathbf{l}'\mathbf{A}\boldsymbol{\beta}/\sigma\sqrt{\mathbf{l}'\mathbf{l}}$ and variance unity, so that

$$(\mathbf{l}'\mathbf{y})^2/\sigma^2\,(\mathbf{l}'\mathbf{l}) \tag{5.2}$$

has the non-central chi-square distribution with one degree of freedom and non-centrality parameter $(\mathbf{l}'\mathbf{A}\boldsymbol{\beta})^2/\sigma^2(\mathbf{l}'\mathbf{l})$. Similarly, if $\mathbf{l}'\mathbf{y}$ and $\mathbf{m}'\mathbf{y}$ are orthogonal linear functions of the observations, then

$$\{(\mathbf{l}'\mathbf{y})^2/\sigma^2\,(\mathbf{l}'\mathbf{l})\} + \{(\mathbf{m}'\mathbf{y})^2/\sigma^2\,(\mathbf{m}'\mathbf{m})\} \tag{5.3}$$

has the non-central chi-square distribution with 2 degrees of freedom.

In vector space language $(\mathbf{l}'\mathbf{y})^2/(\mathbf{l}'\mathbf{l})$ is $\mathbf{y}_0'\mathbf{y}_0$ where $\mathbf{y}_0$ is the projection of $\mathbf{y}$ on $\mathbf{l}$, or on the sub-space spanned by $\mathbf{l}$. Similarly,

$$\{(\mathbf{l}'\mathbf{y})^2/(\mathbf{l}'\mathbf{l})\} + \{(\mathbf{m}'\mathbf{y})^2/(\mathbf{m}'\mathbf{m})\}$$

where $\mathbf{l} \perp \mathbf{m}$, is $\mathbf{y}_0'\mathbf{y}_0$, $\mathbf{y}_0$ now denoting the projection of $\mathbf{y}$ on the subspace spanned by the vectors $\mathbf{l}$ and $\mathbf{m}$. The definitions of sums of squares and degrees of freedom given below are motivated by the above discussion.

5.2 Degrees of Freedom

A linear function $\mathbf{l}'\mathbf{y}$ of the observations ($\mathbf{l} \neq \mathbf{0}$) is said to carry one *degree of freedom*. Each of the observations $y_1, \ldots, y_n$, when regarded as a linear function of the observations, carries one degree of freedom.

A set of linear functions $\mathbf{Ly}$ is said to carry k degrees of freedom, if we can find k linearly independent functions in the set, and no more. In other words, the degrees of freedom carried by the set $\mathbf{Ly}$ equal the rank of the matrix $\mathbf{L}$. Equivalently, k is the dimension of the subspace of linear forms in the variables $y_1, \ldots, y_n$ spanned by the set $\mathbf{Ly}$. Thus, for example, both the sets of functions $y_1 + y_2 + y_3$, $y_1 - y_2 + y_3$, and $y_1 + y_2 + y_3$, $y_1 - y_2 + y_3$, $2y_1 + 2y_3$, carry two degrees of freedom.

Consider the set of all contrasts, i.e. linear functions $l_1 y_1 + \ldots + l_n y_n$ with $l_1 + \ldots + l_n = 0$. This is a subspace of the vector space of all linear functions and has dimension $n - 1$. Thus the set of all contrasts carries $n - 1$ degrees of freedom. Similarly, the set of all contrasts in any k out of the n variables $y_1, \ldots, y_n$ carries $k - 1$ degrees of freedom.

When the functions of the set $\mathbf{Ly}$ are estimates of $\mathbf{\Lambda\beta}$, the degrees of freedom of the set $\mathbf{Ly}$ will also be called the degrees of freedom for the estimates of $\mathbf{\Lambda\beta}$.

Consider the set of all error functions. The degrees of freedom of this set will be called the degrees of freedom belonging to error, or the *error degrees of freedom*, and denoted by n_e. Obviously, $n_e = n - r$, where r is the rank of $\mathbf{A}$.

Similarly, the degrees of freedom of the set of all best estimates (or of the set $\mathbf{A}\hat{\boldsymbol{\beta}}$) will be called the degrees of freedom belonging to estimates, or the *estimate degrees of freedom*. Obviously, the degrees of freedom for estimates are equal to $n - n_e = \text{rank } \mathbf{A}$.

5.3 Sums of Squares

If $\boldsymbol{\alpha}$ is any vector, $\boldsymbol{\alpha}'\boldsymbol{\alpha}$ is the square of the norm of $\boldsymbol{\alpha}$. We shall call it the square of $\boldsymbol{\alpha}$ in short.

Given a linear function $\mathbf{l}'\mathbf{y}$ of the observations, the square of the projection of $\mathbf{y}$ on $\mathbf{l}$ will be called the *sum of squares* (written SS) due to $\mathbf{l}'\mathbf{y}$. Since the function $\mathbf{l}'\mathbf{y}$ has one degree of freedom the SS due to $\mathbf{l}'\mathbf{y}$ is also said to have one degree of freedom.

Similarly, given a set $\mathbf{Ly}$ of linear functions of the observations, the square of the projection of **y** on the row space **R(L)** of **L** will be called the SS due to the set **Ly**, and its degrees of freedom will be the same as those carried by the set **Ly**. Note that the SS due to any set of functions and its degrees of freedom do not depend on the explicit representation **Ly** of the functions. Two sets of functions **Ly** and **My** will yield the same sum of squares if $R(\mathbf{L}) = R(\mathbf{M})$, the degrees of freedom being equal to the dimensions of the row space $R(\mathbf{L})$ or $R(\mathbf{M})$.

If the set **Ly** is the set of estimates of $\mathbf{\Lambda\beta}$ the SS due to the set **Ly** is also called the SS *due to the estimates* of $\mathbf{\Lambda\beta}$.

The SS due to the set of all best estimates (i.e. due to the set $\mathbf{A\hat{\beta}}$) is called the *estimate* SS and has r degrees of freedom. Similarly the SS due to the set of all error functions (i.e. due to the set $y - \mathbf{A\hat{\beta}}$) is called the *error* SS and has $n_e = n - r$ degrees of freedom.

Let $\mathbf{L}_1\mathbf{y}, \ldots, \mathbf{L}_k\mathbf{y}$ be given sets of functions of the observations and $S_1, \ldots, S_k$ be the corresponding sums of squares with $r_1 \ldots r_k$ degrees of freedom respectively. Let **Ly** denote the set of all these functions taken together and S the corresponding sum of squares with r degrees of freedom. If the functions belonging to any two different sets $\mathbf{L}_i, \mathbf{L}_j$ are mutually orthogonal, the row spaces $R(\mathbf{L}_1), \ldots, R(\mathbf{L}_k)$ are mutually orthogonal and the row space $R(\mathbf{L})$, is the sum of the mutually orthogonal spaces $R(\mathbf{L}_1), \ldots, R(\mathbf{L}_k)$. Hence if $\mathbf{y}_0, \ldots, \mathbf{y}_k$ denote the projections of **y** on $R(\mathbf{L})$, $R(\mathbf{L}_1), \ldots, R(\mathbf{L}_k)$ we have

$$\mathbf{y}_0 = \mathbf{y}_1 + \ldots + \mathbf{y}_k \tag{5.4}$$

$$\mathbf{y}_0'\mathbf{y}_0 = \mathbf{y}_1'\mathbf{y}_1 + \ldots + \mathbf{y}_k'\mathbf{y}_k \tag{5.5}$$

so that

$$S = S_1 + \ldots + S_k \tag{5.6}$$

Further, the dimension of $R(\mathbf{L})$ is the sum of the dimensions of $R(\mathbf{L}_1), \ldots, R(\mathbf{L}_k)$, so that

$$r = r_1 + \ldots + r_k \tag{5.7}$$

Thus we see that the sums of squares and the degrees of freedom arising out of mutually orthogonal sets of functions can be added together to give the sum of squares and degrees of freedom for the set of all the functions taken together.

Conversely, if the set **Ly** carrying r degrees of freedom is partitioned into the mutually orthogonal sets $\mathbf{L}_1\mathbf{y}, \ldots, \mathbf{L}_k\mathbf{y}$ carrying $r_1, \ldots, r_k$ degrees of freedom respectively, then the sum of squares S due to the set **Ly** can be written as

$$S = S_1 + \ldots + S_k \tag{5.8}$$

where S_i is the SS due to the set $\mathbf{L}_i\mathbf{y}$ and has r_i degrees of freedom.

Sums of squares arising out of mutually orthogonal sets of functions are called *orthogonal sums of squares*. If $S_1, \ldots, S_k$ are orthogonal sums of squares with $r_1, \ldots, r_k$ degrees of freedom, their sum S is also a sum of squares and has $r = r_1 + \ldots + r_k$ degrees of freedom.

Again, let S with r degrees of freedom be the SS due to the set **Ly**, and S_1 with r_1 degrees of freedom the SS due to the set **My**. If the functions of the set **My** are linear combinations of the functions of the set **Ly**, or in other words the functions **My** belong to the space of linear forms spanned by functions **Ly**, the row space $R(\mathbf{M})$ is a subspace of the row space $R(\mathbf{L})$. In such a case we say that the sum of squares S_1 is a part of the sum of squares S, in the sense that S_1 can be subtracted from S. In fact, the difference $S - S_1$ is the sum of squares due to the set of all those functions which are linear combinations of the functions **Ly** and orthogonal to the functions **My**, and has $r - r_1$ degrees of freedom. This is easily seen from the fact that S_1 and $S - S_1$ are orthogonal sums of squares and their sum S has r degrees of freedom. Thus, the sum of squares $S - S_1$ must have $r - r_1$ degrees of freedom.

5.4 Calculations of Sums of Squares

The projection of **y** on the vector **l** is given by $a\mathbf{l}$ where $a = (\mathbf{l}'\mathbf{y})/(\mathbf{l}'\mathbf{l})$. Hence, the SS due to the linear function $\mathbf{l}'\mathbf{y}$ is $a^2\,(\mathbf{l}'\mathbf{l})$, i.e.

$$(\mathbf{l}'\mathbf{y})^2/(\mathbf{l}'\mathbf{l}) \tag{5.9}$$

It is clear from this (as also from the definition of sums of squares) that the *SS* due to two linear functions $\mathbf{l}'\mathbf{y}$, $\mathbf{m}'\mathbf{y}$ are equal if $\mathbf{m} = k\mathbf{l}$, k being a real number.

Examples: (i) The SS due to the observation y_i (here **l** is a unit vector) is simply y_i^2.

(ii) The SS due to the arithmetic mean of the observations $\bar{y} = (y_1 + \ldots + y_n)/n$ is the same as that due to the grand total

$y_1 + ... + y_n$, and is given by

$$n\bar{y}^2 = (y_1 + ... + y_n)^2/n \tag{5.10}$$

(iii) If $\bar{y}_*$ denotes the arithmetic mean of a subset of k observations $y_{i1}, ..., y_{ik}$, the SS due to $\bar{y}_*$ is $k\bar{y}_*^2$.

An easy rule to remember to obtain the SS due to any linear function is:

$$SS = \frac{(\text{Linear function})^2}{(\text{Sum of the squares of the coefficients})}$$

For example, the SS due to $y_1 + 2y_2 + 3y_3$ is

$$(y_1 + 2y_2 + 3y_3)^2/(1 + 4 + 9)$$
$$= (y_1 + 2y_2 + 3y_3)^2/14$$

Theorem 5.1: The SS due to the set **Ly** is given by

$$S_{\mathbf{L}} = \boldsymbol{\alpha}'\mathbf{Ly} \tag{5.11}$$

where $\boldsymbol{\alpha}$ is any solution of $\mathbf{LL}'\boldsymbol{\alpha} = \mathbf{Ly}$.

Proof: From Theorem 2.7, the projection of **y** on $R(\mathbf{L})$ is given by $\boldsymbol{\alpha}'\mathbf{L}$ where $\boldsymbol{\alpha}$ is any solution of $\mathbf{LL}'\boldsymbol{\alpha} = \mathbf{Ly}$. Hence the SS due to **Ly** is given by

$$S_{\mathbf{L}} = (\boldsymbol{\alpha}'\mathbf{L})(\mathbf{L}'\boldsymbol{\alpha}) = \boldsymbol{\alpha}'\mathbf{LL}'\boldsymbol{\alpha} = \boldsymbol{\alpha}'\mathbf{Ly}$$

The calculation of S_L is simplified if we can find k mutually orthogonal vectors $\mathbf{l}_1, .., \mathbf{l}_k$ spanning $R(\mathbf{L})$, k being the dimension of $R(\mathbf{L})$. Then, the *SS* due to $\mathbf{l}_1'\mathbf{y}, ..., \mathbf{l}_k'\mathbf{y}$ are mutually orthogonal, so that, by the additive property we have

$$S_{\mathbf{L}} = \sum_{i=1}^{k} (\mathbf{l}_i'\mathbf{y})^2/(\mathbf{l}_i'\mathbf{l}_i) \tag{5.12}$$

For example, given n linearly independent linear functions, the SS due to them is the same as the SS due to the orthogonal functions $y_1, y_2, ..., y_n$ (the coefficient vectors of these linear functions are the unit vectors) of R_n. Hence, it is equal to

$$\sum_{i=1}^{k} y_i^2 = \mathbf{y}'\mathbf{y}$$

This SS may also be regarded as the SS due to the set of all linear functions in the variables $y_1, ..., y_n$.

If the linear functions $\mathbf{l}_1'\mathbf{y}, ..., \mathbf{l}_k'\mathbf{y}$ are such that the $\mathbf{l}_i$ are orthonormal, the SS due to them is

$$\sum_{i=1}^{k} (\mathbf{l}_i' \mathbf{y})^2 \tag{5.14}$$

Thus, the sum of squares due to a set of linear functions can always be exhibited as the sum of the squares of linear functions; hence the name sum of squares.

Sometimes, it is not convenient to calculate the SS due to a set Ly by using an orthogonal (or orthonormal) basis for $R(\mathbf{L})$. In such cases the subtraction property of sums of squares can be used. Consider, for example, the sum of squares due to the set of all contrasts in the variables $y_1, \ldots, y_n$. This sum of squares has $n - 1$ degrees of freedom, since the space of all contrasts has dimension $n - 1$, and is a part of the SS due to the set of all linear functions of the variables $y_1, \ldots, y_n$, i.e. of $\sum_{i=1}^{n} y_i^2$, which has n degrees of freedom. Hence, the SS due to all contrasts can be conveniently obtained by subtracting from Σy_i^2 the SS, with one degree of freedom, due to a linear function which is orthogonal to all contrasts, i.e. SS $n\bar{y}^2$ due to the mean. Thus the SS due to all contrasts is

$$\Sigma y_i^2 - n\bar{y}^2 \tag{5.15}$$

Similarly, the SS due to all contrasts in a subset $y_{i_1}, \ldots, y_{i_k}$ of the variables is

$$\sum_{j=1}^{k} y_{i_j}^2 - k\bar{y}_*^2 \tag{5.16}$$

where $\bar{y}_*$ is the arithmetic mean of the variables $y_{i_1}, \ldots, y_{i_k}$. For example, the SS due to the contrasts $y_1 - y_2, y_1 - y_3, y_1 - y_4$ is

$$\sum_{i=1}^{4} y_i^2 - (y_1 + y_2 + y_3 + y_4)^2/4$$

5.5 Distribution of Sums of Squares

Theorem 5.2: Let $S_\mathbf{l}$ denote the SS due to the linear function $\mathbf{l}'\mathbf{y}$. Then

$$E(S_\mathbf{l}) = \sigma^2 + S_{\mathbf{l},\eta} \tag{5.17}$$

where $\boldsymbol{\eta} = E(\mathbf{y}) = \mathbf{A}\boldsymbol{\beta}$, and $S_{\mathbf{l},\eta}$ is obtained from $S_\mathbf{l}$ by replacing $\mathbf{y}$ with $\boldsymbol{\eta}$.

Proof: $$S_\mathbf{l} = (\mathbf{l}'\mathbf{y})^2/(\mathbf{l}'\mathbf{l})$$

Now, $\mathbf{l}'\mathbf{y}$ is normally distributed with mean $\mathbf{l}'\boldsymbol{\eta}$ and variance $(\mathbf{l}'\mathbf{l})\,\sigma^2$, so that,

$$E\,[(\mathbf{l}'\mathbf{y})^2] = \text{Var}\,(\mathbf{l}'\mathbf{y}) + [E(\mathbf{l}'\mathbf{y})]^2 = \sigma^2\,(\mathbf{l}'\mathbf{l}) + (\mathbf{l}'\boldsymbol{\eta})^2$$

Hence,

$$E\,(S_{\mathbf{l}}) = \sigma^2 + (\mathbf{l}'\eta)^2/(\mathbf{l}'\mathbf{l}) = \sigma^2 + S_{\mathbf{l},\,\eta}$$

Corollary: If $\mathbf{l}'\mathbf{y}$ is an error function, with $\mathbf{l} \neq \mathbf{0}$,

$$E\,(S_{\mathbf{l}}) = \sigma^2 \tag{5.18}$$

Theorem 5.3: Let $S_{\mathbf{L}}$ denote the *SS*, with *r* degrees of freedom, due to the set $\mathbf{Ly}$. Then

$$E\,(S_{\mathbf{L}}) = r\sigma^2 + S_{\mathbf{L},\,\eta} \tag{5.19}$$

where $S_{\mathbf{L},\,\eta}$ is obtained from $S_{\mathbf{L}}$ by replacing $\mathbf{y}$ with $\boldsymbol{\eta} = E(\mathbf{y})$.

Proof: Since the set $\mathbf{Ly}$ carries *r* degrees of freedom, we can find orthonormal vectors $\mathbf{l}_1, \ldots, \mathbf{l}_r$ spanning $R(\mathbf{L})$. Then,

$$S_{\mathbf{L}} = \sum_{i=1}^{r} (\mathbf{l}_i'\,\mathbf{y})^2$$

so that,

$$E\,(S_{\mathbf{L}}) = \sum_{i=1}^{r} E\,[(\mathbf{l}_i'\,\mathbf{y})^2] = r\sigma^2 + \sum_{i=1}^{r} (\mathbf{l}_i'\,\boldsymbol{\eta})^2$$

$$= r\sigma^2 + S_{\mathbf{L},\,\eta}$$

Corollary: If the functions of the set $\mathbf{Ly}$ are error functions, then

$$E\,(S_{\mathbf{L}}) = r\sigma^2 \tag{5.20}$$

Theorem 5.4: Let $S_{\mathbf{L}}$ denote the SS, with *r* degrees of freedom, due to the set $\mathbf{Ly}$. Then $S_{\mathbf{L}}/\sigma^2$ has the non-central chi-square distribution with *r* degrees of freedom and non-centrality parameter

$$\delta = S_{\mathbf{L},\,\eta}/\sigma^2 \tag{5.21}$$

Proof: Write

$$S_{\mathbf{L}} = \sum_{i=1}^{r} (\mathbf{l}_i'\,\mathbf{y})^2 = \sum_{i=1}^{r} z_i^2$$

Since the $\mathbf{l}_i$ are orthonormal, the variables z_i are independent normal variables with

$$E(z_i) = \mathbf{l}_i'\,\boldsymbol{\eta},\ \text{Var}\,(z_i) = \sigma^2$$

Hence,

$$S_{\mathbf{L}}/\sigma^2 = \sum_{i=1}^{r} (z_i/\sigma)^2$$

is the sum of the squares of r independent normal variables with variance unity. It follows that $S_{\mathbf{L}}/\sigma^2$ has the non-central chi-square distribution with r degrees of freedom, and non-centrality parameter

$$\delta = \sum_{i=1}^{r} \{E(z_i/\sigma)\}^2 = \sum_{i=1}^{r} (\mathbf{l}_i'\boldsymbol{\eta})^2/\sigma^2 = S_{\mathbf{L},\,\boldsymbol{\eta}}/\sigma^2$$

Corollary: If the functions of the set $\mathbf{Ly}$ are error functions, $S_{\mathbf{L}}/\sigma^2$ has the chi-square distribution with r degrees of freedom.

Theorem 5.5: Let $S_{\mathbf{L}}$ and $S_{\mathbf{M}}$ be orthogonal sums of squares corresponding to the sets $\mathbf{Ly}$, $\mathbf{My}$. Then $S_{\mathbf{L}}$ and $S_{\mathbf{M}}$ are independently distributed.

Proof: Let n_1, n_2 be the degrees of freedom for $S_{\mathbf{L}}$ and $S_{\mathbf{M}}$ respectively. Then we can find orthonormal vectors $\mathbf{l}_1, \ldots, \mathbf{l}_{n_1}$ spanning $R(\mathbf{L})$, and orthonormal vectors $\mathbf{m}_1, \ldots, \mathbf{m}_{n_2}$ spanning $R(\mathbf{M})$. Writing

$$z_i = \mathbf{l}_i'\mathbf{y},\ i = 1, \ldots, n_1$$

$$w_j = \mathbf{m}_j'\mathbf{y},\ j = 1, \ldots, n_2$$

we have

$$S_{\mathbf{L}} = \Sigma z_i^2,\ S_{\mathbf{M}} = \Sigma w_j^2$$

Since $R(\mathbf{M})$ and $R(\mathbf{L})$ are orthogonal subspaces, $\mathbf{l}_i \perp \mathbf{m}_j$ for all i, j. Hence, $z_1, \ldots, z_{n_1}, w_1, \ldots, w_{n_2}$ are mutually independent normal random variables. It follows that $S_{\mathbf{L}}$ and $S_{\mathbf{M}}$ are independently distributed.

Theorem 5.6: If $S_{\mathbf{L}}$ and $S_{\mathbf{M}}$ are orthgonal sums of squares with n_1, n_2 degrees of freedom, and $E(\mathbf{My}) = \mathbf{0}$ (i.e. the functions of the set $\mathbf{My}$ are error functions), then

$$\frac{S_{\mathbf{L}}/n_1}{S_{\mathbf{M}}/n_2}$$

has the non-central F-distribution with n_1, n_2 degrees of freedom and non-centrality parameter

$$\delta = S_{\mathbf{L},\,\boldsymbol{\eta}}/\sigma^2$$

Proof: The proof follows from Theorem 5.5 and the fact that $S_{\mathbf{L}}/\sigma^2$ has the non-central chi-square distribution with n_1 degrees of freedom and non-centrality parameter δ, and S_M/σ^2 has the chi-square distribution with n_2 degrees of freedom.

Corollary: If we also have $E(\mathbf{Ly}) = \mathbf{0}$,

$$\frac{S_L/n_1}{S_M/n_1}$$

has the F-distribution with n_1 and n_2 degrees of freedom.

5.6 Estimate and Error Sums of Squares

The quantity

$$\mathbf{y}'\mathbf{y} = y_1^2 + \ldots + y_n^2 \tag{5.22}$$

is called the *raw sum of squares* (or *crude sum of squares*). It is the SS due to the set of all linear functions and has n degrees of freedom.

The *total sum of squares* is defined to be

$$\sum_{i=1}^{n} (y_i - \bar{y})^2 = \sum_{i=1}^{n} y_i^2 - n\bar{y}^2 \tag{5.23}$$

where $\bar{y}$ is the arithmetic mean of $y_1, \ldots, y_n$. It is the SS due to the set of all contrasts in $y_1, \ldots, y_n$ and has $n - 1$ degrees of freedom. Some authors use the term total SS for the raw sum of squares, and the term total SS adjusted for the mean for what we have called the total SS.

We have already defined in Sec. 5.3 the SS due to estimates and the SS due to error.

Theorem 5.7: The estimate sum of squares is given by

$$\hat{\boldsymbol{\beta}}\mathbf{A}'\mathbf{y} \tag{5.24}$$

where $\hat{\boldsymbol{\beta}}$ is any solution of the normal equations, and has r degrees of freedom, r being the rank of A.

Proof: The estimate sum of squares is the SS due to the set of all best estimates, i.e. due to the set $\mathbf{A}'\mathbf{y}$. Hence it is equal to (by Theorem 5.1) $\hat{\boldsymbol{\beta}}'\mathbf{A}'\mathbf{y}$ where $\hat{\boldsymbol{\beta}}$ is any solution of the equations $\mathbf{A}'\mathbf{A}\boldsymbol{\beta} = \mathbf{A}'\mathbf{y}$.

The degrees of freedom are obviously equal to the rank of A.

Corollary: The SS due to error is given by

$$\mathbf{y}'\mathbf{y} - \hat{\boldsymbol{\beta}}'\mathbf{A}'\mathbf{y} \tag{5.25}$$

and has $n - r$ degrees of freedom.

Theorem 5.8: The error SS is given by

$$(\mathbf{y} - \hat{\boldsymbol{\eta}})'(\mathbf{y} - \hat{\boldsymbol{\eta}}) = \sum_{i=1}^{n} (y_i - \hat{\eta}_i)^2$$

where $\hat{\eta}_i$ is the best estimate of $\eta_i = E(y_i)$.

Proof: If $\hat{\boldsymbol{\beta}}$ is any solution of the normal equations, we have $\hat{\boldsymbol{\eta}} = \mathbf{A}\hat{\boldsymbol{\beta}}$. Thus

$$\begin{aligned}(\mathbf{y} - \hat{\boldsymbol{\eta}})'(\mathbf{y} - \hat{\boldsymbol{\eta}}) &= (\mathbf{y} - \mathbf{A}\hat{\boldsymbol{\beta}})'(\mathbf{y} - \mathbf{A}\hat{\boldsymbol{\beta}}) \\ &= \mathbf{y}'\mathbf{y} - \hat{\boldsymbol{\beta}}'\mathbf{A}'\mathbf{y} - \mathbf{y}'\mathbf{A}\hat{\boldsymbol{\beta}} + \hat{\boldsymbol{\beta}}'\mathbf{A}'\mathbf{A}\hat{\boldsymbol{\beta}} \\ &= \mathbf{y}'\mathbf{y} - \hat{\boldsymbol{\beta}}'\mathbf{A}'\mathbf{y} - \mathbf{y}'\mathbf{A}\hat{\boldsymbol{\beta}} + \mathbf{y}'\mathbf{A}\hat{\boldsymbol{\beta}} \\ &= \mathbf{y}'\mathbf{y} - \hat{\boldsymbol{\beta}}\mathbf{A}'\mathbf{y}\end{aligned}$$

Note that the error sum of squares is the same as the residual sum of squares defined earlier in Sec. 4.7.

Theorem 5.9: Let S^2 and S_e^2 denote the estimate and error sums of squares respectively. Then

(i) $E(S^2) = r\sigma^2 + \boldsymbol{\beta}'\mathbf{A}'\mathbf{A}\boldsymbol{\beta}$ (5.27)

(ii) $E(S_e^2) = (n - r)\,\sigma^2$ (5.28)

(iii) S^2 and S_e^2 are independently distributed.

Proof: (ii) follows from the corrollary to Theorem 5.3, and (iii) from Theorem 5.5. To prove. (i) we have, from Theorem 5.3,

$$E(S^2) = r\sigma^2 + S_\eta^2$$

where S_η^2 is obtained from S^2 by replacing $\mathbf{y}$ with $\boldsymbol{\eta} = E(\mathbf{y}) = \mathbf{A}\boldsymbol{\beta}$. Now,

$$S^2 = \hat{\boldsymbol{\beta}}'\mathbf{A}'\mathbf{y} = \hat{\boldsymbol{\beta}}'\mathbf{A}'\mathbf{A}\hat{\boldsymbol{\beta}} \tag{5.29}$$

where $\hat{\boldsymbol{\beta}}$ is any solution of the normal equations $\mathbf{A}'\mathbf{A}\boldsymbol{\beta} = \mathbf{A}'\mathbf{y}$. The functions $\mathbf{A}\boldsymbol{\beta}$ being estimable

$$E(\mathbf{A}\hat{\boldsymbol{\beta}}) = \mathbf{A}\boldsymbol{\beta} \tag{5.30}$$

Hence, replacing $\mathbf{y}$ by $\boldsymbol{\eta}$ results in replacing $\mathbf{A}\hat{\boldsymbol{\beta}}$ by $\mathbf{A}\boldsymbol{\beta}$, so that

$$S_\eta^2 = \boldsymbol{\beta}'\mathbf{A}'\mathbf{A}\boldsymbol{\beta} \tag{5.31}$$

which completes the proof of (i).

From Theorem 5.9, we get

$$E(S_e^2/n_e) = \sigma^2 \tag{5.32}$$

The quantity S_e^2/n_e is called the *mean square* (MS) *due to error*, or the *error mean square*. We shall denote it by s^2. The error mean square s^2 provides an unbiased estimate of σ^2 and is therefore also called the *least squares estimate* of σ^2 in the linear model.

If $S_{\mathbf{L}}$ is the sum of square due to **Ly** with k degrees of freedom, then $S_{\mathbf{L}}/k$ will be called the *mean square* due to the set **Ly**.

COMPLEMENTS AND PROBLEMS

1. Prove the algebraic identity

$$\Sigma y_i^2 - n\bar{y}^2 = \left(\frac{y_1 - y_2}{2}\right)^2 + \left(\frac{y_1 + y_2 - 2y_3}{6}\right)^2 + \ldots$$
$$+ \left\{\frac{y_1 + y_2 + \ldots + y_{n-1} - (n-1)\,y_n}{n\,(n-1)}\right\}^2$$

(The rhs is the SS due to all constrasts obtained through mutually orthogonal contrasts).

2. The sum of squares due to the linear function $\mathbf{l}'\mathbf{y}$ can be expressed as $(\mathbf{l}'\mathbf{y})^2\,\sigma^2/\text{Var}\,(\mathbf{l}'\mathbf{y})$. This method is useful in cases where $\mathbf{l}'\mathbf{y}$ is not expressed explicitly as a function of $y_1, \ldots, y_n$.

3. If $\mathbf{L}$ is an $m \times n$ matrix of rank m, the sum of squares due to the set of m functions $\mathbf{Ly}$ is $\mathbf{y}'\,\mathbf{L}'\,(\mathbf{LL}')^{-1}\,\mathbf{Ly}$. This follows from Theorem 5.1 by noting $\mathbf{LL}'$ is non-singular.

REFERENCES

The basic ideas and definitions are due to Bose (1949 a).

6

Tests of Linear Hypotheses

6.1 Estimable Linear Hypotheses

In the linear model the parameter vector $\boldsymbol{\beta}$ belongs to the p-dimensional vector space R_p, which we shall call the *parameter space.* If nothing is specified, the vector $\boldsymbol{\beta}$ may lie anywhere in the parameter space.

By the term *hypothesis* we shall mean a statement which restricts $\boldsymbol{\beta}$ to lie in some given subset W of R_p. We thus talk of the hypothesis $\boldsymbol{\beta} \in W$. A test of a given hypothesis is a method to determine, on the basis of the observations, the validity or otherwise of the hypothesis. However, not all hypotheses may be amenable to testing in the linear model. Consequently we are led to confine ourselves to certain types of hypotheses.

A *linear hypothesis* is a hypothesis $\boldsymbol{\beta} \in W$, where W is a subspace of R_p. Since the statement $\boldsymbol{\beta} \in W$ is equivalent to the statement $\boldsymbol{\beta} \perp W^{\perp}$, every linear hypothesis can be described by means of a set of linear equations

$$\boldsymbol{\lambda}_i' \boldsymbol{\beta} = 0, \; i = 1, \ldots k \tag{6.1}$$

It is enough to choose $\boldsymbol{\lambda}_1, \ldots, \boldsymbol{\lambda}_k$ such that they span $W^{\perp}$. We shall always assume that the vectors $\boldsymbol{\lambda}_1, \ldots, \boldsymbol{\lambda}_k$ in (6.1) are linearly independent.

In matrix notation the linear hypothesis (6.1) may be written as

$$\boldsymbol{\Lambda}'\boldsymbol{\beta} = \mathbf{0}, \tag{6.2}$$

where $\boldsymbol{\Lambda} = [\boldsymbol{\lambda}_1 \ldots \boldsymbol{\lambda}_k]$ is a $p \times k$ matrix of rank k; the hypothesis itself may be said to be a *hypothesis of rank* k.

Two hypotheses $\Lambda'\beta = 0$, $\nabla'\beta = 0$ of the same rank are said to be equivalent if each implies the other, or, in other words, both restrict β to lie in the same subspace of R_p.

Theorem 6.1: Two hypotheses $\Lambda'\beta = 0$, $\nabla'\beta = 0$ of the same rank k are equivalent, if and only if, there exists a non-singular $k \times k$ matrix $\mathbf{P}$ such that $\Lambda' = \mathbf{P}\nabla'$.

Proof: Suppose $\Lambda' = \mathbf{P}\nabla'$. Since $\mathbf{P}$ is non-singular, $\Lambda'\beta = 0$ if, and only if, $\nabla'\beta = 0$. Hence, the two hypotheses are equivalent.

Conversely, suppose the two hypotheses are equivalent. Then, $\Lambda'\beta = 0$, if and only if, $\nabla'\beta = 0$, i.e., $\beta \in R(\Lambda')^{\perp}$, if and only if, $\beta \in R(\nabla')^{\perp}$, i.e.

$$R(\Lambda') = R(\nabla') \tag{6.3}$$

Thus, the rows of Λ' are linear combinations of the rows of ∇'. Hence

$$\Lambda' = \mathbf{P}\nabla' \tag{6.4}$$

for some $k \times k$ matrix $\mathbf{P}$. Now, rank Λ' = rank $\nabla' = k$, so that rank $\mathbf{P} = k$, which proves that $\mathbf{P}$ is non-singular.

A linear hypothesis $\lambda_i'\beta = 0$ $(i = 1, \ldots, k)$ will be called an *estimable linear hypothesis* if the linear functions $\lambda_i'\beta$ defining it are all estimable functions. Tests of hypotheses in the linear model are derived for estimable linear hypotheses only.

6.2 The Generalised F-test

Consider the hypothesis

$$\lambda_i'\beta = 0,\ i = 1, \ldots, k \tag{6.5}$$

where $\lambda_1, \ldots, \lambda_k$ are linearly independent and the functions $\lambda_i'\beta$ estimable. Suppose $\mathbf{l}_i'\mathbf{y}$ is the best estimate of $\lambda_i'\beta$ $(i = 1, \ldots, k)$. Let S_H^2 be the sum of squares due to the functions $\mathbf{l}_1'\mathbf{y}, \ldots, \mathbf{l}_k'\mathbf{y}$. Since the functions $\mathbf{l}_i'\mathbf{y}$ are linearly independent, S_H^2 has k degrees of freedom. Let $s_H^2 = S_H^2/k$ be the mean squares due to the functions $\mathbf{l}_i'\mathbf{y}$. Then, assuming $\mathbf{l}_i$'s to be orthogonal,

$$E(s_H^2) = \sigma^2 + \sum_{i=1}^{k} (\mathbf{l}_i'\eta)^2 \tag{6.6}$$

and if the hypothesis is true,

$$E(s_H^2) = \sigma^2 \tag{6.7}$$

since $\mathbf{l}_i'\eta = E(\mathbf{l}_i'\mathbf{y}) = 0$ under the hypothesis.

The mean square s_H^2 and the error mean square s^2 are independently distributed. If the hypothesis is true, the ratio

$$F = s_H^2/s^2 \tag{6.8}$$

has the F-distribution with (k, n_e) degrees of freedom (see corollary to Theorem 5.6). Under the hypothesis we have

$$E(s_H^2) = E(s^2) = \sigma^2 \tag{6.9}$$

and if the hypothesis does not hold,

$$E(s_H^2) > E(s^2) \tag{6.10}$$

Thus we may infer that large values of the ratio s_H^2/s^2 do not support the hypothesis.

Suppose we decide to reject the hypothesis whenever s_H^2/s^2 is greater than a fixed number c. Then, the probability of rejecting the hypothesis when it is true is

$$\Pr\,[s_H^2/s^2 > c]$$

which can be obtained from the upper tail of the F-distribution with (k, n_e) degrees of freedom. If $F_{\alpha,\,k,\,n_e}$ denotes the upper $100\alpha\,\%$ point of the F-distribution with (k, n_e) degrees of freedom, then taking $c = F_{\alpha;\,k,\,n_e}$ we get a test of significance level α. Thus at the 5 % level of significance, we shall reject the hypothesis if

$$s_H^2/s^2 > F_{.05;\,k,\,n_e} \tag{6.11}$$

We call S_H^2 the *hypothesis sum of squares* and s_H^2 the *hypothesis mean square*. The test procedure which rejects the hypothesis when s_H^2/s^2 is greater than a fixed value is called the *generalised F-test.*

It remains to verify that the same hypothesis sum of squares is obtained if the hypothesis (6.5) is replaced by an equivalent hypothesis

$$\boldsymbol{\delta}_i'\,\boldsymbol{\beta} = 0 \quad (i = 1, \ldots, k) \tag{6.12}$$

Writing (6.5) and (6.12) as $\boldsymbol{\Lambda}'\boldsymbol{\beta} = \mathbf{0}$ and $\boldsymbol{\nabla}'\boldsymbol{\beta} = \mathbf{0}$ we have $\boldsymbol{\nabla}' = \mathbf{P}\boldsymbol{\Lambda}'$ for some non-singular matrix $\mathbf{P}$. Let $\mathbf{L}'\mathbf{y}$ denote the best estimates of the functions $\boldsymbol{\Lambda}'\boldsymbol{\beta}$ so that S_H^2 is the SS due to the set $\mathbf{L}'\mathbf{y}$. The best estimates of the functions $\boldsymbol{\nabla}'\boldsymbol{\beta}$ are $\mathbf{PL}'\mathbf{y}$ (Theorem 4.7). $\mathbf{P}$ being non-singular the row spaces $R(\mathbf{L}')$, $R(\mathbf{PL}')$ are identical so that the SS due to the set $\mathbf{PL}'\mathbf{y}$ is the same as the SS due to the set $\mathbf{L}'\mathbf{y}$. Thus the hypothesis sum of squares does not depend on the particular form (6.5) in which the hypothesis is expressed. It is also clear that S_H^2 will be most convenient to calculate if the equations (6.5) are so chosen that the corresponding estimates have orthonormal coefficient vectors $\mathbf{l}_1, \ldots, \mathbf{l}_k$.

6.3 The Generalised *t*-test

In case the hypothesis is described by means of a single equation

$$\boldsymbol{\lambda}'\boldsymbol{\beta} = 0 \tag{6.13}$$

we can, instead of the generalised *F*-test, obtain a test for it based on the *t*-distribution.

If $\mathbf{l}'\mathbf{y}$ is the best estimate of $\boldsymbol{\lambda}'\boldsymbol{\beta}$, then $\mathbf{l}'\mathbf{y}$ is normally distributed with mean $\boldsymbol{\lambda}'\boldsymbol{\beta}$ and variance $\sigma^2(\mathbf{l}'\mathbf{l})$. If the hypothesis is true,

$$(\mathbf{l}'\mathbf{y})/\sigma\sqrt{\mathbf{l}'\mathbf{l}} \tag{6.14}$$

is a standard normal variate (i.e. has mean zero and variance unity).

Now, whether the hypothesis is true or not, S_e^2/σ^2 has the chi-square distribution with n_e degrees of freedom, and $\mathbf{l}'\mathbf{y}$ and S_e^2 are independently distributed. It follows that if the hypothesis is true the random variable

$$t = \frac{(\mathbf{l}'\mathbf{y})/\sigma\sqrt{\mathbf{l}'\mathbf{l}}}{\sqrt{[S^2/n_e\sigma^2]}} = \frac{\mathbf{l}'\mathbf{y}}{s\sqrt{\mathbf{l}'\mathbf{l}}} \tag{6.15}$$

has the *t*-distribution with n_e degrees of freedom.

Under the hypothesis $E(t) = 0$, so that we reject the hypothesis if the value of $|\, t\,|$ is large. The actual procedure, called the *generalized t-test*, is as follows.

Reject the hypothesis if,

$$|\,(\mathbf{l}'\mathbf{y})/s\sqrt{\mathbf{l}'\mathbf{l}}\,| > t_{\alpha/2;\, n_e} \tag{6.16}$$

where $t_{\alpha/2;\, n_e}$ is the upper 100 $(\alpha/2)$ percent point of the *t*-distribution with n_e degrees of freedom. The level of significance of this test is α.

Note that $s^2(\mathbf{l}'\mathbf{l})$ is an unbiased estimate of $\sigma^2(\mathbf{l}'\mathbf{l})$, the variance of $\mathbf{l}'\mathbf{y}$. Thus we may write

$$t = (\mathbf{l}'\mathbf{y})/s\sqrt{\mathbf{l}'\mathbf{l}} = \frac{\mathbf{l}'\mathbf{y}}{\sqrt{[\text{estimate of var}\,(\mathbf{l}'\mathbf{y})]}} \tag{6.17}$$

Comparing this with the well known "*t*"-variable

$$\frac{\bar{x}\sqrt{n}}{\sqrt{[\Sigma(x_i - \bar{x})^2/(n-1)]}} = \frac{\bar{x}}{\sqrt{[\text{estimate of var}\,(\bar{x})}} \tag{6.18}$$

we see the reason for the name generalised *t*-test.

A confidence interval for $\boldsymbol{\lambda}'\boldsymbol{\beta}$ is easily obtained on the basis of the observations $y_1, ..., y_n$. If $\mathbf{l}'\mathbf{y}$ is the best estimate of $\boldsymbol{\lambda}'\boldsymbol{\beta}$ then $\mathbf{l}'\mathbf{y}$ is normally distributed with mean $\boldsymbol{\lambda}'\boldsymbol{\beta}$ and variance $\sigma^2(\mathbf{l}'\mathbf{l})$. Thus

$$z = (\mathbf{l}'\mathbf{y} - \boldsymbol{\lambda}'\boldsymbol{\beta})/\sigma\sqrt{\mathbf{l}'\mathbf{l}} \tag{6.19}$$

is a standard normal variate. Further, S_e^2/σ^2 is a chi-square variable with n_e degrees of freedom, and z and S_e^2 are independently distributed. Thus,

$$\frac{z}{\sqrt{[S_e^2/n_e\sigma^2]}} = \frac{\mathbf{l}'\mathbf{y} - \boldsymbol{\lambda}'\boldsymbol{\beta}}{s\sqrt{\mathbf{l}'\mathbf{l}}} \tag{6.20}$$

has the t-distribution with n_e degrees of freedom. Hence, if $t_{\alpha/2;\, n_e}$ denotes the upper $100(\alpha/2)$ per cent point of the t-distribution with n_e degrees of freedom, we have

$$\Pr\left[\left|\frac{\mathbf{l}'\mathbf{y} - \boldsymbol{\lambda}'\boldsymbol{\beta}}{s\sqrt{\mathbf{l}'\mathbf{l}}}\right| < t_{\alpha/2;\, n_e}\right] = 1 - \alpha \tag{6.21}$$

Thus the confidence interval for $\boldsymbol{\lambda}'\boldsymbol{\beta}$ with confidence coefficient $1 - \alpha$ is given by

$$\mathbf{l}'\mathbf{y} - s\sqrt{\mathbf{l}'\mathbf{l}}\, t_{\alpha/2;\, n_e} < \boldsymbol{\lambda}'\boldsymbol{\beta} < \mathbf{l}'\mathbf{y} + s\sqrt{\mathbf{l}'\mathbf{l}}\, t_{\alpha/2;\, n_e} \tag{6.22}$$

6.4 Examples

(*a*) *Student's "t"*

We have n independent observations from a normal population with mean μ and variance σ^2 and we wish to test the hypothesis $\mu = \mu_0$, μ_0 being a known constant.

The linear model here is

$$y_i = \mu + e_i \qquad i = 1, 2, ..., n \tag{6.23}$$

and there is a single normal equation

$$\sum_{i=1}^{n} y_i = n\mu \tag{6.24}$$

or

$$\bar{y} = (y_1 + ... + y_n)/n = \mu \tag{6.25}$$

The parameter μ is estimable and its best estimate is $\bar{y}$. It follows that $n_e = n - 1$.

The SS due to estimates is simply that corresponding to $\bar{y}$, i.e.

$$n\bar{y}^2$$

and hence

$$S_e^2 = \Sigma y_i^2 - n\bar{y}^2 = \Sigma\,(y_i - \bar{y})^2 \tag{6.26}$$

Hence the appropriate statistic to test the hypothesis $\mu = \mu_0$ is

$$t = \frac{(\bar{y}_0 - \mu_0)/(\sigma/\sqrt{n})}{s/\sigma} = \frac{\sqrt{n}(\bar{y} - \mu_0)}{\sqrt{[\Sigma (y_i - \bar{y})^2/(n-1)]}} \tag{6.27}$$

which is the well known Student's "t".

Note: If $\mu_0 \neq 0$, the hypothesis $\mu = \mu_0$ is not an estimable linear hypothesis as defined earlier. The test for $\mu = \mu_0$ may be easily seen to be the test for the estimable linear hypothesis $\lambda = 0$ in the linear model

$$z_i = \lambda + e_i,\ i = 1, \ldots, n$$

with

$$z_i = x_i - \mu_0,\ \lambda = \mu - \mu_0$$

(See also complements and problems at the end of this chapter).

(*b*) *Fisher's "t"*

We have two independent samples

$$y_1, y_2, \ldots, y_{n_1}$$
$$z_1, z_2, \ldots, z_{n_2}$$

from two normal populations with means μ_1 and μ_2 and common variance σ^2. It is assumed that μ_1, μ_2 and σ^2 are unknown and we wish to test the hypothesis $\mu_1 = \mu_2$

The linear model here is

$$y_i = \mu_1 + e_i,\ i = 1, \ldots, n_1$$
$$z_j = \mu_2 + f_j,\ j = 1, \ldots, n_2 \tag{6.28}$$

e_i and f_j being normal variables with mean zero and variance σ^2. From the normal equations we get

$$\begin{aligned} \bar{y} &= \mu_1 \\ \bar{z} &= \mu_2 \end{aligned} \tag{6.29}$$

Thus both the parameters μ_1 and μ_2 are estimable, their best estimates being $\bar{y}$ and $\bar{z}$. It follows that

$$n_e = n - 2 \tag{6.30}$$

where $n = n_1 + n_2$.

The two estimates $\bar{y}$ and $\bar{z}$ are orthogonal. Hence the SS due to estimates is given by

$$n_1\bar{y}^2 + n_2\bar{z}^2 \tag{6.31}$$

and hence

$$S_e^2 = \Sigma y_i^2 + \Sigma z_j^2 - n_1\bar{y}^2 - n_2\bar{z}^2$$

$$= \sum_i (y_i - \bar{y})^2 + \sum_j (z_j - \bar{z})^2 \qquad (6.32)$$

The hypothesis to be tested can be written as $\mu_1 - \mu_2 = 0$ and the best estimate of $\mu_1 - \mu_2$ is $\bar{y} - \bar{z}$. Also

$$\text{Var}\,(\bar{y} - \bar{z}) = \left(\frac{1}{n_1} + \frac{1}{n_2}\right)\sigma^2$$

Hence the appropriate statistic to test the hypothesis is

$$t = \frac{(\bar{y} - \bar{z}) \Big/ \sigma\sqrt{\dfrac{1}{n_1} + \dfrac{1}{n_2}}}{s/\sigma} = (\bar{y} - \bar{z}) \Big/ s\sqrt{\frac{1}{n_1} + \frac{1}{n_2}} \qquad (6.33)$$

which is the well known Fishers "t" to test the difference of two means.

6.5 Tests of Hypotheses and Least Squares Theory

Suppose the linear hypothesis to be tested is

$$\boldsymbol{\lambda}_1'\boldsymbol{\beta} = \ldots = \boldsymbol{\lambda}_k'\boldsymbol{\beta} = 0 \qquad (6.34)$$

where $\boldsymbol{\lambda}_1'\boldsymbol{\beta}, \ldots, \boldsymbol{\lambda}_k'\boldsymbol{\beta}$ are k independent estimable parametric functions.

We saw that the test of this hypothesis is made by means of the statistic

$$F = \frac{S_H^2/k}{S_e^2/n_e} \qquad (6.35)$$

where S_H^2 is the SS due to the estimates $\mathbf{l}_1'\mathbf{y}, \ldots, \mathbf{l}_k'\mathbf{y}$ of the parametric functions $\boldsymbol{\lambda}_1'\boldsymbol{\beta}, \ldots, \boldsymbol{\lambda}_k'\boldsymbol{\beta}$ respectively, and S_e^2 is the SS due to error. We also saw earlier that

$$S_e^2 = \text{minimum of } (\mathbf{y} - A\boldsymbol{\beta})'(\mathbf{y} - A\boldsymbol{\beta}) \qquad (6.36)$$

the minimum being taken over all $\boldsymbol{\beta} \in R_p$.

If the hypothesis to be tested were true we would have $E(\mathbf{l}_i'y) = \boldsymbol{\lambda}_i'\boldsymbol{\beta} = 0$ for $i = 1, \ldots, k$. Thus the functions $\mathbf{l}_1'\mathbf{y}, \ldots, \mathbf{l}_k'\mathbf{y}$ now belong to error. The effect of the hypothesis is to enlarge the error space. The new SS due to error is given by $S_e^2 + S_H^2$, since S_H^2 and S_e^2 are orthogonal. This would also be the minimum value of $(\mathbf{y} - A\boldsymbol{\beta})'(\mathbf{y} - A\boldsymbol{\beta})$, the minimum now being taken over all values of $\boldsymbol{\beta}$ satisfying the hypothesis. Thus

$$\begin{aligned} S_H^2 = {} & \text{Minimum of } (\mathbf{y} - A\boldsymbol{\beta})'(\mathbf{y} - A\boldsymbol{\beta}) \text{ under the hypothesis} \\ & - \text{Absolute minimum of } (\mathbf{y} - A\boldsymbol{\beta})'(\mathbf{y} - A\boldsymbol{\beta}) \qquad (6.37) \end{aligned}$$

The theory of least squares says that the appropriate SS for testing a linear hypothesis is given by

$$\text{Relative minimum of } (\mathbf{y}-A\boldsymbol{\beta})'(\mathbf{y}-A\boldsymbol{\beta}) \text{ under the hypothesis} \tag{6.38}$$
$$-\text{ Absolute minimum of } (\mathbf{y}-A\boldsymbol{\beta})'(\mathbf{y}-A\boldsymbol{\beta})$$

and we see that this agrees with S_H^2. This method is sometimes more convenient than that of finding S_H^2 from the functions $\mathbf{l}_1'\mathbf{y}, \ldots, \mathbf{l}_k'\mathbf{y}$.

COMPLEMENTS AND PROBLEMS

1. *Linear Regression.* The univariate regression problem can be studied through the linear model

$$y_i = \alpha + \beta x_i + e_i,\ i = 1, \ldots, n$$

Here the y_i are the observations on the random variable y (the dependent variable) corresponding to the values x_i of the independent or explanatory variable x which is assumed to be non-random. The analysis of the model by the method of least squares is considerably simplified if we rewrite it as

$$y_i = \alpha + \beta(x_i - \bar{x}.) + e_i$$

replacing α by $\alpha + \beta\,\bar{x}\cdot$, $\bar{x}.$ denoting, as usual, the arithmetic mean of the values x_i.

With this change the normal equations take the simple form

$$n\alpha = y.$$

$$\beta \sum_i (x_i - \bar{x}.)^2 = \sum_i y_i (x_i - \bar{x}.)$$

giving the usual estimates,

$$\hat{\alpha} = \bar{y}.$$

$$\hat{\beta} = \sum_i y_i (x_i - \bar{x}.)/\sum_i (x_i - \bar{x}.)^2$$

Since both α and β are estimable we have two degrees of freedom for estimates and $n - 2$ for error. The estimates $\hat{\alpha}$ and $\hat{\beta}$ being orthogonal (one is a mean and the other a contrast) the sum of squares for estimates is

$$n\bar{y}.^2 + \sum_i y_i (x_i - \bar{x}.)^2/\Sigma (x_i - \bar{x}.^2)$$

The error sum of squares as well as the tests for the hypotheses $\alpha = \alpha_0$ and $\beta = 0$ are easily obtained.

2. *Multivariate Linear Regression.* The linear model for regression with more than one explanatory variable is

$$y_i = \alpha + \beta_1 x_{1i} + \ldots + \beta_m x_{mi} + e_i,\ i = 1, \ldots, n.$$

Analysis is simplified by rewriting the model as

$$y_i = \alpha + \beta_1 (x_{1i} - \bar{x}_1) + \ldots + \beta_m (x_{mi} - \bar{x}_{m\cdot}) + e_i,\ i = 1, \ldots n$$

where $\bar{x}_{k\cdot}$ is the arithmetic mean of the n values $x_{k1}, x_{k2}, \ldots, x_{kn}$ of the kth explanatory variable x_k. For this model the normal equations can be written as

$$\left[\begin{array}{c:ccc} n & 0 \ldots\ldots & 0 \ldots\ldots & 0 \\ \hdashline \ldots & & & \\ 0 & & \mathbf{S} & \\ \vdots & & & \\ 0 & & & \end{array}\right] \begin{bmatrix} \alpha \\ \ldots \\ \beta \end{bmatrix} = \begin{bmatrix} y. \\ \ldots \\ \xi \end{bmatrix}$$

where $\quad \xi' = (\xi_1, \ldots, \xi_m),\ \xi_k = \Sigma\, y_i\, (x_{ki} - \bar{x}_k),$

$$\beta' = (\beta_1, \ldots, \beta_m),\ \mathbf{S} = (s_{kl})$$

$$s_{kl} = \Sigma\, (x_{ki} - \bar{x}_k)\,(x_{li} - \bar{x}_l)$$

The $m \times m$ matrix $\mathbf{S}$ is usually non-singular so that all the parameters are estimable. The estimates are

$$\hat{\alpha} = \bar{y}$$

$$\hat{\beta} = \mathbf{S}^{-1}\xi$$

The variance-covariance matrix of the estimates is

$$\left[\begin{array}{c:cc} 1/n & 0 \ldots\ldots & 0 \\ \hdashline 0 & & \\ \vdots & & \\ & \mathbf{S}^{-1} & \\ \vdots & & \\ 0 & & \end{array}\right] \sigma^2$$

which shows that $\hat{\beta}_1, \ldots, \hat{\beta}_m$ are orthogonal to $\hat{\alpha}$.

Writing $\mathbf{S}^{-1} = [s^{kl}]$ we have

$$\text{Var}\,(\hat{\beta}_k) = s^{kk}\,\sigma^2$$

$$\text{Cov}\,(\hat{\beta}_k, \hat{\beta}_l) = s^{kl}\,\sigma^2$$

It follows that the SS due $\hat{\beta}_k$ is $(\hat{\beta}_k)^2/s^{kk}$.

The estimate sum of squares is

$$\hat{\alpha}y. + \Sigma\, \hat{\beta}_k\, \xi_k$$

and carries $m+1$ degrees of freedom.

The first term is the SS due to the estimate $\hat{\alpha}$ (1 d.f.). Since $\hat{\alpha}$ is orthogonal to $\hat{\beta}_1, \ldots, \hat{\beta}_m$, the second term is the *SS* due to the estimates $\hat{\beta}_1, \ldots, \hat{\beta}_m$ of the regression parameters, and has m degrees of freedom. The error SS has $n - m - 1$ d.f.

The hypothesis $\beta_1 = \beta_2 = \ldots = \beta_m = 0$ is tested by means of the F-ratio

$$\frac{\Sigma \hat{\beta}_k \xi_k / m}{s^2}$$

which, if the hypothesis is true, has the F-distribution with $(m, n - m - 1)$ degrees of freedom.

The hypotheses $\alpha = a$, $\beta_k = b_k$, $\beta_k = \beta_l$ can be tested by means of the t-statistics $(n - m - 1$ d.f.,)

$$\frac{\bar{y}. - a}{s/\sqrt{n}}, \quad \frac{\beta_k - b_k}{s\sqrt{s^{kk}}} \quad \text{and} \quad \frac{\beta_k - \beta_l}{s\sqrt{(s^{kk} + s^{ll} - 2s^{kl})}}$$

respectively. The hypothesis sum of squares for testing the equality of several regression coefficients is derived more easily by the method described in Sec. 6.5.

3. *Polynomial Regression.* The linear model is

$$y_i = \beta_0 + \beta_1 x_i + \cdots + \beta_m x_i^m + e_i, \; i = 1, \ldots, n$$

The analysis of the polynomial regression model is the same as that for the multivariate regression model taking $x, x^2, \ldots, x^m$ as the m explanatory variables.

4. In the linear regression model

$$y_i = \alpha + \beta x_i + e_i, \; i = 1, \ldots, n,$$

the test for the hypothesis $\alpha + \beta\theta = 0$ (θ known) is derived from the estimate.

$$\hat{\alpha} + \hat{\beta}\theta = \bar{y}. + (\theta - \bar{x}.) [\Sigma y_i (x_i - \bar{x}.)/\Sigma (x_i - \bar{x}.)^2]$$

of $\alpha + \beta\theta$. We have

$$\text{Var}\, [\hat{\alpha} + \hat{\beta}\theta] = \frac{\sigma^2}{n}\left[1 + \frac{n(\theta - \bar{x}.)^2}{\Sigma (x_i - \bar{x}.)^2}\right]$$

The usual t-test of the hypothesis can now be easily obtained.

5. In the linear regression model

$$y_i = \alpha + \beta (x_i - \bar{x}.) + e_i, \; i = 1, \ldots, n$$

the hypothesis that the true regression line is

$$px + qy + r = 0$$

can, by equating the slope and intercept, be written as $\alpha = -\dfrac{p\bar{x}. + r}{q}$, $\beta = -r/q$. Using the estimates $\hat{\alpha} = \bar{y}.$, $\hat{\beta} = \Sigma y_i (x_i - \bar{x}.)/\Sigma (x_i - \bar{x}.)^2$, we find that under the hypothesis,

$$U = \frac{\left[\bar{y}. + \dfrac{p\bar{x}. + r}{q}\right]^2}{\sigma^2/n} + \frac{\left[\dfrac{\Sigma y_i (x_i - \bar{x}.)}{\Sigma (x_i - \bar{x}.)^2} + \dfrac{r}{q}\right]^2}{\sigma^2/\Sigma (x_i - \bar{x}.)^2}$$

has the chi-square distribution with 2 degrees of freedom. At the same time S_e^2/σ^2 has the chi-square distribution with $n - 2$ degrees of freedom. Hence, the ratio

$$\frac{U/2}{S_e^2/(n-2)\,\sigma^2} = \frac{n\,[q\bar{y}. + p\bar{x}. + r]^2\,[\Sigma (x_i - \bar{x}.)^2] + [q\,\Sigma y_i (x_i - \bar{x}.) + p\,\Sigma (x_i - \bar{x}.)^2]^2}{2q^2\,s^2\,[\Sigma (x_i - \bar{x}.)^2]}$$

has the F-distribution with 2 and $n - 2$ degrees of freedom, and can be used to obtain a test of the hypothesis in the usual manner. A more general result is given below,

6. If $\lambda_i' \beta$ $(i = 1, \ldots, k)$ are independent estimable parametric functions, the hypothesis

$$H: \lambda_i' \beta = \alpha_i, \; i = 1, \ldots, k$$

where $\alpha_1, \ldots, \alpha_k$ are known constants, is not an estimable linear hypothesis. The test for H can however be derived as follows. Writing the best estimates of the k parametric functions $\lambda_i' \beta$ as $\mathbf{L}'\mathbf{y}$ and $\alpha' = (\alpha_1, \ldots, \alpha_k)$, it is easy to see that $\mathbf{L}'\mathbf{y}$ has under H the multivariate normal distribution with mean vector α and dispersion matrix $(\mathbf{L}'\mathbf{L})\,\sigma^2$. Thus

$$\frac{U}{\sigma^2} = (\mathbf{L}'\mathbf{y} - \alpha)' \left(\frac{\mathbf{L}'\mathbf{L}}{\sigma^2}\right)^{-1} (\mathbf{L}'\mathbf{y} - \alpha)$$

has the chi-square distribution with k degrees of freedom under H. Thus,

$$F = \frac{U/k}{s^2}$$

has the F-distribution with k and n_e degrees of freedom and provides the usual one-sided (upper tail) F-test of the hypothesis H.

REFERENCES

David and Johnson (1951) discuss the effect of non-normality on the F-test. For the power of the F-test see tables by Tang (1938), Lehmer (1944) and charts by Pearson and Hartley (1951), Fox (1956).

Kolodziejczyk (1935) derived the analysis of variance test as a likelihood ratio test in the non-singular normal equations case. Power of the analysis of variance test is discussed in Hsu (1941), Wald (1942).

Chakrabarti (1962) uses matrix algebra and Cochran's theorem for deriving analysis of variance tests. Seber (1964) uses idempotent matrices.

7

Planning of Experiments

7.1 Introduction

The purpose of an experiment is to test the validity of a given hypothesis. Suppose we wish to test the hypothesis that the time taken to boil a given quantity of water is directly proportional to the volume of water boiled. We could take two pans, one containing double the quantity of water than the other, bring them to boil, and note the time taken for each. If the larger volume takes double the time compared to the smaller volume, we could say that the hypothesis is acceptable. If the result is otherwise, we could say that the hypothesis is rejected. We have thus planned an experiment to test the hypothesis, i.e. described how certain measurements or observations are to be taken, and laid down the rule to decide for or against the hypothesis on the basis of the observations obtained.

A little reflection shows that the experiment described above is not very satisfactory and can be considerably improved. It is certainly not enough to take just two observations—a larger number would be needed before we can place any reliance on our conclusions. A more important shortcoming is that any difference between the boiling times is being attributed entirely to that between the quantities of water boiled ignoring such important factors as the thickness and material of the pans, the nature of the heat source used, the differences in pressure under which the water is boiled in the two pans, etc. The experiment would be improved if, except for the factor under investigation (the quantity of water boiled), all the other factors known to affect the boiling time (the *assignable causes* of variation in the experimental results) were kept constant or *under control.* The observations obtained from such a controlled experiment could then be used to verify the hypothesis of direct proportionality. Any disagreement between the observations and the hypothesis would suggest either that the hypothesis should be discarded or that some other assignable causes have to be looked for. A modified hypothesis can be set up (the modification may be the inclusion of additional factors to be controlled or, a change in the functional relationship assumed) and the experiment replanned. The cycle—hypothesis, experiment, observations, hypothesis, experiment, and so on—continues till a satisfactory conclusion is reached, satisfactory not in any absolute sense but only relative to the standards set for the purposes of the experiment.

An essential feature of the experiment described above is, and this is true of most experiments in physics and chemistry, that the changes in the observations due to the uncontrolled factors are small compared to those resulting from a change in one or more of the factors under control. Once the experimental technique is mastered and the apparatus is working correctly, closely reproducible results are obtained. This fact has two important consequences. First, if the observations change when the factors under control are altered the former may safely be assumed to be due to the latter—one may speak of a "cause and effect" relationship. Second, the hypothesis can usually be given in the form of an exact mathematical relationship between the observations and the assignable causes and because of that is either accepted or rejected—the purpose of the experiment becomes the discovery of "natural laws" governing the phenomenon under study.

In many fields of study, however, the experiments do not possess this feature; the effects under investigation tend to be masked by fluctuations outside the experimenter's control. Repeated observations of the same or similar events do not agree with one another exactly, and the experimenter cannot make the agreement perfect by taking any reasonable precautions. In agricultural field trials, for example, even if the same variety were to be sown on all plots there would be substantial variations in yield from plot to plot or from year to year. In industrial experimentation we find that successive articles produced by a machine are not identical and that, however much care is taken in the adjustment of the machine, the variability in quality of the output cannot be reduced beyond a certain point. In the chemical industry the output from one batch to another will vary considerably even though experiments conducted under laboratory conditions may suggest a close functional relationship between the output and the quantities of raw materials used.

It usually happens, however, that the difference in effects of two controlled factors may be fairly consistent although the absolute effect of each is subject to large variations. In an agricultural trial, for example, it is common for the differences between the yields of two varieties of crop to remain relatively constant even when there is substantial variation from plot to plot, or year to year, in the yields of the individual varieties.

It is with this second type of situation that we shall be concerned with. The attempt will be to devise methods of arranging the experiment so that we may with confidence and accuracy separate the variations in the observations due to variations in the controlled factors from those due to variations in the uncontrolled factors. The aim of the experiments will be to compare the differences in the effects of different controlled factors rather than the determination of the absolute effect of each controlled factor. We have thus two aspects to consider; the *design* of experiments and the *analysis* of the observations resulting from an experiment.

7.2 Nomenclature

Much of the stimulus to statisticians' interest in experiments came originally from agricultural research, and the standard nomenclature still bears evidence of this.

The different factors whose effects are being compared are called *treatments* (from manurial treatments) or *varieties* (from varieties of crop), the experiment itself being called a *field trial, varietal trial* or *treatment trial.* A treatment may be a type of crop or manure, a type of diet in animal feeding experiments, a particular industrial process, a set of environmental conditions like temperature, humidity, etc.

The material to which the treatment is applied is called the *experimental material.* This may be a field on which crops are sown, a batch of animals who are given a prescribed diet, the raw material to which an industrial manufacturing process is applied, and so on.

A *plot* (or *experimental unit*) is the smallest division of the experimental material such that any two units may receive different treatments in the actual experiment. A plot may be an area of land on which a crop is grown, in accordance with the original usage, but it may be a hospital patient, a piece of animal tissue, the site on the body of an animal to receive an injection on a particular occasion, or a machine on which a manufacturing process is to be carried out.

The object of an experiment is to make comparisions between the effects of different treatments, each of which is applied to one or more plots, in terms of measurements or observations made on the separate plots. Any quantitative measure obtained from a plot may be termed a *yield* from that plot under the treatment applied. It is the plot yields that are subjected to analysis.

A group of plots having certain common features is called a *block*. A block may be a collection of neighbouring plots in an agricultural field, a group of animals from the same litter, a collection of machines of the same type, etc.

If a particular treatment is applied to r different plots, we say it has been *replicated* r times. Replication means the repetition of observations, and hence we shall also speak of a whole design being replicated.

7.3 Design

The design of an experiment means:

(i) the set of treatments selected for comparision;

(ii) the specification of the experimental material to which the treatments are to be applied;

(iii) the rule according to which the treatments are allocated to the plots;

(iv) the specification of the measurements, or other records, to be made on each plot.

The relevance of an experiment to the problem under investigation and the trustworthiness of the results depend on (i), (ii), (iii) and (iv). What might be termed the classical theory of experimental design is comprised in (iii).

In an agricultural experiment meant to compare the yields obtained from different varieties of the same crop, say wheat, the experimental material will consist of the field on which crops are sown. The treatments will be the varieties to be compared. The yield will be the amount of grain obtained after harvesting from individual plots into which the field is divided, each plot being sown with one variety. Apart from the varieties, the assignable causes of variation are the differences in soil fertility of the different plots, the amount of irrigation, the type of manure used, the technique of cultivation, etc. The differences in yield from different plots will result from the differences in any of these causes. If two plots are sown with two different varieties and receive two different types of manure, the difference in the yields cannot be ascribed only to varietal differences, as the effect of the different manures also plays a part. In the simplest of such experiments the factors like irrigation, manure, method of cultivation and harvesting, can be kept constant. The differences in soil fertility can, however, not be controlled beyond a certain extent. We say that there is heterogeneity in the experimental material which is beyond the experimenters control. Obviously, the results of the experiment would be more dependable for comparison of the varieties if the two varieties are sown on pairs of plots which are relatively similar in soil fertility. The difference in the yield of the two varieties obtained from such a pair of plots could then be taken as a measure of the difference in performance of the two varieties as to yield. This is the principle of *local control*. We may define local control as a method of allocating the treatments to the experimental units in such a manner that the yields from any two treatments can be compared over pairs of relatively homogeneous plots. Local control is thus a method to overcome the uncertainty in the yields, for comparing treatments, resulting from the hetereo-geneity or non-uniformity of the experimental material to which the treatments are applied.

As we remarked earlier, even if the same variety is sown on plots approximately similar in soil fertility, the yields will differ consider-ably from plot to plot. It is thus insufficient to compare two varieties on the basis of yields obtained from a single pair of uniform plots; not much reliance can be placed on the results from a single pair. If the two varieties are sown on a number of pairs of plots, where the plots in any pair are uniform, but plots from different pairs may vary considerably in fertility, the average of the differences in yields of the two varieties over all the pairs of plots is a more reliable measure of the difference in performance of the two varieties. This is the principle of *replication* in design of experiments. Observations must be repeated a certain number of times and comparisions made only on the basis of the repeated observations. The number of times each variety is replicated in the design will obviously depend on the extent of the variation in individual yields expected to be caused by the hetereogeneity of the experimental material, and the degree of certainty desired in the comparisons.

7.4 Analysis

The term analysis of a design is used for the method by which comparisions of the effects of different treatments are made on the basis of the yields obtained from the experiment. If $y_1, \ldots, y_n$ are the observations obtained from the n experimental units into which the experimental material is divided, we assume that the observation vector **y** represents observations on a random vector **y** (the same symbol is used for both) governed by a linear model

$$\mathbf{y} = \mathbf{A}\boldsymbol{\beta} + \mathbf{e} \tag{7.1}$$

The parameters $\boldsymbol{\beta}$ represent the effect of the hetereogeneity of the experimental material on the one hand, and that of the treatments applied on the other. The elements a_{ij} of the matrix A are determined by the design used. The method of analysis consists in partitioning the variation in the observed yields measured by the raw sum of squares into the sums of squares for estimates and error. That is why the analysis is usually given the name *analysis of variance*. The estimate sum of squares represents the variation in yields resulting from the differences in the effects of the treatments, as also the variation due to other assignable causes. The error sum of squares represents the variation resulting from the uncontrolled factors. We then formulate hypotheses of interest concerning the parameters $\boldsymbol{\beta}$ and test the hypotheses by the use of the general theory developed in Chapters 4 to 6.

The basic assumption is one of additivity of the effects due to different assignable causes including treatments, i.e. the expected yield from any plot is a sum of the effects of the treatment applied and of the other assignable factors operating on that particular plot. For example, the expected value of the yield in a varietal trial in agriculture may be written as

$$\mu + \beta_i + \gamma_j + \tau_k$$

where μ represents the effect common to all the plots in the experiments, τ_k represents the effect of the treatment applied to the particular plot, β_i and γ_j represent the effect of two other factors which are known to affect the yield apart from the treatments. Different plots may have the same or different values for β_i and γ_j depending upon the manner in which the two factors affect the plots.

As a result of the assumption of additivity it is possible to partition the estimate sum of squares further into two orthogonal components of which one represents the variation arising solely from the differencesi n the treatment effects. The latter sum of squares or a part of it can: hen be used to compare the treatments or some of them among themselves.

7.5 Randomisation

Once the numbers of treatments and plots have been fixed, the allocation of treatments to plots can be made in a number of ways. For example, if three treatments *A*, *B* and *C* are to be assigned to 9 plots, divided into 3 blocks of 3 plots each, with each treatment occurring once in each block, a possible allocation is as follows:

Block	I	*A*	*B*	*C*
Block	II	*B*	*C*	*A*
Block	III	*C*	*A*	*B*

The number of such allocations is $6 \times 6 \times 6 = 216$. If we impose a further restriction that each treatment should occur only once in each column (the layout above satisfies this requirement) the possible number of allocations is 12.

The principle of *randomisation* states that in any particular experiment the actual allocation of treatments to plots should be done randomly out of all possible allocations satisfying the constraints imposed, each possible allocation having the same probability of being selected as any other. If the experiment is repeated randomisation should be done afresh. The previous allocation should not be repeated.

Why is randomisation necessary? In the first place, it is the only way of ensuring that comparisons between treatments are not biased by the fact of one treatment being assigned to inherently better plots than another. However honest the experimenter, if he has freedom of choice within the explicit constraints of a design, he is very likely to prejudice results for or against one treatment by the plots that he chooses for it. Of course, any blatant favouritism would be recognised and discarded, but there is abundant evidence that even small subconscious effects may disturb the true objective character of an experiment.

The linear model which forms the basis for the analysis of variance assumes that the observations or yields from different plots are independently distributed. However, this assumption is usually not valid. In an agricultural trial yields from neighbouring plots will be correlated. Similarly, in an industrial process the output from successive runs will tend to be correlated as compared to the output from runs separated by a longer time interval. If randomisation is used, the analysis of the observations based on the assumption of independence is still valid.

Finally, the linear model assumes that the yield from a plot is a sum of the specific effects due to assignable causes and treatments, and an error component which is a normally distributed random variable. The F-test used for testing hypotheses in the linear model is derived on the basis of this assumption of normality. If the error component is not a random but fixed quantity depending on the particular plot, the F-test would no longer be applicable. However, if randomisation has been used for allocating treatments to plots in such a situation, the ratio of the different mean squares would follow approximately an F-distribution, so that the usual tests of hypotheses can still be carried out.

7.6 Missing Plot Technique

Often in experimental work the observations from one or more experimental units may be lost by accident. The experimental design with which the experiment was begun thus gets modified and the method of analysis developed for the original design can not be used. We can, of course, use the general theory to carry out the analysis of variance using only the observations available. However, the analysis is then usually more complicated than it would have been if all the observations were known for the original design. This is specially the case where the original design has some symmetry properties making the analysis easier, and the symmetry is lost due to the missing observations.

The missing plot technique introduced by Yates enables one to carry out the analysis when some observations are missing by using the simplified analysis for the original design, thus rendering it comparatively easier. The technique consists essentially in estimating the missing values and then using the estimated values in the analysis of the original design.

Suppose the experiment had been designed to provide n observations $y_1, \ldots, y_n$ which follow the linear model

$$\mathbf{y} = \mathbf{A}\boldsymbol{\beta} + \mathbf{e} \tag{7.2}$$

Let us write

$$\mathbf{y} = \begin{bmatrix} \mathbf{y}_1 \\ \cdots \\ \mathbf{y}_2 \end{bmatrix} \tag{7.3}$$

where $\mathbf{y}_1' = (y_1, \ldots, y_{n-k})$ are the observations available and $\mathbf{y}_2' = (y_{n-k+1}, \ldots, y_n)$ are the k missing observations. Let the corresponding partitioning of $\mathbf{A}$ and $\mathbf{e}$ be

$$\mathbf{A} = \begin{bmatrix} \mathbf{A}_1 \\ \cdots \\ \mathbf{A}_2 \end{bmatrix} \quad \mathbf{e} = \begin{bmatrix} \mathbf{e}_1 \\ \cdots \\ \mathbf{e}_2 \end{bmatrix} \tag{7.4}$$

Since only $\mathbf{y}_1$ is known our analysis should be based on the model

$$\mathbf{y}_1 = \mathbf{A}_1\boldsymbol{\beta} + \mathbf{e}_1 \tag{7.5}$$

Let us write the original model in the form

$$\begin{aligned} \mathbf{y}_1 &= \mathbf{A}_1\boldsymbol{\beta} + \mathbf{e}_1 \\ \mathbf{y}_2 &= \mathbf{A}_2\boldsymbol{\beta} + \mathbf{e}_2 \end{aligned} \tag{7.6}$$

The normal equations for (7.5) are

$$\mathbf{A}_1'\mathbf{A}_1\boldsymbol{\beta} = \mathbf{A}_1'\mathbf{y}_1 \tag{7.7}$$

and for (7.6) are

$$\mathbf{A}_1'\mathbf{A}_1\boldsymbol{\beta} + \mathbf{A}_2'\mathbf{A}_2\boldsymbol{\beta} = \mathbf{A}_1'\mathbf{y}_1 + \mathbf{A}_2'\mathbf{y}_2 \tag{7.8}$$

Suppose $\boldsymbol{\beta}^*$ is a solution of the normal equations (7.7). If we now choose the unknown

$$\mathbf{y}_2 = \mathbf{A}_2\boldsymbol{\beta}^* \tag{7.9}$$

then $\boldsymbol{\beta}^*$ is also a solution of (7.8). With this value of $\mathbf{y}_2$, the error sum of squares in (7.6) is

$$\begin{aligned} &\mathbf{y}_1'\mathbf{y}_1 + \mathbf{y}_2'\mathbf{y}_2 - \boldsymbol{\beta}^{*\prime}(\mathbf{A}_1'\mathbf{A}_1 + \mathbf{A}_2'\mathbf{A}_2)\boldsymbol{\beta}^* \\ &= (\mathbf{y}_1'\mathbf{y}_1 - \boldsymbol{\beta}^{*\prime}\mathbf{A}_1'\mathbf{A}_1\boldsymbol{\beta}^*) + (\mathbf{y}_2'\mathbf{y}_2 - \boldsymbol{\beta}^{*\prime}\mathbf{A}_2'\mathbf{A}_2\boldsymbol{\beta}^*) \\ &= (\mathbf{y}_1'\mathbf{y}_1 - \boldsymbol{\beta}^{*\prime}\mathbf{A}_1'\mathbf{A}_1\boldsymbol{\beta}^*) \end{aligned}$$

since the second term on the right is zero. Thus we see that the error sum of squares in (7.6) is the same as the error sum of squares in (7.5) provided the values of $\mathbf{y}_2$ are taken as $\mathbf{A}_2\boldsymbol{\beta}^*$.

Now, whatever be the value of $\mathbf{y}_2$ the error SS in (7.6) cannot be less than the error SS in (7.5). This follows from the fact that if $\mathbf{l}'\mathbf{y}$ is an error function in (7.5), then it is also an error function in (7.6). From the above discussion it is clear that the error SS in (7.5), i.e. when some observations are missing, can be obtained according to the following rule.

Denote the missing observations by $\mathbf{y}_2$, obtain the error SS according to the original design, and then choose $\mathbf{y}_2$ so that this error SS is a minimum. The minimum value so obtained is the required SS. The degrees of freedom for error will have to be obtained from the model (7.5). A general rule, to be used when the missing observations leave the estimable functions undisturbed (i.e. functions estimable in the original model are still estimable), is to subtract the number of missing observations from the error degrees of freedom in the original design to get the required error degrees of freedom.

The same method is used for obtaining the sum of squares for testing a given hypothesis when some observations are missing. The missing observations are denoted by $\mathbf{y}_2$, the sum of squares for the hypothesis and error in the original design are added together and their sum minimised with respect to $\mathbf{y}_2$. From the minimum value so obtained we subtract the error sum of squares obtained earlier to get the the hypothesis sum of squares. We are assuming, of course, that the hypothesis is still estimable even though some observations are missing. It follows that the degrees of freedom for the hypothesis sum of squares are the same as they would be in the original design.

7.7 The Dot Notation

In the standard designs to be discussed in the succeeding chapters the yield from a plot will usually be denoted by y_i, y_{ij}, y_{ijk}, etc. where the indices i, j and k are used to specify the particular plot and the treatment applied. The analysis of such designs utilises the totals of the yields summed over one or more indices and the corresponding arithmetic means. The dot notation is a convenient device to represent such totals and the arithmetic means. The total of the yields taken over any index or indices is denoted by replacing such indices by dots and the corresponding means by using a bar over the totals. Thus, if the yields are y_i, $i = 1, \ldots, n$, then

$$y. = y_1 + \dots + y_n = \sum_i y_i \tag{7.10}$$

and

$$\bar{y}. = (y_1 + \dots + y_n)/n = (y./n) \tag{7.11}$$

Similarly, if the yields are y_{ijk} with i varying from 1 to r, j from 1 to s, and k from 1 to t, then

$$y_{i\cdot\cdot} = \sum_{j=1}^{s} \sum_{k=1}^{t} y_{ijk}$$

$$y_{i\cdot k} = \sum_j y_{ijk}$$

$$\bar{y}_{i\cdot\cdot} = y_{i\cdot\cdot}/st$$

$$\bar{y}_{i\cdot k} = y_{i\cdot k}/s$$

REFERENCES

Fisher (1926) laid the basis for the development of the theory of experimental designs and their analysis A geneı al, non-mathematical treatment, of the principles will be found in Cox (1958).

Eisenhart (1947) discusses the basic assumptions underlying the analysis of variance. He also introduces the concepts of fixed and random effect models; these are also discussed in Plackett (1960). The consequences of departures from assumptions are discussed in Cochran (1947) and Box (1954)

Sometimes the transformation of the observations helps where the original observations do not satisfy the basic assumptions of the linear model: see Beall (1942), Curtiss (1943), Bartlett and Kendall (1946), Bartlett (1947), Freeman and Tukey (1950), Tukey (1957), Box and Cox (1964), Draper and Cox (1969).

For a discussion of the advantage of randomisation see Pitman (1937), Welch (1937). Systematic, as opposed to randomised designs, are discussed in Yates (1939) and Cox (1952). The concept of a comparative experiment is due to Anscombe (1948).

The method of dealing with missing values was first discussed in Allan and Wishart (1930). See also Yates (1933 a), Bose and Mahalanobis (1938), Anderson (1946), Pearce (1948), Grundy (1951), Ishii and Kudo (1963).

Application to animal nutrition studies will be found in Dunlop (1933), industrial problems in Davies (1956), in psychology in Kogan (1953), in chemistry and the chemical industry in Bennett and Franklin (1954).

8

The Completely Randomised Design

8.1 Introduction

The simplest of all experimental designs is the *completely randomised design* in which the treatments are assigned randomly to the experimental units. No effort is made to confine treatments to any portion of the experimental material. More specifically, if a particular treatment is to be replicated four times, for example, every group of four units has an equal probability of receiving this treatment.

The completely randomised design is to be used when the experimental material is relatively homogeneous and, consequently, the differences in yield from units receiving different treatments can be assumed to arise mainly due to the differences between the treatments. Such designs have been found most useful in laboratory experiments, in cookery, in certain greenhouse studies, etc. On the other hand, they are seldom used in agricultural field trials.

Suppose n experimental units are available for the comparison of m treatments. We select integers $n_1, n_2, \ldots, n_m$, with $\Sigma n_i = n$. The first treatment is assigned to n_1 units selected with equal probability out of the n units, the second to n_2 units selected with equal probability out of the remaining $n - n_1$ units, and so on. The resulting arrangement will be a completely randomised design for m treatments over n units with the treatments replicated $n_1, n_2, \ldots, n_m$ times respectively. The actual allocation is easily done with the help of tables of random numbers. For example, if there are three treatments with 3, 2 and 5 replications respectively, we number the 10 units in any order. Next, we select a random sequence of 10 two digit or three digit numbers from the tables of random numbers. The ten random numbers so obtained are ranked according to size and the rankings give a random arrangement of the natural numbers from 1 to 10. For example, if the ten random numbers are 09, 90, 73, 75, 54, 08, 28, 53, 91, 89, the smallest is 08 and carries rank 1, the largest, 91 has rank 10. The ranks are

2, 9, 6, 7, 5, 1, 3, 4, 10, 8

Using this random arrangement we assign the first treatment to units 2, 9 and 6; the second to 7 and 5; and the third to the remaining units.

8.2 Normal Equations and Estimates

Let y_{ij} denote the yield from the jth unit receiving the ith treatment. Under the assumption of homogeneity of the experimental material the expected value of y_{ij} may be assumed to depend only on the treatment. The model equations therefore may be taken as

$$y_{ij} = \mu_i + e_{ij} \tag{8.1}$$

with i varying from 1 to m, and j varying from 1 to n_i, where n_i is the number of times the i-th treatment is replicated.

The normal equations are

$$\sum_{j=1}^{n_i} y_{ij} = n_i \mu_i \tag{8.2}$$

or,

$$\bar{y}_{i.} = \mu_i, \; i = 1, \ldots, m \tag{8.3}$$

We see from the normal equations that all the parameters μ_i are estimable, the estimate of μ_i being the mean yield $\bar{y}_i$ of the i-th treatment. It follows that all linear parametric functions are estimable; the estimate of $c_1 \mu_1 + \ldots c_m \mu_m$ is $c_1 \bar{y}_{1.} + \ldots + c_m \bar{y}_{m.}$. In particular the grand mean

$$\bar{y}_{..} = (1/n) \sum_i \sum_j y_{ij} \tag{8.4}$$

where n is total number of observations, is the estimate of $(\Sigma n_i \mu_i)/n$.

8.3 Analysis of Variance

Since all the parameters are estimable there are m degrees of freedom for estimates and consequently $n-m$ for error.

The sum of squares for estimates is that due to the linear functions

$\bar{y}_{1.}, \ldots, \bar{y}_{m.}$ of the observations. These linear functions are orthogonal. Hence

$$\text{SS(est)} = n_1 \bar{y}_{1.}^2 + \ldots + n_m \bar{y}_{m.}^2 \tag{8.5}$$

with m degrees of freedom, and

$$\begin{aligned} \text{SS(err)} &= \sum_i \sum_j y_{ij}^2 - \text{SS(est)} \\ &= \sum_{i,j} (y_{ij} - \bar{y}_{i.})^2 \end{aligned} \tag{8.6}$$

with $n-m$ degrees of freedom.

8.4 Tests of Hypotheses

The main hypothesis of interest is

$$\mu_1 = \mu_2 = \ldots = \mu_m \tag{8.7}$$

i.e. the expected yields are the same for all treatments. The hypothesis is equivalent to

$$\mu_1 - \mu_i = 0 \qquad i = 2, \ldots, m \tag{8.8}$$

The estimate of $\mu_1 - \mu_i$ is $\bar{y}_{1.} - \bar{y}_{i.}$. Hence the sum of squares for testing the hypothesis is the SS due to the $m - 1$ linear functions

$$y_{1.} - \bar{y}_{i.} \qquad i = 2, \ldots, m \tag{8.9}$$

These are linearly independent functions so that the SS due to them has $m - 1$ degrees of freedom.

The estimates $\bar{y}_{1.} - \bar{y}_{i.}$ of $\mu_1 - \mu_i$ being contrasts are orthogonal to $\bar{y}_{..}$. The linear form $\bar{y}_{..}$ is a linear combination of $\bar{y}_{1.}, \ldots, \bar{y}_{m.}$. The same is true for the linear forms $\bar{y}_{1.} - \bar{y}_{i.}$ ($i = 2, \ldots, m$). Hence the SS for testing the hypothesis, which has $m - 1$ degrees of freedom, is easily obtained by subtracting the SS due to $\bar{y}_{..}$, with 1 degree of freedom, from the SS due to $\bar{y}_{1.} \ldots, \bar{y}_{m.}$, with m degrees of freedom. Thus

$$\begin{aligned} \text{S S(Hyp)} &= \sum_{i=1}^{m} n_i \bar{y}_{i.}^2 - n \bar{y}_{1..}^2 \\ &= \sum_i n_i (\bar{y}_{i.} - \bar{y}_{1..})^2 \end{aligned} \tag{8.10}$$

This sum of squares is the weighted sum of the squares of the deviations of the group means $\bar{y}_{i.}$ from the grand mean $\bar{y}_{1..}$ and is therefore usually called the *between groups SS*. Similarly, the SS due to error

$$\sum_i \sum_j (y_{ij} - \bar{y}_{i.})^2 \tag{8.11}$$

being the sum of the squares of the deviations of the individual observations y_{ij} from the respective group means $\bar{y}_{i.}$ is called the *within groups SS*.

The test for the hypothesis is carried out by using the *F*-ratio

$$\frac{\text{SS (Hyp)}/(m-1)}{\text{SS (Err)}/(n-m)} \tag{8.12}$$

which, under the hypothesis, has the *F*-distribution with $(m-1, n-m)$ degrees of freedom. If the *F*-ratio is significant at the chosen level of significance, the hypothesis of equality of treatment means is rejected.

If we are interested in comparing any two specific treatments, say the *i*th and *j*th, we use the generalised *t*-test. The hypothesis is $\mu_i - \mu_j = 0$ and the corresponding *t*-statistic is

$$(\bar{y}_{i\cdot} - \bar{y}_{j\cdot}) \Bigg/ \sqrt{\left[\left\{\frac{\text{SS(err)}}{(n-m)}\right\}\left(\frac{1}{n_i} + \frac{1}{n_j}\right)\right]} \tag{8.13}$$

which has the *t*-distribution with $n - m$ degrees of freedom.

If we wish to test for the equality of a smaller number of treatment means, $\mu_1 = \mu_2 = \mu_3 = \mu_4$ the appropriate SS (Hyp) with 3 d.f. is the SS due to $\bar{y}_{1\cdot} - \bar{y}_{2\cdot}$, $\bar{y}_{1\cdot} - \bar{y}_{3\cdot}$, $\bar{y}_{1\cdot} - \bar{y}_{4\cdot}$. These three independent linear forms belong to the space of linear forms spanned by the mutually orthogonal linear forms $y_{1\cdot}$, $y_{2\cdot}$, $y_{3\cdot}$ and $y_{4\cdot}$. The linear form $(n_1\bar{y}_{1\cdot} + n_2\bar{y}_{2\cdot} + n_3\bar{y}_{3\cdot} + n_4\bar{y}_{4\cdot})/(n_1 + n_2 + n_3 + n_4)$ also belongs to this space and is orthogonal to the three linear forms $\bar{y}_{1\cdot} - \bar{y}_{2\cdot}$, $\bar{y}_{1\cdot} - \bar{y}_{3\cdot}$, $\bar{y}_{1\cdot} - \bar{y}_{4\cdot}$. Hence

$$\text{SS (Hyp)} = \sum_{i=1}^{4} n_i \bar{y}_{i.}^2 - \left(\sum_{i=1}^{4} n_i \bar{y}_{i.}^2\right)/(n_1 + n_2 + n_3 + n_4) \tag{8.14}$$

The usual *F*-test can now be carried out using the *F*-ratio

$$\frac{\text{SS (Hyp)}/3}{\text{SS (Err)}/(n-m)} \tag{8.15}$$

with $(3, n - m)$ degrees of freedom.

8.5 Notes on Computation

(i) The observations are arranged in m columns, the ith column containing the n_i yields y_{ij} for the ith treatment

(ii) The treatment totals $y_{i.}$, the treatment means $\bar{y}_{i.}$ and the correction factors (CF) $n_i \bar{y}_{i.}^2 = y_{i.}^2/n_i$ are then obtained

(iii) Next we obtain the raw SS, the grand mean $\bar{y}_{..}$ and the CF corresponding to the grand mean $n\,\bar{y}^2_{..} = y^2_{..}/n$

(iv) We then calculate the total SS, the SS between groups, and the error SS. The error SS is obtained by subtracting the between groups SS from the total SS

(v) The above results are then presented in the form of an analysis of variance table as shown below:

Table 8.1

Source	df	SS	MS	F
Between groups	$m-1$	S_1^2	$s_1^2 = S_1^2/(m-1)$	s_1^2/s^2
Within groups	$n-m$	S_e^2	$s^2 = S_e^2/(n-m)$	
Total	$n-1$	$\sum_{ij} y_{ij}^2 - n\bar{y}_{..}^2$		

8.6 Example

In an experimental study of the mineral metabolism of pullets, four white Wyandotte pullets of the same strain and hatching were used. During the period of investigation, two pullets (referred to as C_1 and C_2) were given ration C which had a high CaO (calcium) content. The other two pullets (referred to as NC_3 and NC_4) were given ration NC which was generally comparable with ration C apart from the fact that its CaO content was low. In other respects the pullets were treated alike, and an attempt was made to regulate the daily food consumption of the pullets. The table below gives the weights of CaO (in grams) found in the whole eggs laid by pullets.

Table 8.2

C_1	C_2	NC_3	NC_4
2.013	2.005	2.366	2.094
2.195	1.977	2.276	2.152
2.435	2.163	2.147	1.690
2.545	2.088	1.821	1.685
2.542	2.136	1.805	1.354
2.749	2.071	1.758	0.823
2.723	1.895		
2.706	1.870		
2.686	1.824		
2.509			
2.673			
2.721			
2.655			
2.708			

Source: Common, R.H., J. Agri. Sc., Vol. 26, p. 85, 1936.

The basic computations are shown in the following table:

Table 8.3

n_i	14	9	6	6	35	n
$y_{i\cdot}$	35.95	18.029	12.173	9.798	75.950	$y_{\cdot\cdot}$
$\bar{y}_{i\cdot}$	2.5679	2.0032	2.0288	1.6330	2.17	$\bar{y}_{\cdot\cdot}$
$\sum_j y_{ij}^2$	92.8436	36.2333	25.0524	17.2219	171.3512	$\sum_{ij} y_{ij}^2$
CF	92.3145	36.1161	24.6970	16.0001	164.8115	CF
SS	0.5291	0.1172	0.3554	1.2218	6.5397	SS

The raw SS is 171.3512 and after correcting for the grand mean the total SS is 6.5397. The between groups SS is obtained by subtracting the general CF, 164.8115, from the sum of the CF's of the four groups. The within groups SS is obtained by subtraction, and should be equal to the sum of the SS for the four groups, thus providing a check on the calculations. The analysis of variance table is as follows.

Table 8.4

Source	df	SS	MS	F
Between groups	3	4.3162	1.4387	20.065
Within groups	31	2.2235	0.0717	
Total	34	6.5397		

The F value obtained is more than four times the 1 % value. Hence we may conclude that there is a significant variation in the mean CaO contents of the eggs laid by the the four pullets.

We may now wish to examine if the two types of diets are significantly different as to their effect on the calcium content. The relevant hypothesis is

$$\mu_1 + \mu_2 = \mu_3 + \mu_4$$

where μ_1, μ_2, μ_3, μ_4 are the population means for the CaO content of eggs of hens C_1, C_2, NC_3 and NC_4 respectively. This may be done with the help of the t-test. The t-value with 31 degrees of freedom is

$$\frac{2.5679 + 2.0032 - 2.0288 - 1.6330}{s\sqrt{\left[\frac{1}{14}+\frac{1}{9}+\frac{1}{6}+\frac{1}{6}\right]}} = \frac{0.9093}{0.1923} = 4.728$$

which turns out to be highly significant. Alternatively, we may use the F-ratio of MS corresponding to the estimate $\bar{y}_{1\cdot} + \bar{y}_{2\cdot} - \bar{y}_{3\cdot} - \bar{y}_{4\cdot}$ to the error MS.

The SS corresponding to the estimate is

$$\frac{[2.5679 + 2.0032 - 2.0288 - 1.6330]^2}{\left[\frac{1}{14} + \frac{1}{9} + \frac{1}{6} + \frac{1}{6}\right]} = 1.6028,$$

giving the *F*-ratio

$$\frac{1.6028}{0.0717} = 22.354$$

with (1, 31) degrees of freedom, which is highly significant.

Two further hypotheses of interest are $\mu_1 = \mu_2$ and $\mu_3 = \mu_4$. These would test if there are significant differences between the CaO content of eggs of hens receiving the same diet. We could test each hypothesis with the help of the *t*-test. Alternatively, we could use the *F*-test by considering the SS due to the estimates $\bar{y}_{1\cdot} - \bar{y}_{2\cdot}$, $\bar{y}_{3\cdot} - \bar{y}_{4\cdot}$ of $\mu_1 - \mu_2$ and $\mu_3 - \mu_4$ respectively. These two sums of squares are mutually orthogonal and account for two out of the three degrees of freedom for the between groups SS. The third degree of freedom may be accounted for by the SS due to estimate $28\bar{y}_{1\cdot} + 18\bar{y}_{2\cdot} - 23\bar{y}_{3\cdot} - 23_4\bar{y}.$ of the estimable contrast $28\mu_1 + 18\mu_2 - 23\mu_3 - 23\mu_4$. The third estimate is orthogonal to the first two and the SS due to these three estimates together add up to the between groups SS. The between groups SS can now be partitioned as follows.

Table 8.5

Source	df	SS	F
$\mu_1 - \mu_2$	1	1.7469	24.364
$\mu_3 - \mu_4$	1	0.4699	6.554
$28\mu_1 + 18\mu_2 - 23\mu_3 - 23\mu_4$	1	2.0999	
Between groups	3	4.3167	

Both the *F*-values are highly significant; the experiment reveals that there is a significant variation in the calcium content of eggs laid by hens fed on the same type of diet. Thus there is need to carry out further experiments with a larger number of hens.

If there had been equal numbers of observations in each group, the three estimates $\bar{y}_{1\cdot} - \bar{y}_{2\cdot}$, $\bar{y}_{3\cdot} - \bar{y}_{4\cdot}$, $\bar{y}_{1\cdot} + \bar{y}_{2\cdot} - \bar{y}_{3\cdot} - \bar{y}_{4\cdot}$ of the contrasts $\mu_1 - \mu_2$, $\mu_3 - \mu_4$, $\mu_1 + \mu_2 - \mu_3 - \mu_4$ would have been mutually orthogonal. This would have enabled us to partition the between groups SS into the SS corresponding to these estimates, and the the three hypotheses, viz. $\mu_1 = \mu_2$, $\mu_3 = \mu_4$, and $\mu_1 + \mu_2 = \mu_3 + \mu_4$ could have been tested simultaneously. Since in this example there were unequal numbers of observations in the different groups we had to select the odd looking contrast $28\mu_1 + 18\mu_2 - 23\mu_3 - 23\mu_4$ to partition the between groups SS into three orthogonal components.

COMPLEMENTS AND PROBLEMS

1. The expectation of the between groups SS is

$$(m - 1)\,\sigma^2 + \Sigma\, n_i\,(\mu_i - \bar{\mu}_\cdot)^2$$

where

$$\bar{\mu}_\cdot = (\Sigma\, n_i\, \mu_i)/n$$

2. The linear model for the completely randomised design is usually written as

$$y_{ij} = \mu + \tau_i + e_{ij};\ i = 1, ., m;\ j = i, \ldots, n_i$$

where μ is called the general effect common to all the plots and τ_i is called the treatment effect due to the ith treatment.

(a) In this model $a\mu + \Sigma\, t_i\, \tau_i$ is estimable if, and only if, $a = \Sigma\, t_i$. In particular, a linear function in the treatment parameters τ_i is estimable if, and only if, it is a contrast.

(b) If $a\mu + \Sigma\, t_i\, \tau_i$ is estimable, the best estimate is $\Sigma\, t_i\, \bar{y}_{i\cdot}$

(c) The estimate and error sums of squares for this model are the same as those obtained in this chapter.

(d) $\sum_{i,j} (y_{ij} - \mu - \tau_i)^2$ is minimum when $\mu + \tau_i = \bar{y}_{i\cdot}$. Its minimum value under the hypothesis $\tau_i = \ldots = \tau_m$ is obtained by taking $\mu + \tau_i = \bar{y}_{\cdot\cdot}$. Hence, the hypothesis sum of squares is

$$\sum_{i,j} (y_{ij} - \bar{y}_{\cdot\cdot})^2 - \sum_{i,j} (y_{ij} - \bar{y}_{i\cdot})^2 = \Sigma\, n_i\, \bar{y}_{i\cdot}^{\,2} - n\, \bar{y}_{\cdot\cdot}^{\,2}$$

$$= \Sigma\, n_i\, (\bar{y}_{i\cdot} - \bar{y}_{\cdot\cdot})^2$$

(e) If the n_i's are equal, orthogonal parametric functions in the variables τ_i have orthogonal estimates. The same is not true if the n_i's are unequal.

3. *Test for linearity of regression*: The test for linearity of regression is the test for the hypothesis

$$\tau_i = a + bx_i,\ i = 1, \ldots, m$$

in the linear model

$$y_{ij} = \tau_i + e_{ij};\ i = 1, \ldots, m;\ j = 1, \ldots, n_i;\ \Sigma\, n_i = n$$

Here, the x_i are known constants and a and b are unknown parameters. Under the hypothesis the model becomes

$$y_{ij} = a + bx_i + e_{ij}$$

for which the error sum of squares is found to be equal to

$$\Sigma\,\Sigma\, y_{ij}^2 - n\,\bar{y}_{..}^2 - [\Sigma\, y_{i\cdot}\,(x_i - \bar{x}_{\cdot})]^2/\Sigma\, n_i\,(x_i - \bar{x}_{\cdot})^2$$

with $n - 2$ degrees of freedom. Subtracting from this the error sum of squares $\Sigma\,\Sigma\, y_{ij}^2 - \Sigma\, n_i\,\bar{y}_{i\cdot}^2$, with $n - m$ degrees of freedom, of the original model, we get the usual sum of squares

$$\Sigma\, n_i\,\bar{y}_{i\cdot}^2 - n\,\bar{y}_{..}^2 - [\Sigma\, y_{i\cdot}\,(x_i - \bar{x}_{\cdot})]^2/\Sigma\, n_i\,(x_i - \bar{x}_{\cdot})^2$$

with $m - 2 = (n - 2) - (n - m)$ degrees of freedom, for testing the given hypothesis (we have taken $n\,\bar{x}_{\cdot} = \Sigma\, n_i\, x_i$).

Alternatively, the hypothesis may be formulated as

$$\frac{\tau_i - \tau_j}{x_i - x_j} = \frac{\tau_k - \tau_l}{x_k - x_l}, \quad i \neq j\ k \neq l,$$

which can be expressed equivalently in terms of $m - 2$ independent equations

$$\frac{\tau_1 - \tau_2}{x_1 - x_2} = \frac{\tau_1 - \tau_3}{x_1 - x_3} = \ldots = \frac{\tau_1 - \tau_m}{x_1 - x_m}$$

The hypothesis sum of squares is therefore the sum of squares due to the functions

$$\frac{\bar{y}_{1\cdot} - \bar{y}_{2\cdot}}{x_1 - x_2} - \frac{\bar{y}_{1\cdot} - \bar{y}_{j\cdot}}{x_1 - x_j}, \quad j = 3, 4, \ldots, m$$

and has $m - 2$ degrees of freedom. This sum of squares is a part of the sum of squares due to the functions

$$\bar{y}_{1\cdot} - \bar{y}_{j\cdot},\ j = 2, 3, \ldots, m$$

which is equal to $\Sigma\, n_i\,\bar{y}_{i.}^2 - n\,\bar{y}_{..}^2$, and has $m - 1$ degrees of freedom. We have to subtract from it the sum of squares due to a linear function of $\bar{y}_{1\cdot} - \bar{y}_{j\cdot}$ $(j = 2, \ldots, m)$, that is, of $\bar{y}_{1\cdot}, \ldots, \bar{y}_{m\cdot}$, which is orthogonal to the functions

$$\frac{\bar{y}_{1\cdot} - \bar{y}_{2\cdot}}{x_1 - x_2} - \frac{\bar{y}_{1\cdot} - \bar{y}_{j\cdot}}{x_1 - x_j}, \quad j = 3, \ldots, m$$

Such a function is given by

$$n_1\,(x_1 - \bar{x}_{\cdot})\,\bar{y}_{1\cdot} + \ldots + n_m\,(x_m - \bar{x}_{\cdot})\,\bar{y}_{m\cdot} = (x_1 - \bar{x}_{\cdot})\,y_{1\cdot} + \ldots + (x_m - \bar{x}_{\cdot})\,y_{m\cdot}$$

with sum of squares

$$[\Sigma\, y_{i.}\, (x_i - \bar{x}_.)]^2 / \Sigma\, n_i\, (x_i - \bar{x}_.)^2$$

The hypothesis sum of squares is thus seen to be the same as obtained earlier.

REFERENCES

See Eisenhart (1947) for the basic assumptions underlying the analysis, Box (1954) and Cochran (1947) for the consequences of departure from the basic assumptions.

Bartlett (1947) discusses the use of transformation of variables in cases where the basic assumptions are not satisfied.

A number of applications are described in Cochran and Cox (1957).

9

Randomised Block Design

9.1 Introduction

The completely randomised design cannot be used if the experimental material is not homogeneous and there is considerable variation in the characteristics of the different units (or plots) into which the experimental material is divided. The variation in the yields from different plots can no longer be ascribed solely to differences between the treatments applied and is affected in part by the differences between the plots. For example, if two varieties of a crop are sown on plots with a marked difference in soil fertility, the differences in yield will arise not only due to the types of seed used but also from the differences in soil fertility.

In such situations it may sometimes be possible to stratify or group the experimental material into homogeneous subgroups, called *blocks*, so that the plots within a block are relatively similar in comparision to plots belonging to different blocks. The difference in yields from two plots of the same block can then be assumed to result from the difference in the treatments applied to the two plots. If every treatment is applied to one plot in each block, the arrangement is called a *randomised block design*. The blocks used may be days, observers, batches of material, animals, pens, patients, schools, neighbouring plots in a field, etc., provided that these categories do not interact with the treatments. In other words, the randomised block design may be used to control a source of variation in the experimental material and is, therefore, often characterised as a design giving one way control of heterogeneity.

The units belonging to a particular block should be closely

comparable in quality, and a uniform technique should be employed during the course of the experiment for all units in a block. Any changes in technique or in other conditions that may affect the results should be made between blocks. In an agricultural trial, for example, each block consists of a compact group of neighbouring plots which are more likely to be alike in fertility than plots at some distance. Cultivations designed to keep the land clean of weeds will usually be carried out without regard to the blocks, because it may be assumed that the final yields are not affected by the order in which the plots were cultivated. Similarly, the plots will generally be harvested in whatever order is most convenient. If, however, harvesting must be spread over a number of days, it is well to harvest the plots block by block; should rainfall or other factors produce changes in the weight of the crop from day to day.

If v treatments are to be compared we divide the experimental material into b blocks each containing v plots. The treatments are assigned at random to the plots in each block. One of the v treatments is selected at random and assigned to the first plot, then another treatment is selected at random out of the remaining $v-1$ treatments and assigned to the second plot, and so on. This procedure is followed for all the blocks, a fresh random selection being made for each block. As in the case of the completely randomised design randomisation in each block can be done with the help of a table of random numbers.

The randomised block design has been extensively used since its introduction by Fisher, its chief advantages being greater accuracy compared to the completely randomised design, flexibility, and ease of analysis. Its use, however, is not to be recommended when a large number of treatments are to be compared, as it is usually not possible to obtain a large number of relatively similar experimental units. Even if it were possible to obtain a sufficient number of experimental units to form blocks of the requisite size the cost of experimentation may become unduly high. In such cases one has to resort to other designs; for example, the incomplete block designs.

9.2 Normal Equations and Estimates

Let y_{ij} be the yield from the plot receiving the jth treatment in the ith block. Since the experimental material is relatively homogeneous inside each block, the yield y_{ij} may be assumed to depend only on the particular block and treatment. The equations of the linear model may therefore, be written as

$$y_{ij} = \beta_i + \tau_j + e_{ij} \tag{9.1}$$

with i ranging from 1 to b (the number of blocks) and j from 1 to v (the number of treatments). The parameters β_i represent the effect on the yield of the blocks, and τ_j the effect of the treatments.

The normal equations are

$$\begin{aligned} y_{i\cdot} &= v\beta_i + \tau. \quad (i = 1, ..., b) \\ y_{\cdot j} &= \beta. + b\tau_j \quad (j = 1, ..., v) \end{aligned} \tag{9.2}$$

We shall call the totals $y_{i\cdot}$ and $y_{\cdot j}$ the *block* and *treatment totals* respectively. Similarly $\bar{y}_{i\cdot}$ will be called *block means* and $\bar{y}_{\cdot j}$ *treatment means.*

These $b + v$ equations are not independent as we have

$$\sum_{i=1}^{b} y_{i\cdot} = v\beta. + b\tau. = \sum_{j=1} y_{\cdot j} \tag{9.3}$$

Hence, there can be at most $b + v - 1$ independent equations, so that, there are *at most* $b+v-1$ degrees of freedom for estimates.

The normal equations may also be written as

$$\begin{aligned} \bar{y}_{i\cdot} &= \beta_i + \bar{\tau}. \quad (i = 1, ..., b) \\ \bar{y}_{\cdot j} &= \bar{\beta}. + \tau_j \quad (j = 1, ..., v) \end{aligned} \tag{9.4}$$

From the above form of the normal equations we see that the contrasts $\beta_i - \beta_i'$, called *block contrasts,* and the contrasts $\tau_j - \tau_j'$, called *treatment contrasts,* are estimable, the estimates being

$$\bar{y}_{i\cdot} - \bar{y}_{i'\cdot}\, , \bar{y}_{\cdot j} - \bar{y}_{\cdot j'} \tag{9.5}$$

respectively. There are $b - 1$ independent block contrasts and $v - 1$ independent treatment contrasts; the block contrasts are orthogonal to the treatment contrasts. We have thus $v + b - 2$ linearly independent estimates of the block and treatment contrasts. The grand mean $\bar{y}_{\cdot\cdot}$ is the estimate of the estimable function $\bar{\beta}. + \bar{\tau}.$ and is orthogonal to the estimates of the block and treatment contrasts. Having found $v + b - 1$ linearly independent estimates we conclude that there are $v + b - 1$ degrees of freedom for estimates and consequently $(v - 1)(b - 1)$ for error.

The degrees of freedom for estimates can be obtained alternatively as follows. We can show without much difficulty that a parametric function $\sum_i b_i\beta_i + \sum_j t_j\tau_j$ is estimable, if and only if,

$$\sum_{i=1}^{b} b_i = \sum_{j=1}^{v} t_j \tag{9.6}$$

Thus out of the $b + v$ coefficients of an estimable function only $b + v - 1$ coefficients can be chosen arbitarily. Therefore the maximum number of linearly independent estimable parametric functions, which is equal to the degrees of freedom for estimates is $b + v - 1$.

The $b + v - 1$ linearly independent estimable functions can be chosen in many ways. One such set is given by the following:

(i) $b - 1$ independent block contrasts $\sum_i b_i\beta_i$ $(\Sigma b_i = 0)$

(ii) $v - 1$ independent treatment contrasts $\sum_j t_j\tau_j$ $(\Sigma t_j = 0)$

(iii) the function $\bar{\beta}. + \bar{\tau}.$

The corresponding estimates are $\Sigma b_i\bar{y}_{i\cdot}$, $\Sigma t_j\bar{y}_{\cdot j}$ and $\bar{y}_{\cdot\cdot}$. All other estimable functions and estimates are linear combinations of these.

9.3 Analysis of Variance

Every treatment contrast estimate $\Sigma t_j\bar{y}_{\cdot j}$ is orthogonal to every block contrast estimate $\Sigma b_i\bar{y}_{i\cdot}$ and both these are orthogonal to the grand mean $\bar{y}_{\cdot\cdot}$. Hence the SS due to estimates with $b + v - 1$ degrees of freedom can be written as the sum of the SS due to estimates of block contrasts, S_β^2, with $b - 1$ degrees of freedom, the SS due to estimates of treatment contrasts, S_τ^2, with $v - 1$ degrees of freedom, and the SS due to the grand mean $\bar{y}_{\cdot\cdot}$, $bv\bar{y}_{\cdot\cdot}^2$, with one degree of freedom.

The estimates $\Sigma b_i\bar{y}_{i\cdot}$ of block contrasts are orthogonal to $\bar{y}_{\cdot\cdot}$ and both are linear combinations of the block means $\bar{y}_{1\cdot}, \ldots, \bar{y}_{b\cdot}$. Hence,

$$S_\beta^2 = SS \text{ due to } \{\bar{y}_{1\cdot}, \ldots, \bar{y}_{b\cdot}\} - SS \text{ due to } \bar{y}_{\cdot\cdot}$$

$$= \sum_{i=1}^{b} v\bar{y}_{i.}^2 - bv\bar{y}_{\cdot\cdot}^2 = \sum_{i=1}^{b} (y_{i.}^2/v) - (y_{\cdot\cdot}^2/bv) \tag{9.7}$$

Similarly,

$$S_\tau^2 = \sum_{j=1}^{v} b\bar{y}_{.j}^2 - bv\bar{y}_{\cdot\cdot}^2 = \sum_{j=1}^{v} (y_{.j}^2/b) - (y_{\cdot\cdot}^2/bv) \tag{9.8}$$

The SS due to error is, therefore, given by

$$\text{SS (err)} = S_e^2 = \sum_{i,j} y_{ij}^2 - S_\beta^2 - S_\tau^2 - bv\bar{y}_{..}^2$$

$$= \sum_{i,j} (y_{ij} - \bar{y}_{..})^2 - S_\beta^2 - S_\tau^2 \qquad (9.9)$$

and has $(b - 1)(v - 1)$ degrees of freedom.

The analysis of variance table is given below.

Table 9.1

Source	df	SS	MS	F
Between blocks	$b - 1$	S_β^2	$s_\beta^2 = S_\beta^2/(b-1)$	s_β^2/s^2
Between treatments	$v - 1$	S_τ^2	$s_\tau^2 = S_\tau^2/(v-1)$	s_τ^2/s^2
Error	$(b-1)(v-1)$	S_e^2	$s^2 = S_e^2/(b-1)(v-1)$	
Total	$bv - 1$	$\sum_{ij}(y_{ij} - \bar{y}_{..})^2$		

9.4 Tests of Hypotheses

The major hypothesis of interest is the null hypothesis that there are no differences between the treatments, i.e.

$$\tau_1 = \tau_2 = \ldots = \tau_v \qquad (9.10)$$

or,

$$\tau_1 - \tau_2 = \tau_1 - \tau_3 = \ldots = \tau_1 - \tau_v = 0 \qquad (9.11)$$

Since $\tau_1 - \tau_j$ is estimated by $\bar{y}_{.1} - \bar{y}_{.j}$, the SS (hyp) is the SS due to estimates of treatment contrasts, i.e. S_τ^2. Hence, the hypothesis can be tested by means of the *F*-ratio

$$\frac{S_\tau^2/(v-1)}{S_e^2/(b-1)(v-1)} \qquad (9.12)$$

which has, under the null hypothesis, the *F*-distribution with $[(v - 1), (b - 1)(v - 1)]$ degrees of freedom. Other hypotheses of interest concerning comparisons among specific treatments can be tested by means of the usual *t* and *F* tests.

The error MS s^2 provides an estimate of σ^2 with $(b - 1)(v - 1)$ degrees of freedom. Thus the standard error of the difference between any two treatment means say $\bar{y}_{\cdot j} - \bar{y}_{\cdot k}$, is $\sqrt{2s^2/b}$. The ratio

$$(\bar{y}_{\cdot j} - \bar{y}_{\cdot k})/\sqrt{(2s^2/b)} \tag{9.13}$$

has the t-distribution with $(v - 1)(b - 1)$ degrees of freedom. If t_α denotes the $\alpha\%$ value for the t-distribution (i.e. the probability that $|t| > t_\alpha$ is $\alpha\%$), then $\bar{y}_{\cdot j}$, $\bar{y}_{\cdot k}$ will be significantly different at the given level if

$$|(\bar{y}_{\cdot j} - \bar{y}_{\cdot k})| \geqslant t_\alpha \sqrt{(2s^2/b)} \tag{9.14}$$

The quantity $t_\alpha \sqrt{(2s^2/b)}$ is therefore called the *least significant difference*; any two treatment means whose difference exceeds this value are significantly different.

9.5 Notes on Computation

(i) Arrange the observations in a rectangular array with rows corresponding to blocks and columns to treatments

(ii) Calculate the block totals $y_{i.}$, column totals $y_{.j}$ and the grand total $y_{..}$.

(iii) Obtain the raw SS, the general correction factor $y_{..}^2/bv$, and and the total SS

(iv) Obtain the block and treatment SS

$$S_\beta^2 = (\Sigma y_{i.}^2)/v - y_{..}^2/bv$$

$$S_\tau^2 = (\Sigma y_{.j}^2)/b - y_{..}^2/bv$$

Finally obtain the error SS by subtraction and write down the analysis of variance table.

9.6 Example

A randomised block design was used to compare seven varieties of guayule, a Mexican rubber producing plant, with respect to yield of rubber. The varieties are distinguished by the numbers 109, 130, 405, 406, 407, 416 and 593. These were planted in five blocks of seven plots each. One plant was selected at random from each plot and the rubber obtained from it weighed. The layout of the design was as follows; the figures in brackets show the yield (in grams).

Table 9.2

Block	Plot 1	2	3	4	5	6	7
I	407 (2.06)	405 (2.53)	416 (2.96)	109 (1.46)	593 (6.85)	406 (6.65)	130 (4.06)
II	109 (4.07)	593 (5.92)	405 (1.85)	406 (4.06)	416 (4.35)	130 (9.27)	407 (5.00)
III	109 (6.29)	405 (5.20)	130 (6.42)	416 (2.03)	406 (7.77)	407 (2.59)	593 (3.88)
IV	130 (4.43)	109 (6.84)	405 (6.49)	416 (5.41)	593 (6.71)	407 (6.46)	406 (6.12)
V	407 (7.66)	109 (7.35)	406 (8.11)	405 (7.30)	416 (0.48)	130 (6.64)	593 (5.82)

Source: Federer, W.T., Experimental Design, W.T. Macmillan, New York, 1955, p. 12.

The following table gives the yields arranged by variety and block with the treatment and block totals.

Table 9.3

Treatment	Block I	II	III	IV	V	Treatment total
109	1.46	4.07	6.29	6.84	7.35	26.01
130	4.06	9.27	6.42	4.43	6.64	30.82
405	2.53	1.85	5.20	5.49	7.30	23.37
406	6.65	4.06	7.77	6.12	8.11	32.71
407	2.06	5.00	2.59	6.46	7.66	23.77
416	2.96	4.35	2.03	5.41	0.48	15.23
593	6.85	5.92	3.88	6.71	5.82	29.18
Block total	26.57	34.52	34.18	42.46	43.36	181.09

$$\text{Raw SS} = 1096.5773$$

$$\text{CF} = (181.09)^2/35 = 936.9597$$

$$\text{Total SS} = 159.6176$$

$$\text{Between block SS} = \{[(26.57)^2 + .. + (43.36)^2]/7\} - \text{CF}$$
$$= 27.1559$$

$$\text{Between treatments SS} = \{[(26.01)^2 + ... + (29.18)^2]/5\} - \text{CF}$$
$$= 91.2351$$

The analysis of variance table is given below:

Table 9.4

Source	df	SS	MS	F
Blocks	4	27.1559	6.7890	
Varieties	6	41.2266	6.8711	1.8075
Error	24	91.2351	3.8015	
Total	34	159.6176		

5% and 1% values of F (6.24) are 2.51 and 3.67 respectively.

Thus the data do not suggest any significant difference between the varieties as to yield of rubber. But we may formulate other hypotheses of interest concerning the treatments and test them.

In the above case the following facts are known about the seven varieties under investigation. Variety 109 is the only 54± chromosome variety in the group; the remaining are in the 72± category. Varieties 406 and 130 are selections from 593. Varieties 130, 406 and 593 are phenotypically different from the remaining three varieties 405, 407 and 416. The former have round greenish leaves and short branching habit, while the latter have long serrated greyish-green leaves and longer branches.

Keeping these facts in mind we may formulate the following hypotheses:

$$H_1: 6\tau_1 - (\tau_2 + \tau_3 + \tau_4 + \tau_5 + \tau_6 + \tau_7) = 0$$

$$H_2: \tau_2 - \tau_4 = 0$$

$$H_3: \tau_2 + \tau_4 - 2\tau_7 = 0$$

$$H_4: \tau_2 + \tau_4 + \tau_7 - (\tau_3 + \tau_5 + \tau_6) = 0$$

The first hypothesis compares the average yield from 109 with the average yield from the remaining six varieties; H_2 compares the mean yield from 130 and 406; H_3 compares the mean yield from 130 and 406 with the mean yield from 593; H_4 compares the mean yields from the two phenotypically different groups.

The appropriate SS for these four hypotheses are the sums of squares due to the following four linear functions (which, incidentally, are mutually orthogonal).

(i) $6\bar{y}_{\cdot 1} - \sum_{j=2}^{7} \bar{y}_{\cdot j}$

(ii) $\bar{y}_{\cdot 2} - \bar{y}_{\cdot 4}$

(iii) $\bar{y}_{\cdot 2} + \bar{y}_{\cdot 4} - 2\bar{y}_{\cdot 7}$

(iv) $\bar{y}_{\cdot 2} + \bar{y}_{\cdot 4} + \bar{y}_{\cdot 7} - (\bar{y}_{\cdot 3} + \bar{y}_{\cdot 5} + \bar{y}_{\cdot 6})$

The sums of squares due to these functions, each with one degree of freedom, are,

(i) $$\frac{[6(26.01) - 30.82 - 23.37 - 32.71 - 23.77 - 15.23 - 29.18]^2}{5(36 + 1 + 1 + 1 + 1 + 1 + 1)}$$

$$= 0.00457$$

(ii) $$\frac{[30.82 - 32.71]^2}{5(1 + 1)} = 0.35721$$

(iii) $$\frac{[30.82 + 32.71 - 2(29.18)]^2}{5(1 + 1 + 4)} = 0.89096$$

(iv) $$\frac{[30.82 + 32.71 + 29.18 - 23.37 - 23.77 - 15.23]^2}{5(1+1+1+1+1+1)} = 30.68385$$

The four sums of squares are mutually orthogonal and thus account for four out of the six degrees of freedom for estimates of treatment contrasts. The remaining two degrees of freedom may be taken as due to the estimates of the contrasts among the varieties 405, 407,

416, i.e. among τ_3, τ_5, and τ_6. The estimates of these contrasts are orthogonal to the four estimates obtained earlier, and the SS due to them, with two degrees of freedom, is

$$\frac{(23.37)^2 + (23.77)^2 + (15.23)^2}{5} - \frac{(23.37 + 23.77 + 15.23)^2}{15}$$

$$= 9.29008$$

The total of these six sums of squares, $0.00457 + 0.35721 + 0.89096 + 30.68385 + 9.29008 = 41.22667$, with 6 degrees of freedom, agrees with the SS for treatments, thus providing a cheek on the calculations.

The SS due to the estimates of contrasts among τ_3, τ_5 and τ_6 has not been partitioned further as it was not known if any relationship existed among the varieties 405, 407, and 416. If necessary, it can be partitioned further into a sum of two orthogonal SS each carrying one degree of freedom. For example, we may take the mutually orthogonal estimates $\bar{y}_{\cdot 3} - \bar{y}_{\cdot 5}$ and $\bar{y}_{\cdot 3} + \bar{y}_{\cdot 5} - 2\bar{y}_{\cdot 6}$ of the contrasts $\tau_3 - \tau_5$ and $\tau_3 + \tau_5 - 2\tau_6$ respectively. The corressponding SS are

$$\frac{(23.37 - 23.77)^2}{5(1 + 1)} = 0.01600$$

and

$$\frac{[23.37 + 23.77 - 2(15.23)]^2}{5(1 + 1 + 4)} = 9.27408$$

together adding up to 9.29008.

The above partitioning of the treatment sum of squares, with six degrees of freedom, into six orthogonal sums of squares, each having one degree of freedom, can be exhibited in the form of an extended analysis of variance table (Table 9.5).

A considerable part of the treatment sum of squares is accounted for by the SS due to the comparison between the two phenotypically different groups (130, 406, 593) and (405, 407, 416). Consequently, the F-value corresponding to the difference between these groups turns out to be significant at the 1 % level, although the F-test performed earlier did not show that the seven varieties were significantly different. Hence, specific contrasts among the treatments, whenever these are of interest, should always be examined, even if the general treatment mean square is not significant. A word of caution is,

Table 9.5

Source	df	SS	MS	F
Blocks	4	27.1559	6.7890	
Varieties	6	41.2266	6.8711	1.8075
109 vs others	1	0.00457		
130 + 406 vs 593	1	0.89096		
130 vs 406	1	0.35721		
130, 406, 593 vs 405, 407, 416	1	30.68385		8.0715
405 vs 407	1	0.01600		
405, 407 vs 416	1	9.27408		2.4396
Error	24	91.2351	3.8015	
Total	34	159.6176		

however, necessary. Specific treatment contrasts to be examined for significance should relate to hypotheses formulated independently of the actual experimental results. Comparisons suggested subsequently, by a scrutiny of the results of the experiment, are open to suspicion. For example, a comparison of the highest treatment mean with the lowest treatment mean picked out from the results, when the general *F*-test for treatment contrasts does not indicate significance, will often appear to be significant even if there is no difference among the treatments.

9.7 Efficiency of a Randomised Block Design

The purpose of using a randomised block design is to isolate the variation in the yields caused by the heterogeneity of the experimental material from that due to random (or uncontrolled) factors, This is done by grouping the plots into homogeneous blocks, so that the variability in the experimental material appears mainly as the variability between blocks. From the general theory of linear estimation the expected value of the block mean square s_β^2 is given by

$$E\,(s_\beta^2) = \sigma^2 + [v/(b-1)]\,\Sigma\,(\beta_i - \bar{\beta}_.)$$

whereas $E\,(s^2) = \sigma^2$.

Hence, if the block mean square is appreciably larger than the error mean square we could conclude that the design has been effective in achieving its objective. If, on the other hand, s_β^2 is comparable to or less than s^2, we are led to conclude that the design has not controlled the heterogeneity in the experimental material, or that the experimental material is homogeneous. In either case, the block mean square could be pooled up with the error mean square to provide a better estimate

$$\frac{(b-1)s_\beta^2 + (b-1)\,(v-1)\,s^2}{v\,(b-1)} \tag{9.15}$$

of σ^2 based upon a larger number of degrees of freedom.

The randomised block design is usually an improvement over the completely randomised design. The only exception is the case where $s_\beta^2 < s^2$. The relative efficiency of the randomised block design as compared to the completely randomised design is measured by

$$\frac{(b-1)\,s_\beta^2 + (v-1)\,bs^2}{(vb-1)\,s^2} \tag{9.16}$$

and is $\lesseqgtr 1$ according as $s_\beta^2 \lesseqgtr s^2$.

The above measure of efficiency may be derived in various ways. One method is briefly described here. Imagine a randomised block design with b blocks of v plots to each of which a single treatment is applied. We then have $b-1$ degrees of freedom for the block mean square and $b\,(v-1)$ for the error mean square, resulting in an estimate of σ^2 with $b\,(v-1)$ degrees of freedom. If the same experiment is analysed as a completely randomised design, the block mean square and the error mean square get pooled up to yield another estimate of σ^2 with $(vb-1)$ degrees of freedom. The measure of efficiency is the ratio of the relative precision of the two estimates of σ^2. If we utilise now the block and error mean square estimates obtained from the actual randomised block design, the two estimates of σ^2 are, s^2 for the randomised block design, and

$$\frac{(b-1)\,s_\beta^2 + b\,(v-1)\,s^2}{(vb-1)} \tag{9.17}$$

for the completely randomised design, resulting in the measure of efficiency given earlier.

COMPLEMENTS AND PROBLEMS

1. The parametric function $\beta_i + \tau_j$ is estimable and its best estimate is $\bar{y}_{i\cdot} + \bar{y}_{\cdot j} - \bar{y}_{\cdot\cdot}$. The general parametric function $\Sigma\, b_i\, \beta_i + \Sigma\, t_j\, \tau_j$ is estimable if $\Sigma\, b_i = \Sigma\, t_j$ and then its best estimate is $\Sigma\, b_i\, \bar{y}_{i\cdot} + \Sigma\, t_j\, \bar{y}_{\cdot j} - (\Sigma\, b_i)\, \bar{y}_{\cdot\cdot}$
2. In a randomised block design orthogonal treatment contrasts have orthogonal estimates. For example, $\tau_1 - \tau_2$ and $\tau_1 + \tau_2 - 2\tau_3$ have orthogonal estimates $\bar{y}_{\cdot 1} - \bar{y}_{\cdot 2}$ and $\bar{y}_{\cdot 1} + \bar{y}_{\cdot 2} - 2\bar{y}_{\cdot 3}$. The same result holds for orthogonal block contrasts. However, general estimable orthogonal parametric contrasts $\Sigma\, b_i\, \beta_i + \Sigma\, t_j\, \tau_j$, $\Sigma\, b_i'\, \beta_i + \Sigma\, t_j'\, \tau_j$ do not necessarily have orthogonal estimates. For example, the orthogonal contrasts $\beta_1 + \tau_1$ and $\beta_2 + \tau_2$ have estimates $\bar{y}_{1\cdot} + \bar{y}_{\cdot 1} - \bar{y}_{\cdot\cdot}$, and $\bar{y}_{2\cdot} + \bar{y}_{\cdot 2} - \bar{y}_{\cdot\cdot}$ which are not orthogonal.
3. The expectations of the blocks and treatments sums of squares are given by

$$E\,(S_\beta^2) = (b-1)\,\sigma^2 + v\,\Sigma\,(\beta_i - \bar{\beta}_\cdot)^2$$
$$E\,(S_\tau^2) = (v-1)\,\sigma^2 + b\,\Sigma\,(\tau_j - \bar{\tau}_\cdot)^2$$

4. The linear model for the randomised block design is usually taken as

$$y_{ij} = \mu + \beta_i + \tau_j + e_{ij}$$

In this model $a\mu + \Sigma\, b_i\, \beta_i + \Sigma\, t_j\, \tau_j$ is estimable, if and only if, $a = \Sigma\, b_i = \Sigma\, t_j$. In particular $\Sigma\, b_i\, \beta_i$ and $\Sigma\, t_j\, \tau_j$ are estimable, if and only if, they are contrasts. The analysis of variance remains the same as derived in this chapter.

5. *Missing plot technique:* If only one observation is missing, say that on the j-th treatment in the ith block, the estimated missing value, following the method described in Sec. 7.6, comes out to be

$$\frac{b\,y_{i\cdot} + v\,y_{\cdot j} - y_{\cdot\cdot}}{(b-1)\,(v-1)},$$

where $y_{i\cdot}$ is the total of the $v-1$ yields from the block having the missing observation, $y_{\cdot j}$ is the total of the $b-1$ yields from the treatment on which one observation is missing, and $y_{..}$ is the total from all the $bv-1$ available yields.

If more than one observation is missing, we arbitrarily assign values to all but one of the missing values, which is then obtained by using the formula given above. We then use the formula to compute each missing value in turn and continue the procedure till all the values become stabilised. The number of times this iteration procedure has to be carried out depends on how close the initially arbitrarily assigned values are to the computed values finally obtained. Usually three cycles are enough.

6. *Use of controls in RBD:* Suppose we replicate one treatment (called control) r times in each of the b blocks, and replicate v treatments once in each block. The expected yield from the plot getting the jth treatment in the i-th block is $\beta_i + \tau_j$ $(i = 1, \ldots, b; j = 0, 1, \ldots, v)$ with τ_0 corresponding to the control treatment. If the total of the yields from the $v + r$ plots of the i-th block is denoted by B_i, and from the plots getting the j-th treatment by V_j, the normal equations are

$$(v + r)\,\beta_i + r\tau_0 + \tau. = B_i \qquad (\tau. = \tau_1 + \ldots + \tau_v)$$

$$r\beta. + br\,\tau_0 = V_0$$

$$\beta. + b\tau_j \;\; = V_j \qquad (j = 1, \ldots, v)$$

which may be rewriiten as

$$\beta_i + \frac{r\tau_0 + \tau.}{v + r} = \bar{B}_i \qquad \left(\bar{B}_i = \frac{B_i}{v + r}\right)$$

$$\bar{\beta}. + \tau_0 = \bar{V}_0 \qquad \left(\bar{V}_0 = \frac{V_0}{br},\ \bar{\beta}. = \frac{\Sigma\,\beta_i}{b}\right)$$

$$\bar{\beta}. + \tau_j = \bar{V}_j \qquad (j + 1, \ldots, v),\ \bar{V}_j = \frac{V_j}{b}$$

As in an RBD, we find that $\beta_i - \beta_{i'}$ is estimable with estimate $\bar{B}_i - \bar{B}_{i'}$, and $\tau_j - \tau_{j'}$ $(j, j' = 0, \ldots, v)$ is estimable with estimate $\bar{V}_j - \bar{V}_j'$ and the estimates $\widehat{(\beta_i - \beta_{i'})}$ and $\widehat{(\tau_j - \tau_{j'})}$ are orthogonal.

7. The following are some solutions of the normal equations for the randomised block design model of this chapter;

(i) $\hat{\beta}_i = \bar{y}_{i\cdot} - \frac{1}{2}\,\bar{y}_{\cdot\cdot}$, $\quad \hat{\tau}_j = \bar{y}_{\cdot j} - \frac{1}{2}\,\bar{y}_{\cdot\cdot}$

(ii) $\hat{\beta}_i = \bar{y}_i - \bar{y}_{\cdot\cdot}$, $\hat{\tau}_j = \bar{y}_{\cdot j}$

(iii) $\hat{\beta}_i = \bar{y}_{i\cdot}$, $\hat{\tau}_j = \bar{y}_{\cdot j} - \bar{y}_{\cdot\cdot}$

It is easily verified that each of these solutions leads to the same estimaates of estimable functions, and to the same estimate and error sums of squares.

REFERENCES

Randomised block designs were introduced by Fisher (1926). The case of inequality of variances and correlated errors is discussed by Box (1954). For the treatment of missing values see Allan and Wishart (1930), Yates (1933a), and Baten (1952).

The concept of efficiency was introduced by Yates (1935). Experimental comparision of the efficiency of a randomised block design with the completely randomised design is investigated in Cochran (1938).

Kempthorne and Barclay (1953) examine the role of partitioning of error in the analysis of randomised block designs.

A number of interesting applications are discussed in Cochran and Cox (1957).

10

Latin and Graeco-Latin Square Designs

10.1 Introduction

In a randomised block design the effect of heterogeneity of the experimental material is controlled by applying the treatments over compact blocks of relatively homogeneous material. The treatments are allocated randomly to the plots under one restriction: namely, that each treatment must occur in each block. The arrangement enables one to eliminate the effect of one source of variation. In some situations it is possible to divide the experimental material into blocks according to two different criteria. If, for instance, the experimental unit is a cow, blocks may be made up of cows of the same lactation number. The cows may also be divided into groups according to yield in previous lactation. In such situations, where there are two major sources of variation to be controlled, a randomised block design is of no help. In an agricultural field trial, if there is a fertility gradient in one direction only, we can use a randomised block design arranging the blocks at right angles to the direction of the fertility gradient. However, as frequently happens, there may be fertility differences in two directions, in which case a randomised block design would be able to eliminate the effect of the fertility gradient in one direction only. The Latin square design is to be used in such situations where the effect of two major sources of variability in the experimental material is to be eliminated.

10.2 Latin Square Design

An $m \times m$ *Latin square* is an arrangement of m symbols in m rows and m columns such that each symbol occurs once in each row and once in each column. The following is an example of a 4×4 Latin square:

$$\begin{array}{cccc} C & B & A & D \\ B & C & D & A \\ A & D & C & B \\ D & A & B & C \end{array}$$

Any permutation of the rows, columns or symbols results in another Latin square.

In a Latin square design the rows and columns correspond to the division of the experimental material into blocks according to the two sources of variability which have to be controlled. The letters correspond to the treatments which have to be compared. If v treatments are to be compared, we need v^2 experimental units arranged in v rows and v columns, each treatment occurring exactly once in each row and column. In an agricultural field trial the rows and columns may denote the actual subdivision of a square field into rows and columns to correspond to fertility gradients in two perpendicular directions. In general, however, the arrangement of experimental units into blocks according to two different criteria of subdivision of the experimental material need not be a spatial arrangement in the form of a square array represented by the Latin square used. Even in an agricultural experiment the arrangement of plots may be in the form of a continuous strip of land. In this case the rows may be compact blocks of land while the columns specify the order within each block. The actual arrangement of plots and treatments corresponding to the 4×4 Latin square given earlier will be as shown below.

$$C\ B\ A\ D/B\ C\ D\ A/A\ D\ C\ B/D\ A\ B\ C$$

Some examples of the uses of Latin squares in various fields of research may indicate the utility of the design. In a wear-testing machine with four positions the results obtained may vary from position to position, and from run to run. Comparisons between different materials will be more precise if all are tested in the same position and same run. But such an arrangement is clearly impossible.

If one of the two sources of variation is ignored a randomised block design can be used. But to take into account both sources we need a Latin square design. If the rows stand for positions in the wear testing machine and columns for runs, only four runs can be used and only four materials can be tested. If the 4 × 4 Latin square given earlier is used, material A will be tested in the first position on the third run, in the second on the fourth run, and so on.

To take another example, suppose different brands of petrol are to be compared with respect to mileage per gallon achieved in motor cars. Two important sources of variation could be the differences between individual cars and differences in driving habits of the individual drivers. In this case too a Latin square design can be used, taking rows for cars, columns for drivers, and letters for petrol brands. Using the same 4 × 4 Latin square as above the first driver will use petrol C in the first car, B in the second, and so on.

The chief restriction on the utility of the Latin square design is that each treatment has to be replicated as many times as there are treatments to be compared. If the number of treatments is large the number of replications required becomes impractical. Squares larger than 12 × 12 are seldom used, while the most common range is from 5 × 5 to the 8 × 8 square.

10.3 Graeco-Latin Square Design

Consider the two 4 × 4 Latin squares given below

$$\begin{matrix} B & D & A & C \\ A & C & B & D \\ D & B & C & A \\ C & A & D & B \end{matrix} \qquad \begin{matrix} \beta & \delta & \alpha & \gamma \\ \delta & \beta & \gamma & \alpha \\ \gamma & \alpha & \delta & \beta \\ \alpha & \gamma & \beta & \delta \end{matrix}$$

If we superimpose one square on the other we get the arrangement

$$\begin{matrix} B\beta & D\delta & A\alpha & C\gamma \\ A\delta & C\beta & B\gamma & D\alpha \\ D\gamma & B\alpha & C\delta & A\beta \\ C\alpha & A\gamma & D\beta & B\delta \end{matrix}$$

in which each of the letters A, B, C, D occurs once, and only once, with each of the letters α, β, γ, δ. Two Latin squares with the above property, both squares being of the same size, are called *mutually orthogonal* and the arrangement resulting from their superimposition is called a *Graeco-Latin square.* Any permutation of the rows, columns, Latin letters, or Greek letters of a Graeco-Latin square yields another Graeco-Latin square.

With a Graeco-Latin square design we are able to eliminate a third source of variation, or compare an additional set of treatments. If a $v \times v$ Graeco-Latin square design is used the Greek letters may correspond to a third source of variability or to another set of v treatments to be compared. In an agricultural experiment, where the rows and columns correspond to the subdivision of the field into plots, the Latin letters may represent crop varieties and the Greek letters may stand for different manurial treatments. In the wear testing example, suppose, for the sake of illustration, that the observations in the four different positions on a single run are taken by four different operators. It would then be desirable to eliminate the variation arising out of the differences between operators. With the 4×4 Graeco-Latin square given above, with the Greek letters denoting operators, tests would be carried out in the four positions during the first run by the operators β, δ, γ and α on materials *B*, *A*, *D* and *C* respectively. Similarly, in the study on petrol consumption the differences may arise not only due to the different cars and drivers but also due to the time of the day when the test is carried out. This additional source of variation can be eliminated by using a Graeco-Latin square design with the Greek letters now denoting the particular periods of the day when the test is made. Taking the 4×4 square above the first car would be driven by the first driver using petrol *B* during the period β, by the second driver using petrol *D* during the period δ, and so on.

10.4 Randomisation

The ideal randomisation procedure to allocate the treatments to the different units in a Latin square design would be to assign the Latin letters in any way to the treatments, select at random one Latin square out of all possible squares of that size, each square having the same probability of being selected, and then to assign the treatments to the plots in accordance with the arrangement of letters in the selected square. Such a procedure is, however, hardly practicable, as the number of different possible Latin squares increases very rapidly with size. For example, although there are only two different 2×2 squares, the number of different 6×6 squares is 812, 851, 200. A simpler method of randomisation is given in *Statistical Tables for Biological, Agricultural and Medical Research* by Fisher and Yates. The method consists essentially of selecting one of the "typical" squares given in the tables at random and then effecting a random permutation of the rows, columns and letters.

10.5 Normal Equations and Estimates

In a $v \times v$ Latin square, let y_{ij} denote the yield from the plot belonging to the ith row and jth column. We assume that the expected value of the yield y_{ij} depends upon the row and the column to which the plot belongs and the treatment that is applied. The equations of the linear model appropriate to the design are

$$y_{ij} = \rho_i + \kappa_j + \tau_k + e_{ij} \qquad i, j, k = 1, \ldots, v \tag{10.1}$$

The parameter ρ_i will be called the ith row effect, κ_j the j-th column effect, and τ_k the kth treatment effect. The indices i, j, and k vary from 1 to v. However, once i and j are fixed, the value of k is uniquely determined by the particular Latin square used.

Denoting the row and column totals by $y_{i\cdot}$ and $y_{\cdot j}$ respectively, and the treatment totals (i.e. the sum of the y_{ij} over all plots receiving a particular treatment) by t_k, the normal equations are

$$\begin{aligned} y_{i\cdot} &= v\rho_i + \kappa. + \tau. \qquad i = 1, \ldots, v \\ y_{\cdot j} &= \rho. + v\kappa_j + \tau. \qquad j = 1, \ldots, v \\ t_k &= \rho. + \kappa. + v\tau_k \qquad k = 1, \ldots, v \end{aligned} \tag{10.2}$$

The $3v$ normal equations are not linearly independent; the sum of the first v equations (row equations) is equal to the sum of the v column equations, and also equal to the sum of the v treatment equations. Therefore the degrees of freedom for estimates are at most $3v - 2$.

The normal equations may be written as

$$\left.\begin{aligned} \bar{y}_{i\cdot} &= \rho_i + \bar{\kappa}. + \bar{\tau}. \\ \bar{y}_{\cdot j} &= \rho + \kappa_j + \bar{\tau}. \\ \bar{t}_k &= \bar{\rho}. + \bar{\kappa}. + \tau_k \end{aligned}\right\} \tag{10.3}$$

where $\bar{t}_k = (t_k/v)$ is the mean yield from the kth treatment. We see that the row contrasts $\rho_i - \rho_{i'}$, the column contrasts $\kappa_j - \kappa_{j'}$, and the treatment contrasts $\tau_k - \tau_{k'}$ are estimable; the corresponding estimates are $\bar{y}_{i.} - \bar{y}_{i'.}$, $\bar{y}_{.j} - \bar{y}_{.j'}$ and $\bar{t}_k - \bar{t}_{k'}$ respectively. There are $v-1$ linearly independent contrasts of each type and the three types of contrasts are mutually orthogonal. Further, the parametric function $\bar{\rho}. + \bar{\kappa}. + \bar{\tau}.$ is also estimable, its estimate being the grand mean $\bar{y}_{..}$, and is orthogonal to the row, column and treatment contrasts. We have thus found $3(v-1) + 1 = 3v - 2$ linearly independent estimable functions, so that the degrees of freedom for estimates equal $3v - 2$. There are $v^2 - (3v-2) = (v-1)(v-2)$ degrees of freedom for error.

For a Graeco-Latin square design for v treatments the model equations are

$$y_{ijkl} = \rho_i + \kappa_j + \tau_k + \delta_l + e_{ijkl} \tag{10.4}$$

with i, j, k, l varying from 1 to v. The parameters δ_l represent the effect of the additional source of variation apart from the rows and columns, or the effect of the additional set of treatments to be compared. As in a Latin square the values of k and l are uniquely determined, once i and j are fixed, by the particular Graeco-Latin square used.

Denoting by d_l the total of y_{ij} over the plots corresponding to δ_l the normal equations are

$$\begin{aligned}
y_{i.} &= v\rho_i + \kappa. + \tau. + \delta. \qquad & i &= 1, \ldots, v, \\
y_{.j} &= \rho. + v\kappa_j + \tau. + \delta. \qquad & j &= 1, \ldots, v, \\
t_k &= \rho. + \kappa. + v\tau_k + \delta. \qquad & k &= 1, \ldots, v, \\
d_l &= \rho. + \kappa. + \tau. + v\delta_l \qquad & l &= 1, \ldots, v,
\end{aligned} \tag{10.5}$$

The sum of each set of v equations is the same, viz.

$$y_{..} = v\rho. + v\kappa. + v\tau. + v\delta.$$

Hence there are at most $4v - 3$ degrees of freedom for estimates.

As in a Latin square design we find that the contrasts $\rho_i - \rho_{i'}$, $\kappa_j - \kappa_{j'}$, $\tau_k - \tau_{k'}$, $\delta_l - \delta_{l'}$ as well as the function $\bar{\rho}. + \bar{\kappa}. + \bar{\tau}. + \bar{\delta}.$ are estimable. Thus, we have exactly $4v - 3$ degrees of freedom for estimates and $v^2 - (4v - 3) = (v-1)(v-3)$ for error. The estimates corresponding to the estimable functions above are $\bar{y}_{i.} - \bar{y}_{i'.}$, $\bar{y}_{.j} - \bar{y}_{.j'}$, $\bar{t}_k - \bar{t}_{k'}$, $\bar{d}_l - \bar{d}_{l'}$ ($\bar{d}_l = d_l/v$ is the mean yield corresponding to δ_l), and $\bar{y}_{..}$ respectively.

10.6 Analysis of Variance

In a Latin square design the estimates of the row, column and treatment contrasts are mutually orthogonal, and all are orthogonal to the grand mean. Hence the SS due to estimates with $3v-2$ degrees of freedom can be written as the sum of the following four sums of squares:

(i) SS due to estimates $\Sigma\, r_i \bar{y}_{i\cdot}$ of the row contrasts $\Sigma\, r_i \rho_i$ $(\Sigma\, r_i = 0)$,

$$S_\rho^2 = [(\Sigma\, y_{i.}^2)/v] - [y_{..}^2/v^2] \tag{10.6}$$

with $v-1$ degrees of freedom.

(ii) SS due to estimates $\Sigma\, q_j \bar{y}_{\cdot j}$ of the column contrasts $\Sigma\, q_j \kappa_j$ $(\Sigma\, q_j = 0)$,

$$S_k^2 = [(\Sigma\, y_{.j}^2)/v] - [y_{..}^2/v^2] \tag{10.7}$$

with $v-1$ degrees of freedom.

(iii) SS due to estimates $\Sigma\, a_k \bar{t}_k$ of the treatment contrasts $\Sigma\, a_k \tau_k$ $(\Sigma\, a_k = 0)$,

$$S_\tau^2 = [(\Sigma\, t_k^2)/v] - [y_{..}^2/v^2] \tag{10.8}$$

with $v-1$ degrees of freedom.

(iv) SS due to the grand mean, $y_{..}^2/v^2$, with one degree of freedom.

The SS due to error has $(v-1)(v-2)$ degrees of freedom and can be obtained by subtraction. We have

$$\begin{aligned} \text{SS (Err)} = S_e^2 &= \Sigma\, y_{ij}^2 - S_\rho^2 - S_k^2 - S_\tau^2 - (y_{..}^2/v^2) \\ &= \Sigma\, (y_{ij} - \bar{y}_{..})^2 - S_\rho^2 - S_k^2 - S_\tau^2 \end{aligned} \tag{10.9}$$

The analysis of variance table is given below.

Table 10.1

Source	df	SS	MS	F
Between rows	$v-1$	S_ρ^2	s_ρ^2	
Between columns	$v-1$	S_κ^2	s_κ^2	
Between treatments	$v-1$	S_τ^2	s_τ^2	s_τ^2/s^2
Error	$(v-1)(v-2)$	S_e^2	s^2	
Total	v^2-1	$\sum_{ij}(y_{ij} - \bar{y}_{..})^2$		

Similarly, in the Graeco-Latin square design the SS due to estimates with $4v - 3$ degrees of freedom is partitioned into five mutually orthogonal components. Four of these are the same as for the Latin square; the fifth is the SS due to estimates $\Sigma b_l \bar{d}_l$ of the contrasts $\Sigma b_l \delta_l$ $(\Sigma b_l = 0)$,

$$S_\delta^2 = [(\Sigma d_l^2)/v] - [y_{..}^2/v^2] \tag{10.10}$$

with $v - 1$ degrees of freedom.

The SS due to error is now

$$\text{SS (Err)} = S_e^2 = \Sigma (y_{ij} - \bar{y}_{..})^2 - S_\rho^2 - S_\kappa^2 - S_\tau^2 - S_\delta^2 \tag{10.11}$$

and has $(v - 1)(v - 3)$ degrees of freedom. The analysis of variance table is as follows.

Table 10.2

Source	df	SS	MS	F
Between rows	$v - 1$	S_ρ^2	s_ρ^2	
Between columns	$v - 1$	S_κ^2	s_κ^2	
Between treatments (Latin)	$v - 1$	S_τ^2	s_τ^2	s_τ^2/s^2
Between treatments (Greek)	$v - 1$	S_δ^2	s_δ^2	s_δ^2/s^2
Error	$(v - 1)(v - 3)$	S_e^2	s^2	
Total	$v^2 - 1$	$\sum_{ij} (y_{ij} - \bar{y}_{..})^2$		

10.7 Tests of Hypotheses

Tests of hypotheses in Latin and Graeco-Latin squares are carried out as in a randomised block design. In Latin square designs the F-ratio s_τ^2/s^2 with $[(v - 1), (v - 1)(v - 2)]$ degrees of freedom provides the test for differences among the treatments. Further partitioning of the treatment sum of squares can be done if some specific treatment contrasts are to be examined.

In the Graeco-Latin square design the test for differences among the first set of treatments is carried out by means of the F-ratio s_τ^2/s^2 with $[v-1, (v-1)(v-3)]$ degrees of freedom. If there is a second set of treatments to be compared we use the F-ratio s_δ^2/s^2 which too has $[v-1, (v-1)(v-3)]$ degrees of freedom. The sums of squares for the two sets of treatments may be further partitioned if necessary to test specific contrasts.

10.8 Notes on Computation

(i) Calculate first the row totals $y_{i.}$, the column totals $y_{.j}$, the treatment totals t_k, d_l, and the grand total $y_{..}$

(ii) Obtain the raw SS, the general correction factor $y_{..}^2/v^2$ and the total SS

(iii) Obtain the following sums of squares

$$S_\rho^2 = [(\Sigma y_{i.}^2)/v] - (y_{..}^2/v^2)$$
$$S_\kappa^2 = [(\Sigma y_{.j}^2)/v] - (y_{..}^2/v^2)$$
$$S_\tau^2 = [(\Sigma t_\kappa^2)/v] - (y_{..}^2/v^2)$$
$$S_\delta^2 = [(\Sigma d_l^2)/v] - (y_{..}^2/v^2)$$

(iv) Obtain the error SS by subtraction and complete the analysis of variance table.

10.9 Example

A fumigation experiment was carried out on a strip of land at Rothamsted. Preliminary examination had shown it to be infested with wireworms and the efficacy of the following fumigants was to be compared:

(i) Chlorodinitrobenzene (N) — 2 cwt. per acre
(ii) Chloropicrin (P) — 2 cwt. per acre
(iii) o- and p-dichlorobenzene (K) — 5 cwt. per acre
(iv) Sodium cyanide and anhydrous magnesium sulphate (M) — 7.5 cwt. per acre

In addition it was decided to have some plots without any fumigants to serve as control. This we denote by O (no treatment).

The land was ploughed and cultivated for sowing with sugarbeet. The experimental plots (40′ × 18′) were arranged in a Latin square design but with blocks laid end to end. The four treatments and the control were each replicated five times. The fumigants, diluted with sand, were sprinkled in the bottom of the plough furrow, and the next furrow slice covered the fumigant.

Sampling was done nine weeks after fumigation. The soil samples (9″ × 9″ × 5″) were taken in each third of every plot across the beet rows. In addition, one plot in each block was sampled in between the rows.

Table 10.3 gives the total count of wire-worms per plot and the layout of the Latin square design used.

Table 10.3

P	6	*O*	3	*N*	29	*K*	8	*M*	17
M	8	*K*	13	*O*	18	*N*	12	*P*	16
O	16	*M*	12	*K*	7	*P*	10	*N*	28
N	14	*P*	11	*M*	13	*O*	22	*K*	7
K	7	*N*	26	*P*	24	*M*	14	*O*	20

Source: Landell, W.R.S., *Annals of Applied Biology*, Vol. 25, p. 341, 1938.

The basic calculations are as follows:

Row totals: 63, 67, 73, 67, 91

Column totals: 51, 65, 91, 66, 88

Treatment totals: 67(*P*), 64(*M*), 109(*N*), 42(*K*), 79(10)

Grand Total: 361

$$\text{Raw SS} = 6445.00$$

$$\text{CF} = (361)^2/25 = 5212.84$$

$$\text{Total SS} = 1232.16$$

$$S_P^2 = \{[(63)^2 + (67)^2 + \dots + (91)^2]/5\} - 5212.84 = 98.56$$

$$S_K^2 = \{[(51)^2 + (65)^2 + \dots + (88)^2]/5\} - 5212.84 = 228.56$$

$$S_T^2 = \{[(67)^2 + (64)^2 + \dots + (79)^2]/5\} - 5212.84 = 481.36$$

The analysis of variance table is given in Table 10.4.

Table 10.4

Source	df	SS	MS	F
Rows	4	98.56	24.6400	
Columns	4	228.56	57.1400	1.6184
Treatments	4	481.36	120.3400	3.4084
Error	12	423.68	35.3067	
Total	24	1232.16		

The 5% and 1% values of F with (4, 12) degrees of freedom are 3.26 and 5.41 respectively. The treatment MS gives an F-value of 3.4084; there is thus some evidence of differences between the effects of fumigants.

We may now compare the effect of the fumigants against the control (no treatment); the relevant hypothesis is

$$4\tau_0 - (\tau_P + \tau_M + \tau_N + \tau_K) = 0 \tag{10.12}$$

The SS for testing this hypothesis is the SS due to the estimate

$$4\bar{t}_0 - (\bar{t}_P + \bar{t}_M + \bar{t}_N + \bar{t}_K) \tag{10.13}$$

and is equal to

$$\frac{[4(79) - (67 + 64 + 109 + 42)]^2}{5(16 + 1 + 1 + 1 + 1)} = 11.56$$

The F-value for testing the hypothesis is $(11.56)/(35.3067) < 1$. Thus, the fumigants seem to have no appreciable effect on wire-worm infestation.

The difference between the effect of the control and the average effect of the four fumigants is estimated by

$$\bar{t}_0 - [(\bar{t}_P + \bar{t}_M + \bar{t}_N + \bar{t}_K)/4] \tag{10.14}$$

The variance of this estimate is

$$(\sigma^2/5) - (1/16)\,(4\sigma^2/5) = \sigma^2/4 \tag{10.15}$$

Using the 5% value of t with 12 degrees of freedom, which is equal to 2.179, the 95% confidence interval for the difference between the effect of the control and the average of the effects of the four fumigants is

$$[(79/5) - (1/4)\,(67+64+109+42)\,(1/5)] \pm (2.179)\,\sqrt{(35.3067/4)}$$

$$= 1.70 \pm 6.4738 = (-4.77,\ 8.17)$$

As was to be expected, the difference not being significant, the confidence interval covers the value zero.

The four degrees of freedom for treatments may be split up into four orthogonal contrasts. If we take $4\tau_0 - (\tau_P + \tau_M + \tau_N + \tau_K)$ as one of these, the other three may be taken as $\tau_P - \tau_M$, $\tau_N - \tau_K$, and $(\tau_P + \tau_M) - (\tau_N + \tau_K)$. The estimates of these four contrasts are mutually orthogonal. The SS due to the first has already been calculated and is 11.56. The SS due to the other three are:

$$\frac{(67-64)^2}{5(1+1)} = 0.9$$

$$\frac{(109-42)^2}{5(1+1)} = 448.9$$

$$\frac{(67+64-109-42)^2}{5(1+1+1+1)} = 20.0$$

The four sums of squares, with one degree of freedom each, add up to

$$11.56 + 0.9 + 448.9 + 20.0 = 481.36$$

i.e. the treatment sum of squares. We notice that the major portion of the treatment sum of squares is accounted for by the difference between the treatments N and K.

10.10 Efficiency of the Latin Square Design

As in the randomised block design, the effectiveness of the Latin square design in eliminating the variation due to the rows and columns is indicated by the row and column mean squares s_P^2 and s_K^2. If these are large compared to the error mean square the design may be considered to have achieved its objective. In the example considered in Sec. 10.9 we find that s_ρ^2 is less than s^2, and s_κ^2 though larger than s^2 is not significantly different. Thus, we may conclude that in this case the Latin square design has not been of much use. Similar considerations apply to the Graeco-Latin square design.

The efficiency of a Latin square design compared to a completely randomised design is measured by

$$\epsilon_1 = \frac{(v-1)\,s_\rho^2 + (v-1)\,s_\kappa^2 + (v-1)^2\,s^2}{(v^2-1)\,s^2}$$

The efficiency relative to a randomised block design is

$$\epsilon_2 = \frac{(v-1)\,s_\kappa^2 - (v-1)^2\,s^2}{v\,(v-1)\,s^2}$$

if the rows are regarded as blocks; and is

$$\epsilon_3 = \frac{(v-1)\,s_\rho^2 - (v-1)^2\,s^2}{v\,(v-1)\,s^2}$$

if the columns are regarded as blocks. We note that $\epsilon_2 \gtrless 1$ according as $s_\kappa^2 \gtrless s^2$, $\epsilon_3 \gtrless 1$ according as $s_\rho^2 \gtrless s^2$, and $\epsilon_1 \gtrless 1$ according as $s_\rho^2 + s_\kappa^2 \gtrless 2s^2$.

COMPLEMENTS AND PROBLEMS

1. For the Latin square design we have the following results.

(a) The parametric function $\rho_i + \kappa_j + \tau_k$ is estimable and its best estimate is $\bar{y}_{i\cdot} + \bar{y}_{\cdot j} + \bar{t}_k - 2\,\bar{y}_{\cdot\cdot}$.

(b) The general parametric function $\Sigma\, r_i\, \rho_i + \Sigma\, k_j\, \kappa_j + \Sigma\, s_k\, \tau_k$ is estimable if, and only if, $\Sigma\, r_i = \Sigma\, k_j = \Sigma\, s_k$, and then its best estimate is $\Sigma\, r_i\, \bar{y}_{i\cdot} + \Sigma\, k_j\, \bar{y}_{\cdot j} + \Sigma\, s_k\, \bar{t}_k - 2\,(\Sigma\, r_i)\,\bar{y}_{\cdot\cdot}$.

(c) Orthogonal treatment contrasts have orthogonal estimates. The same result holds for orthogonal row or column contrasts. However, general estimable orthogonal parametric functions do not necessarily have orthogonal estimates.

(d) $E\,(S_\rho^2) = (v-1)\,\sigma^2 + v\,\Sigma\,(\rho_i - \bar{\rho}_\cdot)^2$

(e) $E\,(S_\kappa^2) = (v-1)\,\sigma^2 + v\,\Sigma\,(\kappa_j - \bar{\kappa}_\cdot)^2$

(f) $E\,(S_\tau^2) = (v-1)\,\sigma^2 + v\,\Sigma\,(\tau_k - \bar{\tau}_\cdot)^2$

2. *Missing plot technique:* If only one observation is missing, say that in the ith row and jth column and having the kth treatment, the estimated missing value, following the method described in Sec. 7.6, is given by

$$\frac{v\,(y_{i\cdot} + y_{\cdot j} + t_k) - 2\,y_{\cdot\cdot}}{(v - 1)\,(v - 2)}$$

where $y_{i\cdot}$ is the total of the $v - 1$ yields from the i-th row, $y_{\cdot j}$ the total of the $v - 1$ yields from the j-th column, t_k is the total of the $v - 1$ yields from the plots getting the k-th treatment, and $y_{\cdot\cdot}$ is the total from the $v^2 - 1$ available yields.

If more than observation is missing, we arbitrarily assign values to all but one of them which is then determined by means of the above formula. The same formula is then used to compute each missing value in turn and the procedure continued till all the values get stabilised. The number of times this iteration procedure has to be carried out depends on how close the values arbitrarily assigned initially are to the computed values finally obtained.

3. If we write the linear model for the Latin Square design as

$$y_{ij} = \mu + \rho_i + \kappa_j + \tau_{k(ij)} + e_{ij}$$

the parametric function $a\mu + \Sigma\, r_i\, \rho_i + \Sigma\, k_j\, \kappa_j + \Sigma\, t_k\, \tau_k$ is estimable if, and only if, $a = \Sigma\, r_i = \Sigma\, k_j = \Sigma\, t_k$. In particular $\Sigma\, r_i\, \rho_i$, $\Sigma\, k_j\, \kappa_j$, and $\Sigma\, t_k\, \tau_k$ are estimable if, and only if, they are contrasts. The analysis of variance for this model is the same as that derived in this chapter.

REFERENCES

Latin square designs were introduced by Fisher (1926). Their use in field experiments is discussed by Yates (1933b). The concept of efficiency was introduced by Yates (1935).

For a discussion of missing values see Delury (1946), and Yates and Hale (1939). Freeman (1964) considers the case where further treatments are added to a Latin-square design. Application of Latin-Square designs to psychology is found in Grant (1948).

For a design more general than a Graeco-Latin square design see Pothoff (1962).

Construction of mutually orthogonal Latin squares is described in detail in Raghavarao (1971).

Instructive examples are found in Cochran and Cox (1957).

11

Incomplete Block Designs: General Theory

11.1 Introduction

In a randomised block or Latin square design the experimental units are grouped into blocks, or into rows and columns, and every treatment occurs once in each group. If the number of treatments to be compared is large, such designs are not suitable. This is due to the fact that sufficient numbers of homogeneous experimental units may not be available to form a block large enough to accomodate all the treatments. In biological work on animals, for example, it will be desirable, if at all possible, to compare treatments within litters, but the litter size will depend on the particular species and will often be such that it is impossible to include all treatments within a litter. Similar considerations apply to greenhouse pot experiments, where the block is restricted to the width of the bench; to experiments on plant virus diseases, where the block consists of a small number of leaves on each plant; to cookery experiments where only a limited number of stoves are available; and to experiments on the control of fruit pests, where the tree constitutes the experimental unit. Even if a sufficient number of homogeneous experimental units is available to include all treatments inside a block, their use may make the cost of the experiment prohibitive.

Incomplete block designs have been introduced to meet such situations; as their name implies, each block receives only some, and not all, of the treatments to be compared. Smaller block size enables one to eliminate the effect of heterogeneity of the experimental material to a greater extent than would be possible with designs using larger blocks. An additional advantage is the reduction in experimental costs. In this chapter we shall develop the general analysis for incomplete block designs. Two special types of incomplete block designs will be discussed in Chapters 12 and 13.

11.2 Normal Equations

Suppose v treatments have to be compared and b blocks are available. Let k_i denote the number of plots in the ith block, and r_j the number of plots receiving the jth treatment. If N is the total number of plots we have

$$N = r_1 + \ldots + r_v = k_1 + \ldots + k_b$$

Each treatment may occur more than once in each block, or may not occur at all. If n_{ij} denotes the number of times the jth treatment occurs in the ith block we have

$$\sum_j n_{ij} = k_i,\ i = 1, \ldots, b$$

and,

$$\sum_i n_{ij} = r_j,\ j = 1, \ldots, v$$

Let y_{ijm} denote the yield from the mth replicate of the jth treatment in the ith block. The model equations are assumed to be

$$y_{ijm} = \beta_i + \tau_j + e_{ijm} \tag{11.1}$$

where β_i represents the effect of the ith block, and τ_j the effect of the jth treatment.

Let $B_1, \ldots, B_b$ denote the *block totals* ($B_i = \sum_{jm} y_{ijm}$), $V_1, \ldots, V_v$ the *treatment totals* ($V_j = \sum_{im} y_{ijm}$), and Y the *grand total* ($Y = \sum_{ijm} y_{ijm}$). The normal equations are

$$k_i\,\beta_i + n_{i1}\,\tau_1 + \ldots + n_{iv}\,\tau_v = B_i, \quad i = 1, \ldots, b \tag{11.2}$$

and

$$n_{1j}\,\beta_1 + \ldots + n_{bj}\,\beta_b + r_j\,\tau_j = V_j, \quad j = 1, \ldots, v \tag{11.3}$$

The $b + v$ equations are not linearly independent as the sum of the b equations (11.2) equals the sum of the v equations (11.3). Thus, there are at most $b + v - 1$ degrees of freedom for estimates. We shall see later that the degrees of freedom for estimates may be less than $b + v - 1$; they are equal to $b + v - 1$ only in special cases.

11.3 Adjusted Treatment Totals

Using equations (11.2) to eliminate the block parameters β_i from equations (11.3) we get the following v equations

$$V_j = n_{1j}\,[(B_1 - n_{11}\,\tau_1 - \ldots - n_{1v}\,\tau_v)/k_1] + \ldots$$
$$+ n_{bj}\,[(B_b - n_{b1}\,\tau_1 - \ldots - n_{bv}\tau_v)/k_b] + r_j\,\tau_j$$

i.e.,

$$V_j - (n_{1j}\,B_1/k_1) - \ldots - (n_{bj}\,B_b/k_b) = \tau_1\,[-(n_{11}\,n_{1j})/k_1 - \ldots$$
$$- (n_{b1}\,n_{bj})/k_b] + \ldots + \tau_j\,[r_j - (n_{1j}^2/k_1) - \ldots$$
$$-(n_{bj}^2/k_b)] + \ldots + \tau_v\,[-(n_{1v}\,n_{1j}/k) - \ldots - (n_{bv}\,n_{bj}/k_b)]$$
$$j = 1, \ldots, v. \qquad (11.4)$$

The quantities

$$Q_j = V_j - \left\{\frac{n_{1j}\,B_1}{k_1} + \ldots + \frac{n_{bj}\,B_b}{k_b}\right\}, \quad j = 1, \ldots, v \qquad (11.5)$$

are called the *adjusted treatment totals* (adjusted for block effects). There are k_i plots in the ith block, and hence B_i/k_i may be called the average yield per plot from the ith block. Thus, $n_{ij}\,B_i/k_i$ may be considered as the average contribution to the jth treatment total from the ith block. The adjusted treatment total Q_j is obtained by removing from the jth treatment total V_j the sum of the average contributions of the b blocks.

We write equations (11.4) as

$$Q_j = c_{j1}\,\tau_1 + \ldots + c_{jv}\,\tau_v, \quad j = 1, \ldots, v \qquad (11.6)$$

where

$$c_{jj} = r_j - (n_{1j}^2/k_1) - \ldots - (n_{bj}^2/k_b)$$

and

$$c_{jj'} = -(n_{1j}\,n_{1j'}/k_1) - \ldots - (n_{bj}\,n_{bj'}/k_b), \quad j \neq j'$$

The $v \times v$ matrix $\mathbf{C} = [c_{ij}]$ is called the *C-matrix* of the incomplete block design. The C-matrix is symmetric and its row and column sums are zero; it follows that the C-matrix is singular. Using the C-matrix, equations (11.6) can be written as

$$\mathbf{Q} = \mathbf{C}\,\boldsymbol{\tau} \qquad (11.7)$$

where $\mathbf{Q}' = (Q_1, \ldots, Q_v)$ and $\boldsymbol{\tau}' = (\tau_1, \ldots, \tau_v)$. We shall call (11.7) the *reduced normal equations.* The normal equations (11.2) and (11.3) of the model are equivalent to (11.2) and (11.7).

Theorem 11.1: The adjusted treatment totals Q_j are orthogonal to the block totals B_i.

Proof: It is enough to prove that Cov $(B_i, Q_j) = 0$ for all i and j. Now

$$\text{Cov}\,(B_i, Q_j) = \text{Cov}\,[B_i, V_j - \sum_i (n_{ij}/k_i)\, B_i]$$

$$= \text{Cov}\,(B_i, V_j) - (n_{ij}/k_i)\,\text{Var}\,(B_i)$$

since the block totals are mutually orthogonal. As B_i and V_j have n_{ij} observations in common and the observations are mutually independent, we have

$$\text{Cov}\,(B_i, V_j) = n_{ij}\,\sigma^2$$

Also

$$\text{Var}\,(B_i) = k_i\,\sigma^2$$

Hence,

$$\text{Cov}\,(B_i, Q_j) = n_{ij}\,\sigma^2 - (n_{ij}/k_i)\,k_i\,\sigma^2 = 0$$

Theorem 11.2: For the adjusted treatment totals Q_j we have

(i) $E(\mathbf{Q}) = \mathbf{C}\boldsymbol{\tau}$

(ii) $D(\mathbf{Q}) = \sigma^2\mathbf{C}$

Proof: (i) follows from the fact that equations (11.7) are obtained by carrying out linear operations on the normal equations. Further,

$$\text{Var}(Q_j) = \text{Var}\,[V_j - \sum_i (n_{ij}/k_i)\,B_i]$$

$$= \text{Var}\,(V_j) + \sum_i (n_{ij}/k_i)^2\,\text{Var}\,(B_i) - 2\sum_i (n_{ij}/k_i)\,\text{Cov}\,(V_j, B_i)$$

$$= r_j\,\sigma^2 + \sum_i (n_{ij}/k_i)^2\,k_i\,\sigma^2 - 2\sum_i (n_{ij}/k_i)\,n_{ij}\,\sigma^2$$

$$= r_j\,\sigma^2 - \sum_i (n_{ij}^2/k_i)\,\sigma^2 = c_{jj}\,\sigma^2 \tag{11.8}$$

Also,

$$\text{Cov}\,(Q_j, Q_{j'}) = \text{Cov}\,[V_j - \sum_i (n_{ij}/k_i)\,B_i, V_{j'} - \sum_i (n_{ij'}/k_i)\,B_i]$$

$$= -\sum_i (n_{ij'}/k_i)\,\text{Cov}\,(V_j, B_i) - \sum_i (n_{ij}/k_i)\,\text{Cov}\,(B_i, V_{j'})$$

$$+ \sum_i (n_{ij}\,n_{ij'}/k^2)\,\text{Var}\,(B_i)$$

$$= -\sum_i (n_{ij'} n_{ij}/k_i)\,\sigma^2 - \sum_i (n_{ij} n_{ij'}/k_i)\,\sigma^2 + \sum_i (n_{ij} n_{ij'}/k_i^2)\, k_i\,\sigma^2$$

$$= c_{jj'}\,\sigma^2 \tag{11.9}$$

The proof of (ii) i s contained in (11.8) and (11.9).

The adjusted treatment totals are best estimates of their expected values. We have seen that, unlike the treatment totals V_j, the adjusted treatment totals are orthogonal to the block totals. However, though the block totals B_i are mutually orthogonal (the same is true for the treatment totals V_j), the adjusted treatment totals Q_j are not linearly independent. In fact we have

$$\sum_j Q_j = \Sigma\,[V_j - \sum_i (n_{ij}/k_i)\, B_i] = Y - \sum_i\sum_j (n_{ij}/k_i)\, B_i$$

$$= Y - \sum_i B_i = 0$$

This is not the only linear relation among the Q_j's; there may be more than one such relation, as we shall see later, depending on the arrangement of the treatments in the incomplete block.

10.4 Estimates

From the normal equations (11.2) and (11.3) we see that a linear parametric function is estimable, if and only if, it is of the form

$$\sum_j q_j\, E(V_j) + \sum_i p_i\, E(B_i) \tag{11.10}$$

for some constants p_i, q_j, i.e., of the form

$$\sum_j q_j\,(r_j\,\tau_j + \sum_i n_{ij}\,\beta_i) + \sum_i p_i\,(k_i\,\beta_i + \sum_i n_{ij}\,\tau_j)$$

i.e.,

$$\sum_j (q_j\, r_j + \sum_i p_i\, n_{ij})\,\tau_j + \sum_i (p_i\, k_i + \sum_j q_j\, n_{ij})\,\beta_i \tag{11.11}$$

Theorem 11.3: A necessary condition for $\sum_i b_i\beta_i + \sum_j t_j\tau_j$ to be estimable is $\sum_i b_i = \sum_j t_j$.

Proof: If the function is estimable there exist constants p_i and q_j such that $b_i = p_i\, k_i + \sum_j q_j\, n_{ij}$, and $t_j = q_j\, r_j + \sum_i p_i\, n_{ij}$.

Hence,

$$\sum_j t_j = \sum_j q_j\, r_j + \sum_j\sum_i p_i\, n_{ij}$$

$$= \sum_j q_j r_j + \sum_i p_i (\sum_j n_{ij})$$

$$= \sum_j q_j r_j + \sum_i p_i k_i = \sum_i b_i$$

Corollary: A necessary condition for the function $\sum_j t_j \tau_j$ to be estimable is $\sum_j t_j = 0$, i.e. it must be a contrast. (We shall see later that the condition is not sufficient.)

We now examine in detail the problem of estimating linear parametric functions $\sum_j t_j \tau_j$ involving the treatment parameters alone.

Theorem 11.4: The best estimate of an estimable function $\sum_j t_j \tau_j$ is a linear function of the adjusted treatment totals.

Proof: It follows from (11.11) that

$$\sum_j t_j \tau_j = \sum_j (q_j r_j + \sum_i p_i n_{ij}) \tau_j + \sum_i (p_i k_i + \sum_j q_j n_{ij}) \beta_i \quad (11.13)$$

for some constants p_i, q_j; and from (11.10) that its best estimate is

$$\sum_j q_j V_j + \sum_i p_i B_i \quad (11.14)$$

But we must have

$$p_i k_i + \sum_j q_j n_{ij} = 0, \; i = 1, \ldots, b$$

Hence the best estimate is

$$\sum_j q_j V_j + \sum_i (-\sum_j q_j n_{ij}/k_i) B_i$$

i.e.,

$$\sum_j q_j [V_j - \sum_i (n_{ij}/k_i) B_i] = \sum_j q_j Q_j \quad (11.15)$$

Theorem 11.5: The parametric function $\mathbf{t}'\boldsymbol{\tau} = \sum_j t_j \tau_j$ is estimable, if and only if, $\mathbf{t}' \in R(\mathbf{C})$, i.e., $\mathbf{t}' = \mathbf{q}'\mathbf{C}$ for some $\mathbf{q}$.

Proof: Suppose $\mathbf{t}'\boldsymbol{\tau}$ is estimable and let $\mathbf{q}'\mathbf{Q}$ be its best estimate. Then we have

$$E(\mathbf{q}'\mathbf{Q}) = \mathbf{q}'\mathbf{C}\boldsymbol{\tau} = \mathbf{t}'\boldsymbol{\tau}$$

for all values of $\boldsymbol{\tau}$. Hence $\mathbf{t}' = \mathbf{q}'\mathbf{C}$. Conversely, if $\mathbf{t}' = \mathbf{q}'\mathbf{C}$, $\mathbf{q}'\mathbf{Q}$ is the estimate of $\mathbf{t}'\boldsymbol{\tau}$ and $\mathbf{t}'\boldsymbol{\tau}$ is estimable.

Corollary: The maximum number of linearly independent estimable functions $\mathbf{t}'\boldsymbol{\tau}$ equals the rank of the C-matrix.

Theorem 11.6: If $\mathbf{t}'\boldsymbol{\tau}$ is estimable, its best estimate is $\mathbf{t}'\hat{\boldsymbol{\tau}}$, where $\hat{\boldsymbol{\tau}}$ is any solution of the reduced normal equations.

Proof: Since $\mathbf{t}'\boldsymbol{\tau}$ is estimable, we have $\mathbf{t}' = \mathbf{q}'\mathbf{C}$ so that its best estimate is $\mathbf{q}'\mathbf{Q}$. If $\hat{\boldsymbol{\tau}}$ is any solution of the reduced normal equations, we have $\mathbf{C}\hat{\boldsymbol{\tau}} = \mathbf{Q}$, so that

$$\mathbf{t}'\hat{\boldsymbol{\tau}} = \mathbf{q}'\mathbf{C}\hat{\boldsymbol{\tau}} = \mathbf{q}'\mathbf{Q}$$

Note that the reduced normal equations, having been obtained from the normal equations of the incomplete block model, are consistent, so that a solution $\hat{\boldsymbol{\tau}}$ always exists.

The above discussion shows that for finding estimable functions involving treatment perameters only, and their estimates, it is enough to consider equations (11.7). They provide all the information relating to functions of the form $\mathbf{t}'\boldsymbol{\tau}$, and the complete set of normal equations (11.2) and (11.3) need not be used. It is for this reason that we have called equations (11.7) the reduced normal equations.

Suppose $\mathbf{t}'\boldsymbol{\tau}$ is estimable and its best estimate is $\mathbf{q}'\mathbf{Q}$. Then,

$$\text{Var}\,(\mathbf{q}'\mathbf{Q}) = (\mathbf{q}'\mathbf{C}\mathbf{q})\,\sigma^2$$

But, we have $E\,(\mathbf{q}'\mathbf{Q}) = \mathbf{q}'\mathbf{C}\boldsymbol{\tau} = \mathbf{t}'\boldsymbol{\tau}$ for all $\boldsymbol{\tau}$, so that $\mathbf{q}'\mathbf{C} = \mathbf{t}'$.

Using this result we can express the variance of the estimate $\mathbf{q}'\mathbf{Q}$ of $\mathbf{t}'\boldsymbol{\tau}$ as

$$\text{Var}\,(\mathbf{q}'\mathbf{Q}) = (\mathbf{t}'\mathbf{q})\,\sigma^2$$

Thus the variance is easily obtained if both $\mathbf{t}$ and $\mathbf{q}$ are known. In particular, if $\mathbf{q}' = (q_1, \ldots, q_v)$ the variance of the estimate of any contrast of the type $\tau_j - \tau_{j'}$ is

$$(q_j - q_{j'})\,\sigma^2$$

Thus, to obtain the variance of the estimate of the contrast $\tau_j - \tau_{j'}$ we need only calculate the coefficients of Q_j, $Q_{j'}$ in the estimate $\mathbf{q}'\mathbf{Q}$ of $\tau_j - \tau_{j'}$, coefficients of the other terms need not be calculated.

11.5 Analysis of Variance

Since the best estimates are linear combinations of the block and treatment totals B_i and V_j, the degrees of freedom for estimates equal the degrees of freedom carried by the set $\{B_i, V_j\}$ of the $b + v$

block and treatmcnt totals. The normal equations (11.2) and (11.3) are equivalent to the set of equations (11.2) and (11.7); hence the stimate degrees of freedom are also equal to the degrees of freedom carried by the set $\{B_i, Q_j\}$.

Since the block totals B_i are orthogonal to the adjusted treatment totals Q_j, the degrees of freedom of the set $\{B_i, Q_j\}$ are equal to the sum of the degrees of freedom carried by the sets $\{B_i\}$ and $\{Q_j\}$. The block totals B_i being mutually orthogonal, the set $\{B_i\}$ carries b degrees of freedom. The adjusted treatment totals are not linearly independent, since $\sum_i Q_j = 0$. Thus, the set $\{Q_j\}$ has at most $v - 1$ degrees of freedom. Since each Q_j estimates a contrast $t'\tau$ and the maximum number of linearly independent estimable contrasts of the type $\mathbf{t}'\boldsymbol{\tau}$ equals the rank of the C-matrix, the degrees of freedom carried by the set $\{Q_j\}$ are equal to rank (C), say $v - m$ $(m \geqslant 1)$.

Thus we have $b + v - m$ degrees of freedom for estimates. The SS due to estimates is the sum of the SS due to the set $\{B_i\}$ with b degrees of freedom, and the SS due to the $\{Q_j\}$ with $v - m$ degrees of freedom. The SS due to a total B_i is $B_i{}^2/k_i$, and hence the SS due to the set B_i is

$$\sum_i (B_i^2/k_i) \quad \text{with } b \text{ d.f.} \tag{11.16}$$

The grand mean Y/N, which is the best estimate of $\sum_i k_i\beta_i + \sum_j r_j\tau_j$, is orthogonal to any contrast among the block means B_i/k_i.

Hence the sum of squares

$$S_B^2 = \sum_i (B_i^2/k_i) - (Y^2/N) \tag{11.17}$$

with $b - 1$ degrees of freedom, is the SS due to contrasts among the block means, we call it the *unadjusted block SS.* (It is also called the *SS for blocks ignoring treatments.*)

It remains to find the SS due to the set $\{Q_j\}$ which we denote by $S_{V/B}^2$ to distinguish it from S_V^2 which is defined later. Writing

$$\mathbf{Q} = \mathbf{By} \tag{11.18}$$

we get (see Th.5.1) $S_{V/B}^2 = \hat{\boldsymbol{\alpha}}'\mathbf{By} = \hat{\boldsymbol{\alpha}}\mathbf{Q}$, where $\hat{\boldsymbol{\alpha}}$ is any solution of the equations,

$$\mathbf{BB}'\boldsymbol{\alpha} = \mathbf{By} = \mathbf{Q} \tag{11.19}$$

But we have

$$\mathbf{D(Q)} = \mathbf{D(By)} = \mathbf{BD(y)\,B'} = \mathbf{BB'}\sigma^2 = \mathbf{C}\sigma^2$$

Thus, $\mathbf{BB'} = \mathbf{C}$ and it follows that

$$S^2_{V/B} = \hat{\boldsymbol{\tau}}'\mathbf{Q} \tag{11.20}$$

where $\hat{\tau}$ is any solution of the reduced normal equations. The degrees of freedom are $v - m$. We call $S^2_{V/B}$ the *adjusted treatment SS*. (It is also called the *SS for treatments eliminating blocks.*)

The error SS is now given by

$$\text{SS(Err)} = S_e^2 = \sum_{ijm} y_{ijm}^2 - (Y^2/N) - S_B^2 - S^2_{V/B} \tag{12.21}$$

with $N - (b + v - m)$ degrees of freedom.

Instead of adjusting the treatment totals by eliminating the block parameters we could also eliminate the treatment parameters from the equation (11.2), using (11.3), to get the *adjusted block totals*

$$B_i - \sum_j (n_{ij}/r_j)\, V_j, \qquad i = 1, \ldots, b \tag{11.22}$$

These are orthogonal to the treatment totals V_j and the estimate sum of squares can also be partioned into the *unadjusted treatment SS* (also called *SS for treatments ignoring blocks*)

$$S_V^2 = \sum_j (V_j^2/r_j) - (Y^2/N) \tag{11.23}$$

with $v - 1$ degrees of freedom, the *adjusted block SS* (also called *SS for blocks eliminating treatments*)

$$S^2_{B/V} = \text{SS due to the set of adjusted block totals}$$

$$= S_B^2 + S^2_{V/B} - S_V^2 \tag{11.24}$$

with $b - m$ degrees of freedom, and the SS due to the grand mean, Y^2/N, with one degree of freedom. The error SS then comes out to be the same as obtained earlier.

As in the case of $S^2_{V/B}$, the adjusted block SS, $S^2_{B/V}$, could also have been obtained by solving the reduced normal equations for the adjusted block totals. However, $S^2_{B/V}$ is obtained more easily by using the relation

$$S_B^2 + S^2_{V/B} = S_V^2 + S^2_{B/V} \tag{11.25}$$

The error *SS* obtained above is, for reasons to appear later, called the *intrablock error*.

The analysis of variance is usually presented in the form given in Table 11.1 on page 138.

11.6 Example

Suppose four treatments (1, 2, 3, 4) are laid out in five blocks (I, II, III, IV, V) as below:

I: 1, 1, 2

II: 1, 2, 2

III : 3, 4

IV : 3, 4

V : 3, 4

The adjusted treatment totals are

$$Q_1 = V_1 - (2/3)\, B_1 - (1/3)\, B_2$$
$$Q_2 = V_2 - (1/3)\, B_1 - (2/3)\, B_2$$
$$Q_3 = V_3 - (B_3 + B_4 + B_5)/2$$
$$Q_4 = V_4 - (B_3 + B_4 + B_5)/2$$

so that we have $Q_1 + Q_2 = Q_3 + Q_4 = 0$. The reduced normal equations are most easily written by equating each Q_j to its expected value.

We have

$$\begin{aligned} E\,(Q_1) &= E\,(V_1) - (2/3)\, E\,(B_1) - (1/3)\, E\,(B_2) \\ &= (2\beta_1+2\tau_1+\beta_2+\tau_1) - (2/3)\,(3\beta_1+2\tau_1+\tau_2) \\ &\qquad\qquad - (1/3)\,(3\beta_2+\tau_1+2\tau_2) \\ &= (4/3)\,\tau_1 - (4/3)\,\tau_2 \end{aligned}$$

Similarly, we obtain $E\,(Q_2)$, $E\,(Q_3)$ and $E\,(Q_4)$ to get the reduced normal equations below:

$$\begin{bmatrix} 4/3 & -4/3 & 0 & 0 \\ -4/3 & 4/3 & 0 & 0 \\ 0 & 0 & 3/2 & -3/2 \\ 0 & 1 & -3/2 & 3/2 \end{bmatrix} \tau = \mathbf{Q}$$

Table 11.1

Source	df	SS	SS	df	Source
Blocks (unadj.)	$b - 1$	S_B^2	$S_{B/V}^2$	$b - m$	Blocks (adj.)
Treatments (adj.)	$v - m$	$S_{V/B}^2$	$S_V{}^2$	$v - 1$	Treatments (unadj.)
Error	$N - b - v + m$	S_e^2	S_e^2	$N - b - v + m$	Error
Total	$N - 1$	$\Sigma\, y_{ijk}^2 - \dfrac{Y^2}{N}$	$\Sigma\, y_{ijk}^2 - \dfrac{Y^2}{N}$	$N - 1$	Total

Rank of the *C*-matrix is 2; only two linearly independent treatment contrasts are estimable. If we take these as $\tau_1 - \tau_2$, $\tau_3 - \tau_4$, their best estimate are (3/4) Q_1 and (2/3) Q_3. We see that the contrasts $\tau_2 - \tau_3$, $\tau_2 - \tau_4$, $\tau_1 - \tau_3$ and $\tau_1 - \tau_4$ are not estimable.

The variances of the estimates of $\tau_1 - \tau_2$ and $\tau_3 - \tau_4$ are

$$\text{Var}\,(3Q_1/4) = (9/16)\,\text{Var}\,(Q_1) = (9/16)\,(4\sigma^2/3) = 3\sigma^2/4$$

and

$$\text{Var}\,(2Q_3/3) = (4/9)\,\text{Var}\,(Q_3) = (4/9)\,(3\sigma^2/2) = 2\sigma^2/3$$

11.7 Connectedness and Balance

We saw in Section 11.6 that all treatment contrasts may not be estimable in an incomplete block design. An incomplete block design in which all treatment contrasts are estimable is called a *connected design*. The following result is an immediate consequence of Theorem 11.5.

Theorem 11.7: An incomplete block design is connected if, and only if, the rank of the *C*-matrix is $v - 1$.

It is clear that if treatments j and j' occur in the same block, the contrast $\tau_j - \tau_{j'}$ is estimable. Further, if treatments j and j' occur together in a block, and j' and j'' occur together in another block, then $\tau_j - \tau_{j'}$, $\tau_{j'} - \tau_{j''}$, and hence $\tau_j - \tau_{j''}$, are estimable. These considerations lead us to another definition of a connected design: a design will be called a connected design if, given any two treatments θ and ϕ, it is possible to construct a chain of treatments $\theta = \theta_0, \theta_1, \ldots, \theta_n = \phi$ such that every consecutive pair of treatments in the chain occurs together in a block. The two definitions of connectedness can be shown to be equivalent.

A design which is not connected breaks up into *connected components*. In the example of Sec. 11.6 the design is not connected and has two connected components consisting of blocks I, II and blocks III, IV, and V. Treatment contrasts in a connected component are estimable. A general treatment contrast is estimable if it can be expressed as a linear combination of treatment contrasts belonging to connected components.

We have seen (Sec. 11.6) that the estimates of treatment contrasts of the type $\tau_j - \tau_{j'}$, may have unequal variances in an incomplete block design. Thus some treatment comparisons will be made with greater precision than others. In experimenting it is desirable to estimate the differences between any pair of treatments with the same precision. A design in which all estimable treatment contrasts of the type $\tau_j - \tau_{j'}$ are estimated with the same precision is called a *balanced design*. Randomised block and Latin square designs are balanced as well as connected.

COMPLEMENTS AND PROBLEMS

1. Each adjusted treatment total Q_j is a contrast in the observations y_{ijk}.
2. If $\mathbf{l}_i' \mathbf{Q}$ are the estimates of the k linearly independent estimable parametric functions $\mathbf{t}_i' \boldsymbol{\tau}$ $(i = 1, \ldots, k)$, then the vectors $\mathbf{l}_i$ are linearly independent.

3 (a) A connected design is balanced if, and only if, all the $v-1$ non-zero characteristic roots of its C-matrix are equal.

(b) If the $v-1$ non-zero characteristic roots of the C-matrix of a design are all equal to θ, then each elementary contrast $\tau_i - \tau_j$ is estimated with a variance $2\sigma^2/\theta$ (see Raghavarao (1971)).

4. Denoting by $\mathbf{N} = (n_{ij})$, where n_{ij} is the number of times the ith treatment occurs in the jth block, the *incidence matrix* of the incomplete block design we have

(i) $\mathbf{C} = \text{diag}\,[r_1, \ldots, r_v] - \mathbf{N}\left\{\text{diag}\left[\frac{1}{k_1}, \ldots, \frac{1}{k_b}\right]\right\}\mathbf{N}'$

(ii) $\mathbf{Q} = \mathbf{V} - \mathbf{N}\left\{\text{diag}\left[\frac{1}{k_1}, \ldots, \frac{1}{k_b}\right]\right\}\mathbf{B}$

($\mathbf{Q}$, $\mathbf{V}$, $\mathbf{B}$ are the vectors of adjusted treatment totals, treatment totals, and block totals respectively).

5. If the linear model for the general incomplete block design is taken as

$$y_{ijm} = \mu + \beta_i + \tau_j + e_{ijm}$$

the normal equations are

$$N\mu + \Sigma\, k_i\, \beta_i + \Sigma\, r_j\, \tau_j = Y$$

$$k_i\mu + k_i\, \beta_i + \sum_j n_{ij}\, \tau_j = B_i$$

$$r_j\mu + \sum_i n_{ij}\, \beta_i + r_j\, \tau_j = V_j$$

The number of linearly independent normal equations, and hence the degrees of freedom for estimates, are, as before, at most $b + v - 1$. Eliminating μ and β_i we get the reduced normal equations, which are the same as obtained in this chapter. The analysis of variance is unaffected by the change in the model equations.

6. The SS due to the linear function $\mathbf{q}'\mathbf{Q}$ is $(\mathbf{q}'\mathbf{Q})^2/(\mathbf{q}'\,\mathbf{C}\,\mathbf{q})$. In particular, if $\tau_i - \tau_j$ is estimable, the SS for the hypothesis $\tau_i = \tau_j$ is given by $(Q_i - Q_j)^2/(c_{ii} + c_{jj} - 2\,c_{ij})$, where c_{ij} are the elements of $\mathbf{C}$.

REFERENCES

Incomplete block designs were introduced by Yates (1936). For the general method of analysis see Rao (1947), Tocher (1952), Chakrabarti (1962).

Kempthorne (1953) discusses the use of blocks of two plots each; Das and Kulkarni (1966) describe some incomplete block designs for use in bio-assays.

A matrix inversion method of solving the reduced equations is given in John (1965) and Shah (1959).

For a discussion of balance see Pearce (1960), Kshirsagar (1966) and Raghavarao (1971).

12

Balanced Incomplete Block Design

12.1 Introduction

The *balanced incomplete block design* (BIBD) is an incomplete block design in which the b blocks have the same number k of plots each, and every treatment is replicated r times in the design. Each treatment occurs at most once in a block, i.e. $n_{ij} = 0$ or 1; further, every pair of treatments occurs together in λ of the b blocks. We may denote such a design by $(b, k; v, r; \lambda)$. An example of a (10, 3; 6, 5; 2) BIBD is given below:

I:	1	2	5	VI:	2	3	4
II:	1	2	6	VII:	2	3	5
III:	1	3	4	VIII:	2	4	6
IV:	1	3	6	IX:	3	5	6
V:	1	4	5	X:	4	5	6

with Roman numerals representing blocks and Arabic numerals standing for treatments.

The parameters b, k, v, r, and λ of a BIBD cannot be chosen arbitrarily. They must satisfy the following relations:

(i) $$bk = vr$$

(ii) $$\lambda (v - 1) = r (k - 1) \tag{12.1}$$

(iii) $$b \geqslant v \text{ (and hence } r \geqslant k)$$

Even if the parameters satisfy these relations it is not always possible to arrange the treatments in blocks to get the corresponding design. In fact, necessary and sufficient conditions to be satisfied by the parameters for the existence of a BIBD are not known; (12.1) states some necessary conditions. The construction of these designs, i.e. actual arrangement of treatments into blocks, presents interesting problems in combinatorial mathematics, and a considerable amount of literature is available on the subject. Although such a design can be found for any number of treatments, v, and any block size, k, most of these are not of much interest, since they require too many replications. Tables are available giving all designs involving at most 20 replications and their method of construction (for details see Fisher and Yates, *Statistical Tables*, Oliver and Boyd, London, and D. Raghavarao, *Constructions and Combinatorial Problems in Design of Experiments*, Wiley, New York).

A BIBD is called *symmetrical* if the number of blocks equals the number of treatments. This implies that the block size k must be equal to the number of replications r of each treatment. In a symmetrical BIBD, any two blocks have λ treatments in common.

Since the BIBD is an incomplete block design, and since every treatment can occur at most once in a block, we must have $v \geqslant k$. If $v = k$, each treatment occurs once in every block and the design reduces to a randomised block design.

Thus in a BIBD we can always assume $v > k$. In the same way, it can be shown that $\lambda < r$. Obviously, $\lambda \leqslant r$, and if $\lambda = r$, then from $\lambda (v - 1) = r (k - 1)$ we see that $v = k$, so that the design is again a randomised block design.

12.2 Intrablock Analysis

We shall follow the notation, now in standard use, first introduced by Yates. As in Chapter 11, the block totals, treatment totals, adjusted treatment totals, and the grand total will be denoted by B_i, V_j, Q_j and Y respectively. We denote by T_j the total of all blocks containing the jth treatment, and by W_j the following:

$$W_j = (v - k) V_j - (v - 1) T_j + (k - 1) Y, \ (\Sigma W_j = 0) \quad (12.2)$$

Then the adjusted treatment totals Q_j are given by

$$Q_j = V_j - (T_j/k) \quad (12.3)$$

or

$$k Q_j = k V_j - T_j \quad (12.4)$$

The elements of the C-matrix are

$$c_{jj} = r(1 - 1/k)$$
$$c_{jj'} = -\lambda/k \quad (j \neq j')$$

from which we get the reduced normal equations as

$$Q_1 = r\left(1 - \frac{1}{k}\right)\tau_1 - \frac{\lambda}{k}\tau_2 - \ldots - \frac{\lambda}{k}\tau_v$$
$$Q_2 = -\frac{\lambda}{k}\tau_1 + r\left(1 - \frac{1}{k}\right)\tau_2 - \ldots - \frac{\lambda}{k}\tau_v \qquad (12.5)$$
$$\cdots\cdots\cdots\cdots\cdots\cdots\cdots\cdots\cdots\cdots$$
$$Q_v = -\frac{\lambda}{k}\tau_1 - \frac{\lambda}{k}\tau_2 - \ldots + r\left(1 - \frac{1}{k}\right)\tau_v.$$

These can be written as

$$Q_j = r\left(1 - \frac{1}{k}\right)\tau_j - \frac{\lambda}{k}(\tau. - \tau_j) \qquad (12.6)$$

or, using (12.1), as

$$Q_j = \frac{\lambda v}{k}\tau_j - \frac{\lambda}{k}\tau. \qquad j = 1, \ldots, v \qquad (12.7)$$

A solution of these equations is

$$\hat{\tau}_j = (k/\lambda v)\, Q_j \qquad (12.8)$$

with $\sum_j \hat{\tau}_j = 0$.

We find from (12.7) that all treatment contrast are estimable: the estimate of $\tau_j - \tau_{j'}$ is

$$(k/\lambda v)(Q_j - Q_{j'}) \qquad (12.9)$$

The variance of this estimate is

$$(k^2/\lambda^2 v^2)\,[\text{Var}(Q_j) + \text{Var}(Q_{j'}) - 2\,\text{Cov}(Q_j, Q_{j'})]$$
$$= (k^2/\lambda^2 v^2)(c_{jj} + c_{j'j'} - 2\,c_{jj'})\,\sigma^2$$
$$= (k^2/\lambda^2 v^2)\left[2r\left(1 - \frac{1}{k}\right) + \frac{2\lambda}{k}\right]\sigma^2$$
$$= (2k/\lambda v)\,\sigma^2 \qquad (12.10)$$

It follows from (12.9) and (12.10) that the BIBD is a connected and balanced design.

Using the solution (12.9) we find that the adjusted treatment SS is given by

$$S^2_{V/B} = \sum_j \hat{\tau}_j Q_j = (k/\lambda v) \sum Q_j^2 \tag{12.11}$$

and has $v - 1$ degrees of freedom. The unadjusted block and treatment SS are given by

$$S^2_B = [(\sum_i B_i^2)/k] - (Y^2/bk) \tag{12.12}$$

$$S^2_V = [(\sum_j V_j^2)/r] - (Y^2/vr) \tag{12.13}$$

The adjusted block SS, $S^2_{B/V}$, and the error SS are obtained by subtraction The analysis of variance table is given on p. 146.

Tests of hypothesis can now be carried out. The major hypothesis of interest is

$$\tau_1 = \tau_2 = \ldots = \tau_v$$

or,

$$\tau_1 - \tau_2 = \tau_1 - \tau_3 = \ldots = \tau_1 - \tau_v = 0 \tag{12.14}$$

As before, the SS for testing this hypothesis is, from (12.9), the SS due to the set of functions

$$(k/\lambda v)(Q_1 - Q_j), \quad j = 2, \ldots, v \tag{12.15}$$

Since the contrasts (12.14) are linearly independent the SS for testing the hypoth sis has $v - 1$ degrees of freedom. We therefore conclude that the SS due to the set (12.15) is the same as the SS due to the set $\{Q_j\}$. In other words, the SS for testing the hypothesis is the adjusted treatment SS, $S^2_{V/B}$. Hence the test is made by using the F-ratio

$$\frac{S^2_{V/B}/(v-1)}{s^2} \tag{12.16}$$

of the adjusted treatment *MS* to the error *MS*, which follows the F-distribution with $(v - 1, bk - v - b + 1)$ degrees of freedom.

Comparisons between any two specific treatments can be made by means of the t-test. Using (12.9) and (12.10) we see that to test the hypothesis $\tau_j = \tau_{j'}$ we have to use the statistic

$$\frac{(k/\lambda v)(Q_j - Q_{j'})}{s\sqrt{(2k/\lambda v)}} = \frac{Q_j - Q_{j'}}{s\sqrt{(2\lambda v/k)}} \tag{12.17}$$

which is distributed as t with $bk - v - b + 1$ degrees of freedom.

By a similar argument it can be shown that to test the hypothesis of equality of the block parameters we use the F-ratio.

Table 12.1

Source	df	SS	SS	df	Source
Blocks (unadj.)	$b-1$	S_B^2	$S_{B\|V}^2$	$b-1$	Blocks (adj.)
Treatments (adj.)	$v-1$	$S_{V\|B}^2$	S_V^2	$v-1$	Treatments (unadj.)
Error	$bk-v-b+1$	S_e^2	S_e^2	$bk-v-b+1$	Error
Total	$bk-1$	$\mathbf{y}'\mathbf{y}-Y^2/bk$	$\mathbf{y}'\mathbf{y}-Y^2/bk$	$bk-1$	Total

$$\frac{S^2_{B|V}/(b-1)}{s^2} \tag{12.17}$$

with $(b-1, bk-v-b+1)$ degrees of freedom.

In a BIBD the variance of the estimate of $\tau_j - \tau_{j'}$ is $(2k/\lambda v)\,\sigma^2$ while in a randomised block design with the same number r of replications for each treatment, the variance of the estimate of $\tau_j - \tau_{j'}$ is $2\sigma^2/r$. Hence, the efficiency of a BIBD compared to a randomised block design may be measured by

$$(2\sigma^2/r) \div (2k/\lambda v)\,\sigma^2,$$

i.e. by $\lambda v/kr$. The quantity $\lambda v/kr$ is called the efficiency of the BIBD and denoted by E.

From (12.1) we have $rk - \lambda v = r - \lambda > 0$; hence, $E < 1$. The difference between any two treatments is estimated, in a BIBD, with variance $(2k/\lambda v)\,\sigma^2 = 2\sigma^2/rE$ as compared to the variance $2\sigma^2/r$ in a randomised block design. Thus, we may say that a BIBD with r replications for each treatment is as efficient as a randomised block with only rE replications per treatment.

12.3 Interblock Analysis

The estimates of the treatment differences $\tau_j - \tau_{j'}$ have been obtained above by means of the adjusted treatment totals, i.e. by the elimination of the block parameters β_i from the normal equations for the treatment parameters τ_j. The elimination of block parameters really implies the formation of contrasts of the yields inside a block; those are then used to build up the estimates of treatment contrasts, hence the name intrablock analysis. However, in case of incomplete block designs, the block totals may also be expected to provide some information on the treatments, since the sets of treatments occurring in different blocks are different. The appropriate technique for the utilisation of block totals to estimate treatment differences, first introduced by Yates, is called interblock analysis. The method is applicable to the case where the number of blocks is larger than the number of treatments.

For interblock analysis we assume that the yield y_{ij} from the jth treatment in the ith block is given by

$$y_{ij} = b_i + \tau_j + e_{ij} \tag{12.18}$$

where the e_{ij} are independent normal variables with mean zero and variance σ_e^2, the b_i are independent normal variables with mean 0 and variance σ_b^2, the e_{ij} and b_i are independently distributed, and τ_j are parameters representing the effects of the different treatments. If the b_i were also to be regarded as fixed parameters, then the best estimates of the treatment contrasts are provided, by least squares theory, by the intrablock analysis already discussed. It is only when the b_i are considered as random variables that one can utilise the information provided by the block totals. This idea will be discussed in more detail later (Sec. 12.8).

The observations y_{ij} are no longer independent; we have

$$\text{Cov}\,(y_{ij}, y_{i'j'}) = \begin{cases} \sigma_b^2 \text{ if } i = i' \\ 0 \text{ if } i \neq i' \end{cases} \tag{12.19}$$

although they all have the same variance $\sigma_b^2 + \sigma_e^2$. Thus, the general theory is no longer applicable. We could use the theory relating to correlated observations (Sec. 4.8); we shall follow instead another method and work with the block totals B_i in place of the yields y_{ij}. We have

$$E\,(B_i) = \sum_{j(i)} \tau_j$$

$$\text{Var}\,(B_i) = k^2\,\sigma_b^2 + k\,\sigma_e^2$$

$$\text{Cov}\,(B_i, B_{i'}) = 0 \tag{12.20}$$

where $\sum_{j(i)}$ denotes summation over the treatments occurring in the ith block. The general theory can thus be applied to the block totals B_i. The normal equations for the τ_j are easily obtained by the addition rule (see Sec. 4.5). To get the equation corresponding to τ_1 we add over all the blocks containing the first treatment and get

$$T_1 = r\tau_1 + \lambda\,(\tau. - \tau_1)$$

The first treatment occurs in r blocks to give the term $r\tau_1$; every other treatment occurs in λ blocks with the first to give the term $\lambda\,(\tau. - \tau_1)$. Thus, the normal equations are

$$T_j = r\tau_j + \lambda\,(\tau. - \tau_j), \quad j = 1, \ldots, v \tag{12.21}$$

Adding these v equations, we get

$$\sum_j T_j = r\tau. + \lambda\,(v\tau. - \tau.) = rk\tau.$$

But $\sum_j T_j = kY$ since each block total occurs in all those T_j which correspond to the treatments occurring in that block, i.e. k times. Hence, the normal equations can be solved giving the solution

$$\tau.^* = Y/r$$

$$\tau_j^* = (T_j - \lambda\tau.^*)/(r - \lambda) = (T_j - \lambda Y/r)/(r - \lambda) \qquad (12.22)$$

We have, therefore, v degrees of freedom for estimates and $b - v$ for the SS due to error, which we shall call the *residual*. The SS due to estimates is

$$\sum_j \tau_j^* T_j = [\sum_j T_j^2 - (\lambda k/r)\, Y^2]/(r - \lambda)$$

The SS due to estimates can be partitioned into the SS due to estimates of the contrasts $\tau_j - \tau_{j'}$ with $v - 1$ degrees of freedom, and the SS due to the total Y with one degree of freedom. Since $Y = \sum_{i=1}^{b} B_i$, the SS due to Y is Y^2/b. Hence SS due to estimates of the contrasts $\tau_j - \tau_{j'}$ is

$$\{[\sum_j T_j^2 - (\lambda k/r)\, Y^2]/(r - \lambda)\} - (Y^2/b) = [\sum_j T_j^2 - (k^2 Y^2/v)]/(r - \lambda)$$

$$= [\sum_j T_j^2 - (T.^2/v)]/(r - \lambda) \qquad (12.23)$$

The residual SS can now be obtained by subtraction. The analysis of variance with the block totals B_i as variables is given in Table 12.2.

Table 12.2

Source	df	SS
Between treatments	$v - 1$	$[\Sigma T_j^2 - (k^2 Y^2/v)]/(r - \lambda)$
Residual	$b - v$	$\Sigma B_i^2 - [\Sigma T_j^2/(r - \lambda)]$ $+ [v\,(k-1)/b\,(v-k)]\, Y^2$
Total	$b - 1$	$k\, S_B^2 = \Sigma B_i^2 - (Y^2/b)$

Writing the block totals B_i in terms of the yields y_{ij} as

$$B_i = \mathbf{n}_i' \mathbf{y} \tag{12.24}$$

we see that

$$\mathbf{n}_i' \mathbf{n}_i = k, \quad \mathbf{n}_i' \mathbf{n}_{i'} = 0 \ (i \neq i') \tag{12.25}$$

Take two orthognal linear functions $\sum_i l_i B_i$, $\sum_i m_i B_i$ of the block totals, i.e. for which $\Sigma\, l_i m_i = 0$. Let

$$\sum_i l_i B_i = \sum_i l_i (\mathbf{n}_i' \mathbf{y}) = \boldsymbol{\lambda}' \mathbf{y}$$

$$\sum_i m_i B_i = \sum_i m_i (\mathbf{n}_i' \mathbf{y}) = \boldsymbol{\mu}' \mathbf{y}$$

then

$$\boldsymbol{\lambda}' \boldsymbol{\mu} = k \sum_i l_i m_i = 0$$

Thus, the functions $\sum_i l_i B_i$, $\sum_i m_i B_i$ are again orthogonal when considered as linear functions of the yields. Further, the SS due to the function $\sum_i l_i B_i$ considered as a function of the block totals is

$$(\sum_i l_i B_i)^2 / \sum_i l_i^2$$

and the SS due to $\sum_i l_i B_i$ considered as a function $\boldsymbol{\lambda}'\mathbf{y}$ of the yields is

$$(\sum_i l_i B_i)^2 / (\boldsymbol{\lambda}'\boldsymbol{\lambda}) = (\sum_i l_i B_i)^2 / (k \sum_i l_i^2)$$

Now, the total SS using the block totals as variables is

$$\Sigma\, B_i^2 - (Y^2/b) = k S_B^2 \tag{12.27}$$

where S_B^2 is the unadjusted block SS in the intrablock analysis. Thus we find that the unadjusted block SS can be further partitioned into orthogonal components as follows.

Table 12.3

Source	df	SS
Varietal component	$v-1$	$[\Sigma\, T_j^2 - (k^2 Y^2/v)]/k\,(r-\lambda)$
Remainder	$b-v$	R
Total	$b-1$	S_B^2

In the above table the residual is the SS due to those linear functions $\boldsymbol{\lambda}'\mathbf{y}$ which can be expressed as $\sum_i l_i B_i$ and are such that

$$E(\boldsymbol{\lambda}'\mathbf{y}) = E(\sum_i l_i B_i) = \sum_i l_i (\sum_{j(i)} \tau_j) = \sum_j a_j \tau_j = 0$$

i.e. for which

$$a_j = \sum_{i(j)} l_i = 0 \tag{12.28}$$

for all j, $\sum_{i(j)}$ denoting summation over all blocks containing the jth treatment. If C_i denote the adjusted block totals, and D_i the sum of the treatment totals of the treatments occurring in the ith block, we have $C_i = B_i - (D_i/r)$.

Hence

$$\sum_i l_i B_i = \sum_i l_i C_i + [\sum_i (l_i D_i)/r] = \sum_i l_i C_i + [\sum_j (a_j V_j)/r]$$

If $E(\sum_i l_i B_i) = 0$, each $a_j = 0$, and therefore

$$\sum_i l_i B_i = \sum l_i C_i$$

In other words, every linear function $\boldsymbol{\lambda}'\mathbf{y}$ belonging to the residual can be expressed as a linear function of the adjusted block totals, so that the residual SS is also a part of the adjusted block SS.

We may, therefore, partition the adjusted block sum of squares $S^2_{B|V}$ with $(b - 1)$ degree of freedom into the above residual R with $b - v$ degrees of freedom, and the difference (which we again call the varietal component) with $v - 1$ degrees of freedom. The varietal component is

$$\begin{aligned} S^2_{B/V} - R &= (S^2_B - R) + S^2_{V/B} - S^2_V \\ &= \{[(\sum_j T_j^2) - (T_.^2/v)]/k(r - \lambda)\} + [(\sum_j Q_j^2)/(k/\lambda v)] \\ &\quad - [(\sum_j V_j^2)/r - (Y^2/vr)] \end{aligned}$$

which after simplification reduces to

$$(\sum_j W_j^2)/rv\,(v - k)\,(k - 1) \tag{12.29}$$

The complete analysis of variance, including both intra and inter-block analysis, is usually presented in the form given in Table 12.4.

Table 12.4

Source	df	SS	SS	df	Source
Blocks (unadj.)					Blocks (adj.)
Varietal Component	$v-1$	$\Sigma (T_j-\bar{T}.)^2/k\ (r-\lambda)$	$\Sigma W_j^2/rv\ (v-k)\ (k-1)$	$v-1$	Varietal component
Remainder	$b-v$	R	R	$b-v$	Remainder
Total for blocks	$b-1$	S_B^2	$S_{B/V}^2$	$b-1$	Total for blocks
Treatments (adj.)	$v-1$	$S_{V/B}^2$	S_V^2	$v-1$	Treatment (unadj.)
Intrablock error	$bk-v-b+1$	S_e^2	S_e^2	$bk-v-b+1$	Intrablock error
Total	$bk-1$	$\Sigma y_{ij}^2-Y^2/bk$	$\Sigma y_{ij}^2-Y^2/bk$	$bk-1$	Total

12.4 Distributions in Interblock Analysis

We have found that the two sums of squares for blocks, S_B^2 and $S_{B/V}^2$, can be further partitioned into orthogonal components consisting of a varietal component and a residual. However, these components, though orthogonal, can no longer be considered to be distributed independently. The general theory of distribution of sums of squares, applies only to the case of independent observations. In the interblock analysis the observations y_{ij} are correlated and hence the problem of distribution of the different sums of squares must be examined afresh.

Theorem 12.1: The adjusted treatment totals Q_j, and the intrablock error sum of squares have the same distribution in the interblock analysis model as obtained earlier for intrablock analysis.

Proof: Fixing the values of the random variables b_i, say $b_i = b_i^*$, the conditional distribution of the adjusted treatment totals Q_j and the error sum of squares are the same as obtained in intrablock analysis. Since the conditional distributions do not involve the fixed value b_i^* of the random variables b_i, the result follows:

Corollary: (i) $E(Q_j) = (\lambda v/k)(\tau_j - \bar{\tau}.)$

(ii) $\text{Var}(Q_j) = r(1 - 1/k)\sigma_e^2$

(iii) $\text{Cov}(Q_j, Q_j') = -(\lambda/k)\sigma_e^2$

(iv) $E(s^2) = \sigma_e^2$

where s^2 denotes the intrablock error mean square.

Theorem 12.2: $E(S_{V/B}^2) = (v-1)\sigma_e^2 + (\lambda v/k)\Sigma(\tau_j - \bar{\tau}.)^2$

Proof: We have

$$S_{V/B}^2 = (k/\lambda v)\Sigma Q_j^2$$

with $v - 1$ degrees of freedom. By Theorem 12.1 the Q_j follow a multivariate normal distribution. Hence, from Theorem 5.3 we have

$$E(S_{V/B}^2) = (v-1)\sigma_e^2 + (k/\lambda v)\sum_j \{E(Q_j)\}^2$$

$$= (v-1)\sigma_e^2 + (\lambda v/k)\sum_j(\tau_j - \bar{\tau}.)^2 \qquad (12.30)$$

Theorem 12.3:

$$\text{(i) Cov }(B_i, V_j) = \begin{cases} k\sigma_b^2 + \sigma_e^2 & \text{if } n_{ij} = 1 \\ 0 & \text{if } n_{ij} = 0 \end{cases}$$

$$\text{(ii) Cov }(B_i, T_j) = \begin{cases} k^2\sigma_b^2 + k\sigma_e^2 & \text{if } n_{ij} = 1 \\ 0 & \text{if } n_{ij} = 0 \end{cases} \tag{12.31}$$

(iii) Cov $(B_i, Q_j) = 0$

Proof: We have

$$\text{Cov }(y_{ij}, y_{ij'}) = \sigma_b^2, \; j \neq j'$$

$$\text{Cov }(y_{ij}, y_{i'k}) = 0, \; i \neq i'$$

$$\text{Var }(y_{ij}) = \sigma_b^2 + \sigma_e^2$$

If $n_{ij} = 0$, none of the observations in V_j comes from the ith block and hence from the results above

$$\text{Cov }(B_i, V_j) = 0$$

On the other hand, if $n_{ij} = 1$, there is one observation y_{ij} in V_j which comes from the ith block. Hence

$$\begin{aligned} \text{Cov }(B_i, V_j) &= \sum_{j' \neq j} \text{Cov }(y_{ij'}, y_{ij}) + \text{Var }(y_{ij}) \\ &= (k-1)\,\sigma_b^2 + \sigma_b^2 + \sigma_e^2 \\ &= k\sigma_b^2 + \sigma_e^2 \end{aligned}$$

where the summation $\sum_{j' \neq j}$ is over all the plots of the ith block not containing the jth treatment.

(ii) Since $T_j = \sum_{i'(j)} B_{i'}$, where the summation is over all the blocks containing the jth treatment, we have, if $n_{ij} = 1$,

$$\begin{aligned} \text{Cov }(B_i, T_j) &= \sum_{i'(j)} \text{Cov }(B_i, B_{i'}) \\ &= \text{Var }(B_i) \\ &= k^2\sigma_b^2 + k\sigma_e^2 \end{aligned}$$

and, if $n_{ij} = 0$, Cov $(B_i, T_j) = 0$.

$$\begin{aligned} \text{(iii) Cov }(B_i, Q_j) &= \text{Cov }[B_i, V_j - (T_j/k)] \\ &= \text{Cov }(B_i, V_j) - (1/k)\text{ Cov }(B_i, T_j) \\ &= 0 \end{aligned}$$

Theorem 12.4:

(i) $E(S_B^2) = (b-1)(k\sigma_b^2 + \sigma_e^2) + [(r-\lambda)/k] \sum_j (\tau_j - \bar{\tau}.)^2$

(ii) $E(S_V^2) = (v-k)\,\sigma_b^2 + (v-1)\,\sigma_e^2 + r \sum_j (\tau_j - \bar{\tau}.)^2$

(iii) $E(S_{V|B}^2) = v\,(r-1)\,\sigma_b^2 + (b-1)\,\sigma_e^2$

Proof: (i) In the analysis of variance using block totals, to which the general theory is applicable, the total SS with $b-1$ degrees of freedom is $\Sigma B_i^2 - (Y^2/b) = kS_B^2$. The common variance of the B_i is $k^2\sigma_b^2 + k\sigma_e^2$, $E(B_i) = \sum_{j(i)} \tau_j$, and $E(Y) = r\tau_{..}$

Hence, by Theorem 5.3, we have

$$E(kS_B^2) = (b-1)(k^2\sigma_b^2 + k\sigma_e^2) + \sum_i (\sum_{j(i)} \tau_j)^2 - [(r\tau.)^2/b]$$

$$= (b-1)(k^2\sigma_b^2 + k\sigma_e^2) + r\Sigma\tau_j^2 + 2\lambda\Sigma\tau_j\tau_{j'} - [(r^2\tau_.^2)/b]$$

$$= (b-1)(k^2\sigma_b^2 + k\sigma_e^2) + r\Sigma\tau_j^2$$

$$+ \lambda(\tau.^2 - \Sigma\tau_j^2) - [(r^2/b)\,\tau.^2]$$

$$= (b-1)(k^2\sigma_b^2 + k\sigma_e^2) + (r-\lambda)\,\Sigma\,(\tau_j - \bar{\tau}.^2)$$

so that

$$E(S_B^2) = (b-1)(k\sigma_b^2 + \sigma_e^2) + [(r-\lambda)/k]\,\Sigma(\tau_j. - \bar{\tau}.)^2$$

(ii) $$E(S_V^2) = E[(\Sigma V_j^2/r) - (Y^2/vr)]$$

But

$$E(V_j^2) = \text{Var}(V_j) + [E(V_j)]^2$$

$$= r(\sigma_b^2 + \sigma_e^2) + r^2\tau_j^2$$

and

$$E(Y^2) = \text{Var}(Y) + [E(Y)]^2$$

$$= \text{Var}(\sum_i B_i) + [E(Y)]^2$$

$$= b(k^2\sigma_b^2 + k\sigma_e^2) + (r\tau.)^2$$

Hence

$$E(S_V^2) = (1/r) \sum_j [r(\sigma_b^2 + \sigma_e^2) + r^2\tau_j^2] - (1/vr)\,[b(k^2\sigma_b^2 + k\sigma_e^2) + (r\tau.)^2]$$

$$= v\,(\sigma_b^2 + \sigma_e^2) + r\,\Sigma\,\tau_j^2 - (k\sigma_b^2 + \sigma_e^2) - (r/v)\,\tau^2$$

$$= (v-k)\,\sigma_b^2 + (v-1)\,\sigma_e^2 + r\sum_j (\tau_j - \bar{\tau}.)^2$$

(iii) $$E\,(S^2_{B|V}) = E\,(S^2_B) + E\,(S^2_{V|B}) - E\,(S^2_V)$$

$$= v\,(r-1)\,\sigma_b^2 + (b-1)\,\sigma_e^2$$

12.5 Interblock Estimation of Treatment Contrasts

We have seen that in the interblock analysis model $E\,(Q_j) = (\lambda v/k)$ $(\tau_j - \bar{\tau}.)$ and $E\,(T_j) = (r-\lambda)\,\tau_j + \lambda v\bar{\tau}..$ Hence the contrast $\tau_j - \tau_{j'}$ can be estimated by $(k/\lambda v)\,(Q_j - Q_{j'})$ and $(T_j - T_{j'})/(r-\lambda)$. The first of these two estimates is the same as that obtained in the intrablock analysis, and will, therefore, be called *the intrablock* estimate of $\tau_j - \tau_{j'}$. The second will be called the *interblock estimate.* If we write $\hat{\tau}_j = (k/\lambda v)\,Q_j$, $\tau_j^* = T_j/(r-\lambda)$, a general treatment contrast $\sum_j t_j\,\tau_j (\Sigma\, t_j = 0)$ has the intrablock estimate

$$(k/\lambda v)\sum_j t_j\,Q_j = \sum_j t_j\,\hat{\tau}_j \qquad (12.33)$$

and the interblock estimate

$$(\sum_j t_j\,T_j)/(r-\lambda) = \sum_j t_j\,\tau_j^* \qquad (12.34)$$

Since Cov $(B_i, Q_j) = 0$, these two estimates are independently distributed, and a better estimate of the contrast $\Sigma\, t_j\tau_j$ is obtained by pooling them, i.e. by taking a weighted average of the two estimates with the weights being inversely proportional to the variances.

Now,

$$\text{Var}\,(\Sigma\, t_j\hat{\tau}_j) = (k/\lambda v)^2\, \text{Var}\,(\Sigma\, t_j\,Q_j)$$

$$= (k/\lambda v)^2\,[\sum_j t_j^2\,\text{Var}\,(Q_j) + 2\sum_{j\neq j'} t_j\,t_{j'}\,\text{Cov}\,(Q_j, Q_{j'})]$$

$$= (k/\lambda v)^2\,[r\,(1 - 1/k)\,\sigma_e^2\,\Sigma\, t_j^2 - (\lambda/k)\,\sigma_e^2\,\{t.^2 - \Sigma\, t_j^2\}]$$

$$= (k/\lambda v)\,\sigma_e^2\,\Sigma\, t_j^2 \qquad (12.35)$$

Similarly,

$$\text{Var}\,(\Sigma\, t_j\,\tau_j^*) = \text{Var}\,(\Sigma\, t_j\,T_j)/(r-\lambda)^2$$

$$= [\Sigma\, t_j^2\,\text{Var}\,(T_j) + 2\sum_{j\neq j'} t_j\,t_{j'}\,\text{Cov}\,(T_j, T_{j'})]/(r-\lambda)^2$$

$$= [r\,(k^2\,\sigma_b^2 + k\sigma_e^2)\,\Sigma\, t_j^2 + \lambda\,(k^2\,\sigma_b^2 + k\sigma_e^2) + (t_.^2 - \Sigma\, t_j^2)]/(r-\lambda)^2$$

$$= (k^2\,\sigma_b^2 + k\sigma_e^2)\,(\Sigma\, t_j^2)/(r-\lambda) = [\sigma_f^2/(r-\lambda)]\,\Sigma\, t_j^2 \qquad (12.36)$$

where we have put $k^2\,\sigma_b^2 + k\sigma_e^2 = \sigma_f^2$.

Thus the weights are proportional to $\lambda v/k\sigma_e^2$ and $(r-\lambda)/\sigma_f^2$. Writing $w = 1/\sigma_e^2$ and $w' = k/\sigma_f^2$, the weights are proportional to $w\lambda v$ and $w'\,(r-\lambda)$, so that the *pooled estimate* is

$$\frac{(w\lambda v)\,\Sigma\, t_j\,\hat{\tau}_j + w'\,(r-\lambda)\,\Sigma\, t_j\,\tau_j^*}{w\lambda v + w'\,(r-\lambda)}$$

$$= \Sigma\, t_j[w\lambda v\,\hat{\tau}_j + w'\,(r-\lambda)\,\tau_j^*]/[w\lambda v + w'\,(r-\lambda)]$$

$$= \sum_j t_j\,\tau_j^{**} \qquad (12.37)$$

where

$$\tau_j^{**} = [w\lambda v\,\hat{\tau}_j + w'\,(r-\lambda)\,\tau_j^*]/[w\lambda v + w'\,(r-\lambda)] \qquad (12.38)$$

We have

$$w\lambda v\,\hat{\tau}_j + w'\,(r-\lambda)\,\tau_j^* = wkQ_j + w'\,T_j$$

and

$$w\lambda v + (r-\lambda)\,w' = [wvr\,(k-1)/(v-1)] + [r - \{r(k-1)/(v-1)\}]\,w'$$

$$= [wrv\,(k-1) + w'r\,(v-k)]/(v-1)$$

Hence,

$$\tau_j^{**} = [(v-1)\,(wkQ_j + w'T_j)]/[wrv\,(k-1) + w'r\,(v-k)]$$

$$= (v-1)\,[w\,(kV_j - T_j) + w'T_j]/r\,[wv\,(k-1) + w'\,(v-k)]$$

$$= [wk\,(v-1)\,V_j + (w-w')\{W_j - (v-k)\,V_j$$

$$- (k-1)\,Y]/r\,[wv\,(k-1) + w'\,(v-k)]$$

$$= [\{wv\,(k-1) + w'\,(v-k)\}\,V_j + (w-w')\{W_j$$

$$- (k-1)\,Y\}]/r\,[wv\,(k-1) + w'\,(v-k)]$$

$$= [V_j + \mu\,W_j - \mu\,(k-1)\,Y]/r \qquad (12.39)$$

where

$$\mu = (w-w')/[wv\,(k-1) + w'\,(v-k)] \qquad (12.40)$$

Since $\sum_j t_j = 0$, the pooled estimate of the contrast $\sum_j t_j\,\tau_j$ can be written as

$$\sum_j t_j \tau_j^{**} = \sum_j t_j (V_j + \mu W_j)/r \tag{12.41}$$

To sum up, we have, for a general contrast $\Sigma\, t_j\, \tau_j$, an intrablock estimate given by (12.33), an interblock estimate given by (12.34), and the pooled estimate given by (12.41). The variances of these three estimates are

$$(k/\lambda v)\, \sigma_e^2\, (\Sigma\, t_j^2) = (k/\lambda v w)\, (\Sigma\, t_j^2)$$

$$[\sigma_f^2/(r-\lambda)]\, (\Sigma\, t_j^2) = [k/w'\,(r-\lambda)]\, (\Sigma\, t_j^2)$$

and

$$[k/\{\lambda v w + w'\,(r-\lambda)\}]\, (\Sigma\, t_j^2)$$

respectively.

12.6 Estimating the Weights w and w'

The best estimate obtained in Sec. 12.5 by pooling the intra-and-inter block estimates involves the unknown weights w and w'. This is a situation one faces while analysing a linear model with correlated observations (see Sec. 4.8). In practice, we replace the weights w and w' by their estimates obtained from the data provided by the experiment. The resulting pooled estimate is, of course, no longer the best but only approximately so.

We have $w = 1/\sigma_e^2$ and the intrablock error mean square s^2 is an unbiased estimate of σ_e^2 with $(vr - v - b + 1)$ degrees of freedom. Hence we take

$$\hat{w} = 1/s^2 \tag{12.42}$$

as the estimate of w. To obtain the estimate of $w' = 1/(k\sigma_e^2 + \sigma_e^2)$ we need first an unbiased estimate of σ_b^2. For this we use $S_{B|V}^2$, the SS due to blocks eliminating treatments. Writing $s_b^2 = S_{B|V}^2/(b-1)$, we have

$$E(s_b^2) = \sigma_e^2 + [v\,(r-1)/(b-1)]\, \sigma_b^2 \tag{12.43}$$

and hence

$$E\,[(s_b^2 - s^2)\,(b-1)/v\,(r-1)] = \sigma_b^2$$

We take

$$(s_b^2 - s^2)\,(b-1)/v\,(r-1) \tag{12.44}$$

as the estimate of σ_b^2. The estimate of $k\sigma_b^2 + \sigma_e^2$ is then given by

$$[k\,(b-1)\,(s_b^2 - s^2)/v\,(r-1)] + s^2 = [k\,(b-1)\, s_b^2 - (v-k)\, s^2]/v\,(r-1) \tag{12.45}$$

Finally, we get the estimate $\hat{w}'$ of $\hat{w}$ as

$$\hat{w}' = v(r-1)/[k(b-1)s_b^2 - (v-k)s^2] \tag{12.46}$$

Using $\hat{w}$ and $\hat{w}'$ we get the estimate $\hat{\mu}$ of μ using (12.40). The approximate best pooled estimate of the contrast $\Sigma\, t_j\tau_j$ is

$$\Sigma t_j (V_j + \hat{\mu} W_j)/r \tag{12.47}$$

and its variance may approximately be estimated by

$$(\Sigma\, t_j^2)/[\lambda v \hat{w} + (r-\lambda)\,\hat{w}'] \tag{12.48}$$

In the analysis of variance using block totals we had found that the expectation of the residual R, with $b - v$ degrees of freedom, was $(b-v)(k^2\sigma_b^2 + \sigma_e^2)$. Hence we could have used R as well to obtain an unbiased estimate of σ_b^2 and then of w'. However, we prefer to use $S_{B/V}^2$, instead of R, as it is based on a larger number of degrees of freedom.

12.7 Notes on Computation

(i) Calculate the block totals B_i and the treatment total V_j.

(ii) Obtain the raw SS, the general correction factor Y^2/bk, the total SS, and S_B^2.

(iii) Write down in tabular form for each treatment (with the rows of the table corresponding to treatments) the quantities V_j, T_j, kQ_j, W_j and $Y_j = V_j + \hat{\mu}\, W_j$, in that order. The V_j, T_j, $k\,Q_j$ and W_j columns are filled first and are used to carry out the intra- and interblock analysis of variance. From this analysis the estimates $\hat{w}$, $\hat{w}'$ and $\hat{\mu}$ are obtained and the Y_j column then filled using $\hat{\mu}$. At the foot of each column enter the total of that column and below it the divisors r, $k(r-\lambda)$, $k\lambda v$, $vr(v-k)(k-1)$ in the columns corresponding to V_j, T_j, Q_j and W_j respectively. The totals and the divisors are used to obtain the different sums of squares

$$S_V^2 = (\Sigma V_j^2)/r] - [(\Sigma V_j)^2/bk]$$

$$S_{V/B}^2 = (\Sigma k Q_j)^2/k\lambda v$$

$$\text{Varietal component (of } S_B^2) = [\Sigma T_j^2 - (\Sigma T_j)^2/v]/k(r-\lambda)$$

$$\text{Varietal component (of } S_{B/V}^2) = \Sigma W_j^2/vr(v-k)(k-1)$$

The totals of the k Q_j and W_j columns should add up to zero, providing a check on the calculations.

(iv) Obtain the analysis of variance table by entering the values of S_B^2, $S_{V/B}^2$, S_V^2 and the varietal components, and obtain $S_{B/V}^2$ and the intrablock error SS by subtraction. Obtain next the residual in one part by subtraction and enter it in the other part. All the subtotals should now ageee and this provides a check on the calculations.

12.8 Recovery of Interblock Information

In the discussion relating to interblock analysis we assumed that the block effects were random variables. This assumption is appropriate if the blocks can be regarded as a random sample from a large population of blocks. For example, if we were to compare different makes of automobile tyres for durability, the natural block would be a car and the plots, the four wheels of the car. If the cars in a particular experiment were selected at random from a large number of cars we could treat the block effects as random variables.

If the block effects are regarded as constants, the best estimate of the treatment contrast is, by the general theory of least squares, the intrablock estimate. If, however, the block effects are regarded as random variables the estimate can be improved by utilising the information provided by the block totals. Since the treatments occurring in different blocks are not all the same, one can expect the differences between the block totals to provide some information about the differences between the treatments. The procedure to get the improved est ate which, as described earlier, consists in obtaining the interblo timates τ_j^* and then the pooled estimates τ_j^{**}, is called recovery of interblock information. The gain in information may be small in some cases and appreciable in others. Since the additional work involved is not much, and the resulting gain cannot in any case be assessed until the interblock analysis is performed, it would appear best to follow this method of analysis in all cases.

The increase in precision using interblock analysis as compared to intrablock analysis is measured by

[1/Var (pooled estimate) ÷ 1/Var (intra-block estimate)] − 1

$$= \frac{\lambda v w + w'(r-\lambda)}{\lambda v w} - 1 = \frac{w'(r-\lambda)}{\lambda v w} \qquad (12.49)$$

and may be estimated by the quantity,

$$\frac{\hat{w}'(r-\lambda)}{\lambda v \hat{w}} \tag{12.50}$$

One usually foregoes the recovery calculation if the estimated increase is small, say less than 10% (in particular, if it is negative).

Though $w > w'$ it may happen that the estimates $\hat{w}$ and $\hat{w}'$ do not satisfy this relation. If the adjusted block mean square s_b^2 is less than the intrablock error mean square s^2 then $\hat{w}'$ will be greater than $\hat{w}$. It may also happen that the estimate $\hat{w}'$ of w' is negative. In such cases one usually takes $\hat{w} = \hat{w}'$, and the pooled estimate of the contrast $\Sigma t_j \tau_j$ becomes $(\Sigma t_j V_j)/r$, as $\hat{\mu}$ is now equal to zero.

12.9 Example

A varietal trial involving 6 varieties of wheat crop was conducted in an agricultural college in India in a balanced incomplete block design with 10 blocks of 3 plots each. Each variety was replicated 5 times and each pair of varieties occurred together in 2 blocks. Each plot was 1/20th of an acre in area and the yield was measured in kilograms. (The actual field trial was conducted in a symmetric BIBD with $v = b = 13$, $r = k = 4$, $\lambda = 1$. We have adapted the data from that experiment to obtain a BIBD as described above.)

The layout plan and the yields from the 6 varieties are given in Table 12.5.

The block totals B_i and the treatment totals V_j are shown as marginal totals in the table. For this data we have

$$\text{Raw SS} = 111569$$

$$\text{CF} = (1809)^2/30 = 109082.7$$

$$\text{Total SS} = 2486.3$$

We next obtain V_j, T_j, kQ_j and W_j. These are given in Table 12.6.

From Table 12.5 we obtain the the following sums of squares

$$S_B^2 = [(173)^2 + (180)^2 + \ldots + (170)^2]/3 - \text{CF}$$

$$= 328649/3 - \text{CF} = 466.96$$

$$S_V^2 = [(351)^2 + (300)^2 + \ldots + (279)^2]/5 - \text{CF}$$

$$= 549125/5 - \text{CF} = 742.30$$

Table 12.5

Blocks	Varieties 1	2	3	4	5	6	Total
1	69	54	50				173
2	77	65		38			180
3	72		45		54		171
4	63			60		39	162
5	70				65	54	189
6		65	68			67	200
7		57		60	62		179
8		59			65	63	187
9			75	62	61		198
10			59	55		56	170
Total	351	300	297	275	307	279	1809

Table 12.6

Variety	V_j	T_j	$kQ_j = 3V_j - T_j$	$W_j = 3V_j - 5T_j + 2Y$	$Y_j = V_j + \hat{\mu} W_j$
1	351	875	178	296	360.59
2	300	919	−19	−77	297.51
3	297	912	−21	−51	295.35
4	275	889	−64	−2	274.94
5	307	924	−3	−81	304.38
6	279	908	−71	−85	276.25
Total	1809	5427	0	0	1809.02
Divisor	$r = 5$	$k(r - \lambda) = 9$	$k\lambda v = 36$	$vr(v-k)(k-1) = 180$	

The remaining sums of squares for intrablock and interblock analysis of variance are obtained from Table 12.6. We get

$$S^2_{V/B} = [(178)^2 + (-19)^2 + \dots + (-71)^2]/36$$
$$= 41632/36 = 1156.44$$

The varietal components are given by

$$[(875)^2 + (919)^2 + \dots + (908)^2 - (1809)^2/6]/9 = 196.61$$

and

$$[(296)^2 + (-77)^2 + \dots + (-85)^2]/180 = 610.73$$

The analysis of variance is shown in Table 12.7: (a) refers to the analysis in which S^2_B is taken as the block sum of squares, and (b) refers to the analysis in which it is $S^2_{B|V}$.

The estimates of the weights w and w' and μ are:

$$\hat{w} = 1/57.53 = 0.0174$$

$$\hat{w}' = 6\,(5-1)/[3\ \ (10-1)\,(97.90) - (6-3)\,(57.53)] = 0.0097$$

$$\hat{\mu} = (0.0174 - 0.0097)/[6\,(3-1)\,(0.0174) + (6-3)\,(0.0097)]$$
$$= 0.0324$$

With this value of $\hat{\mu}$ we obtain the values of Y_j; these have been entered in the last column of Table 12.6.

Using intrablock analysis only, the F-ratio for testing varietal differences is

$$231.288/57.53 = 4.02$$

which is almost equal to 4.56, the 1% value of F with (5.15) degrees of freedom. Thus, we may conclude that the six varieties differ significantly as to their yields.

The difference in yields of two different varieties is estimated by

$$k\,(Q_j - Q_{j'})/(\lambda v) = (Q_j - Q_{j'})/4$$

The variance of any such estimate is

$$2k\sigma_e^2/(\lambda v) = \sigma_e^2/2$$

and hence the standard error of the estimate is

$$\sqrt{(57.53)/2} = \sqrt{28.765} = 5.3633$$

For 15 degrees of freedom, the 5% value of t is 2.131; hence, the

Table 12.7

Source	df	SS(a)	MS(a)	SS(b)	MS(b)
Blocks					
Varietal component	5	196.61		610.73	
Remainder	4	270.35		270.35	
Total for Blocks	9	466.96		881.08	97.90
Treatments	5	1156.44	231.288	742.30	
Error	15	862.90	57.53	862.90	57.53
Total	29	2486.30		2486.28	

least significant difference at the 5% level of significance is

$$(2.131)\,(5.3633) = 11.436$$

The estimates of the differences in yield due to different varieties are given below, those significant at the 5% level being marked with an asterisk.

	2	3	4	5	6
1	49.25*	49.75*	60.5*	45.25*	62.25*
2		0.50	11.25	− 4.00	13.00*
3			10.75	− 4.50	12.50*
4				− 15.25*	1.75
5					17.00*

With recovery of interblock information the differences in yield from two varieties are estimates by

$$(Y_j - Y_{j'})/r = (Y_j - Y_{j'})/5$$

The variance of this estimates is

$$2k/[\lambda v\omega + (r - \lambda)\,\omega']$$

and hence the standard error of the estimate of the difference in yield from two varieties is

$$\sqrt{6/[12(0.0174) + 3\,(0.0097)]} = \sqrt{6/0.2397}$$

$$= \sqrt{25.2207} = 5.0220$$

Comparing this value with the standard error obtained in intra-block analysis (5.3633) we see that not very much is gained by the use of interblock information. The actual gain in information (ignoring errors due to inaccurate weighting) may be measured by

$$(28.765/25.2207) - 1 = 0.14$$

i.e. 14%.

Using the 5% value of t with 15 degrees of freedom (2.131), we see that the difference in yields due to two different varieties will be significant if

$$(Y_j - Y_{j'})/5 \geqslant (2.131)\,(5.0220) = 10.702$$

However, the test is only an approximate one. After recovery of interblock information the estimates of the difference in yields from two different varieties are as follows; significant differences are again marked with an asterisk.

	2	3	4	5	6
1	12.62*	13.05*	17.13*	11.24*	16.87*
2		0.43	4.51	− 1.37	4.25
3			4.08	− 1.81	3.82
4				− 5.89	0.26
5					5.63

From intrablock analysis we may conclude that varieties 4 and 6 are not much different and give an inferior yield as compared to the others, varieties 2, 3 and 5 do not differ significantly from each other, and variety 1 is superior to all the others. With recovery of interblock information the picture changes slightly; variety 1 is still superior to all the others, but varieties 2 to 6 do not differ significantly from each other.

COMPLEMENTS AND PROBLEMS

1 (*a*) *Complement of a BIBD:* Consider a BIBD with parameters $(b, k; v, r; \lambda)$. Replace each block by another block containing those treatments which are not included in the original block. This gives us another block design which is again a BIBD with parameters $(b, v-k; v, b-r; b-2r+\lambda)$. The new design is called the complement of the original BIBD.

(*b*) *Residual of a BIBD:* Consider a symmetrical BIBD with parameters $(b, k; b, k; \lambda)$. Fix one of the blocks. From each of the remaining blocks delete all those treatments which it has in common with the fixed block. The remaining blocks after deletion of the common treatments form a BIBD with parameters $(b-1, k-\lambda; b-k, k; \lambda)$, called the residual design.

(*c*) *Derived design of a BIBD:* If instead of the procedure followed to obtain the residual design we retain in each of the remaining blocks only those treatments which it has in common with the fixed block we get another

BIBD with parameters $(b-1, \lambda; k, k-1; \lambda-1)$, called the derived design.

2. *Resolvable BIBD:* A BIBD is called resolvable if the blocks can be divided into r groups such that each group contains each of the v treatments exactly once. More generally, it may be possible to divide the blocks into c groups of b/c blocks each such that inside a group each of the v treatments occurs r/c times.

Assuming that the blocks inside a group are more homogeneous, the model equations for intrablock analysis can be written as

$$y_{ijl} = \gamma_i + \beta_{ij} + \tau_{l(ij)} + e_{ijl}, \qquad i = 1, \ldots, c$$
$$j = 1, \ldots, b/c$$
$$l = 1, \ldots, k$$

Here, y_{ijl} is the yield from the lth plot in the jth block of the ith group, γ_i is the group effect, β_{ij} is the effect of the jth block in the ith group, $\tau_{l(ij)}$ is the effect due to the treatment applied to the lth plot of the jth block in the ith group, and e_{ijl} is the random (error) component.

Denoting the totals of the yields of the ith group by G_i, of the blocks by B_{ij}, and varieties by V_l, the normal equations for intrablock analysis are

$$G_i = (bk/c)\,\gamma_i + k \sum_j \beta_{ij} + (r/c) \sum_l \tau_l, \quad i = 1, \ldots, c$$

$$B_{ij} = k\,\gamma_i + kB_{ij} + \sum_{l(ij)} \tau_l, \qquad j = 1, \ldots, b/c$$

$$V_l = (r/c)\,\gamma. + \sum_{ij(l)} \beta_{ij} + r\,\tau_l, \qquad l = 1, \ldots, v$$

where $\sum_{l(ij)} \tau_l$ denotes the sum over the varieties occuring in the jth block of the ith group, and $\sum_{ij(l)} \beta_{ij}$ denotes the sum over all the blocks containing the lth variety. (Note the difference in the use of the index l in the model and normal equations.)

The usual intrablock and interblock analysis for a BIBD is modified only to the extent that the unadjusted and adjusted block sums of squares, with $b-1$ degrees of freedom each, are expressed as sums of two orthogonal components, the between groups sum of squares $S_G^2 = (c/bk) \sum G_i^2 - (Y^2/bk)$, with $c-1$ degrees of freedom, and the between blocks within groups sum of squares

$$S_{B(G)}^2 = S_B^2 - S_G^2$$

or

$$S_{B(G)/V}^2 = S_{B/V}^2 - S_G^2$$

with $b-c$ degrees of freedom. The remainder sums of squares in interblock analysis are now obtained by subtracting the varietal components from $S^2_{B(G)}$ or $S^2_{B(G)|V}$.

The intra- and interblock estimates, and hence the pooled estimates, are unaffected. The estimates of the weights ω and ω' will now change since σ_b^2 can no longer be estimated by using $S^2_{B|V}$, as $E\,(S^2_{B|V})$ now involves γ_l also. Instead we use $S^2_{B(G)|V}$ to estimate σ_b^2 giving the estimates of ω and ω' as

$$\hat{\omega} = \frac{1}{s_e^2},\ \hat{\omega}' = \frac{v\,(r-1) - k\,(c-1)}{k\,(b-c)\,s_b^2 - (v-k)\,s_e^2} = \frac{r-1}{r\,s_b^2 - s_e^2} \text{ (if } r = c)$$

3. (a) In a BIBD the C-matrix can be written as

$$\mathbf{C} = \frac{\lambda v}{k}\,\mathbf{I}_v - \frac{\lambda}{k}\,\mathbf{E}_{v\,v}$$

where $\mathbf{I}_n$ is the identity matrix of order n and $\mathbf{E}_{r\,s}$ denotes an $r \times s$ matrix all of whose elements are equal to unity.

(b) $$\text{Var}\,(\mathbf{t}'\,\mathbf{Q}) = (\mathbf{t}'\,\mathbf{C}\,\mathbf{t})\,\sigma^2 = \left[\frac{\lambda v}{k}\,\mathbf{t}'\mathbf{t} - \frac{\lambda}{k}\,\mathbf{t}'\mathbf{E}_{vv}\,\mathbf{t}\right]\sigma^2$$

$$= \frac{\lambda v}{k}\,(\mathbf{t}'\,\mathbf{t})\,\sigma^2 \text{ (if } \mathbf{t} \text{ is a contrast)}$$

(c) $$\text{Cov}\,(\mathbf{t}'\mathbf{Q}, \mathbf{s}'\mathbf{Q}) = \left[\frac{\lambda v}{k}\,\mathbf{t}'\mathbf{s} - \frac{\lambda}{k}\,\mathbf{t}'\mathbf{E}_{vv}\,\mathbf{s}\right]\sigma^2$$

(d) $\mathbf{t}'\mathbf{Q}$ and $\mathbf{s}'\mathbf{Q}$ are orthogonal, if and only if,

$$v\,(t_1\,s_1 + \ldots + t_v\,s_v) = (t_1 + \ldots + t_v)\,(s_1 + \ldots + s_v)$$

In particular Q_1 is orthogonal to $Q_1 + (v-1)\,Q_2$; $Q_1 + Q_2$ is orthogonal to $3Q_1 - Q_2 + Q_3 + \ldots + Q_v$.

If, in analogy with the covariance of a sample from a bivariate population, we define

$$\text{Cov}\,(\mathbf{t}, \mathbf{s}) = \frac{1}{v}\,(t_1\,s_1 + \ldots + t_v s_v) - \frac{(t_1 + \ldots + t_v)}{v}\,\frac{(s_1 + \ldots + s_v)}{v}$$

then Cov $(\mathbf{t}'\,\mathbf{Q}, \mathbf{s}'\,\mathbf{Q}) = 0$ if, and only if, Cov $(\mathbf{t}, \mathbf{s}) = 0$.

(e) If $\mathbf{t}$ or $\mathbf{s}$ is a contrast Cov $(\mathbf{t}'\mathbf{Q}, \mathbf{s}'\mathbf{Q})$ equals $\left(\frac{\lambda v}{k}\,\mathbf{t}'\mathbf{s}\right)\sigma^2$, so that, $\mathbf{t}'\mathbf{Q}$ and $\mathbf{s}'\mathbf{Q}$ are orthogonal, if and only if, $\mathbf{t}$ is orthogonal to $\mathbf{s}$.

(f) SS due to $\mathbf{t}'\mathbf{Q} = (\mathbf{t}'\,\mathbf{Q})^2/(\lambda v/k)\,\mathbf{t}'\mathbf{t}$ if $\mathbf{t}$ is a contrast.

(g) SS due to all contrasts among Q_1, Q_2, Q_3

$$= \text{SS due to } Q_1 - Q_2,\ Q_1 - Q_3$$

$$= \text{SS due to } Q_1 - Q_2,\ Q_1 + Q_2 - 2\,Q_3$$

$$= \frac{(Q_1 - Q_2)^2}{2\lambda v/k} + \frac{(Q_1 + Q_2 - 2Q_3)^2}{6\lambda v/k}$$

$$= \frac{k}{\lambda v}\left[\sum_1^3 Q_j^2 - \frac{(Q_1 + Q_2 + Q_3)^2}{3}\right]$$

(h) Generalising the resu lt in (g) above we get another derivation of

$$S^2_{V/B} = \frac{k}{\lambda v}\sum_1^v Q_j^2$$

as the SS due to all contrasts in $Q_1, Q_2, \ldots, Q_v$.

(i) The SS obtained in (*g*) above is the SS due to the hypothesis $\tau_1 = \tau_2 = \tau_3$.

4. *Incidence matrix.* The matrix $\mathbf{N} = (n_{ij})$, where n_{ij} is the number of times the ith treatment occurs in the jth block, is called the incidence matrix of an incomplete block design.

For a BIBD, n_{ij} is equal to 0 or 1, and we have the following results.

(a) $\mathbf{NN}' = (r - \lambda)\,\mathbf{I} + \lambda \mathbf{E}_{vv}$

(b) $\mathbf{C} = r\mathbf{I} - \frac{1}{k}\,\mathbf{NN}'$

(c) Rank $(\mathbf{NN}') = v$

(d) If the BIBD is symmetric we also have

$$\mathbf{N}'\mathbf{N} = (r - \lambda)\,\mathbf{I} + \lambda\,\mathbf{E}_{vv}$$

so that, any two blocks have λ treatments in common.

(e) In a symmetric BIBD, $(r - \lambda)$ is a perfect square if v is even. This follows easily from the fact that the determinant of the matrix $\mathbf{N}'\mathbf{N}$ is

$$r^2\,(r - \lambda)^{v-1}$$

5. $v =$ Rank $(\mathbf{NN}') =$ Rank $\mathbf{N} \leqslant b$.

The result $b \geqslant v$, known as Fisher's inequality, can be improved to $b \geqslant v + r - k$ by noting that $(v - k)\,(r - k) \geqslant 0$.

6. In a randomised block design the adjusted treatment sum of squares is equal to the unadjusted treatment sum of squares.

7. If the linear model for inter-block analysis in a BIBD is taken as

$$y_{ij} = \mu + b_i + \tau_j + e_{ij}$$

we get

$$E(B_i) = k\mu + \sum_{j(i)} \tau_j$$

$$\text{Var}\,(B_i) = k^2\,\sigma_b^2 + k\,\sigma_e^2$$

and,

$$\text{Cov}\,(B_i, B_{i'}) = 0$$

The normal equations for the linear model using the block totals B_i as variables now become

$$bk^2\mu + kr\tau. = kY$$

$$kr\mu + (r - \lambda)\,\tau_j + \lambda\tau. = T_j, \quad j = 1, \ldots v$$

The number of linearly independent equations, and hence the degrees of freedom for estimates, are at most v.

A solution of the normal equations is

$$\mu^* = Y/bk, \; \tau_j^* = \left(T_j - \frac{rY}{b}\right)\Big/(r-\lambda)$$

so that the sum of squares for estimate is

$$\mu^* = (kY) + \Sigma\, \tau_j^*\, T_j = \left[\Sigma\, T_j^2 - \frac{T_{\cdot}^2}{v}\right]\Big/(r-\lambda) + \frac{Y^2}{b}$$

All contrasts in τ_j are estimable, the estimate of the contrast $\Sigma\, t_j \tau_j$ being $\Sigma\, t_j\, \tau_j^* = \Sigma\, t_j\, T_j/(r-\lambda)$. The grand total Y is also an estimate which is orthogonal to the estimates $\Sigma\, t_j\, T_j/(r-\lambda)$. Thus, there are v degrees of freedom for estimates. It is now easily verified that the analysis of variance with the block totals B_i as variables is the same as that obtained earlier in this chapter.

REFERENCES

Balanced incomplete blocks were introduced by Yates (1936). Fisher's inequality was obtained by Fisher (1940), and another proof given by Bose (1949 b).

For related combinational problems see Raghavarao (1971). These designs were used to obtain designs for two-way control of heterogeneity by Shrikhande (1951) and Agrawal (1966). For the case where a double grouping of blocks into replications is possible, see Robinson (1966).

The case of missing plots is discussed in Cornish (1940).

Recovery of interblock information was invented by Yates (1940). For further discussion see Nair (1944), Sprott (1956), Roy and Shah (1962), Shah (1964). Seshadri (1966), Stein (1966), Chakrabarti (1962) Cunningham and Henderson (1966).

A Bayesian approach to analysis will be found in Tiao and Draper (1968).

13

Partially Balanced Incomplete Block Design

13.1 Introduction

The balanced incomplete block designs studied in the previous chapter have the important property that they are the most efficient among all connected incomplete block designs in which each block has the same number of plots and each treatment is replicated the same number of times. However, balanced incomplete block designs do not always exist, and for certain numbers of treatments they exist only with an inordinately large number of replicates. For example, with eight treatments, to be arranged in blocks of three plots each, at least 56 blocks are required. Consequently, every treatment has to be replicated at least 21 times. The actual arrangement consists of putting in each block one of the 56 combinations of 8 treatments taken 3 at a time.

Partially balanced incomplete block (PBIB) designs were introduced to meet this difficulty. With their use the number of replications for each treatment can be made much smaller compared to a BIB design. However, the design though connected is no longer balanced; all treatment contrasts of the type $\tau_j - \tau_{j'}$ are not estimated with the same variance.

13.2 Association Schemes

The concept of an *association scheme* is needed for the definition of a PBIB design. Given v symbols $1, 2, \ldots, v$, we say we have an association scheme with m associate classes if the following conditions are satisfied.

(i) Any two symbols are either 1st, 2nd, . . ., or mth associates, the relation of association being symmetrical; that is, if the symbol α is the ith associate of the symbol β, then β is the ith associate of α.

(ii) Each symbol α has n_i ith associates, the number n_i being independent of α ($i = 1, \ldots, m$).

(iii) If α and β are ith associates, then the number of symbols that are jth associates of α, and kth associates of β, is p^i_{jk} and is independent of the pair of ith associates α and β.

A simple example of an association scheme with three associate classes is derived from the following arrangement of six symbols

1	2	3
4	5	6

With respect to each symbol the two other symbols in the same row are the first associates, the one other symbol in the same column is the second associate, and the remaining two symbols are the third asssociates. Thus 2, 3 are the first associates, 4 the second associate, and 5, 6 the third associates of 1. We verify easily that the result is a 3-class association scheme with $n_1 = 2$, $n_2 = 1$, $n_3 = 2$. This method can be used to generate 3-class association schemes for $m \times n$ symbols by arranging them in m rows and n columns. Such schemes are called *rectangular association schemes.*

Another simple example, of a 2-class association scheme, is the so-called *triangular association scheme* obtained by arranging $v = n(n - 1)/2$ symbols in n rows and columns as follows:

(i) The positions in the principal diagonal are left blank.

(ii) The $n(n - 1)/2$ positions above the principal diagonal are filled by the numbers $1, \ldots, v$ corresponding to the symbols.

(iii) The positions below the principal diagonal are filled so as to maintain symmetry about the principal diagonal.

The symbols entering in the same row or column with i are the first associates of i, and the rest are the second associates. For example, taking $v = 6$ we have the arrangement

×	1	2	3
1	×	4	5
2	4	×	6
3	5	6	×

The first and second associates are as follows:

Treatment	First associates	Second associates
1	2, 3, 4, 5	6
2	1, 3, 4, 6	5
3	1, 2, 5, 6	4
4	1, 2, 5, 6	3
5	1, 3, 4, 6	2
6	2, 3, 4, 5	1

13.3 PBIB Design

Let the v treatments follow an m-class association scheme. The arrangement of the v treatments in b blocks of k plots each will be called a PBIB design with m associate classes if every treatment is replicated r times, and any two ith associates occur together in λ_i blocks ($i = 1, \ldots, m$). The BIB design may be called a PBIB design with one associate class. We note that some of the λ_i may be zero.

The parameters $b, k, v, r, \lambda_1, \ldots, \lambda_m, n_1, \ldots, n_m$ are usually called the *parameters of the first kind*, and the p^i_{jk} the *parameters of the second kind*. Note that $n_1, \ldots, n_m$, and p^i_{jk} are obtained from the association scheme used for the design. As in a BIB design, the parameters of a PBIB design cannot be chosen arbitrarily; they must satisfy the following relations

$$bk = vr, \quad \sum_{i=1}^{m} n_i = v - 1$$

$$\sum_{i=1}^{m} n_i \lambda_i = r(k-1)$$

$$n_k p^k_{ij} = n_i p^i_{jk} = n_j p^j_{ki} \tag{13.1}$$

$$\sum_{k=1}^{m} p^i_{jk} = \begin{cases} n_j - 1 & \text{if } i = j \\ n_j & \text{if } i \neq j \end{cases}$$

From the above conditions it follows that there are only $m(m^2-1)/6$ independent parameters of the second kind.

From the triangular association scheme with six symbols

×	1	2	3
1	×	4	5
2	4	×	6
3	5	6	×

we can get the following PBIB designs

(i)

(1 2 3)
(1 4 5)
(2 4 6)
(3 5 6)

with $b = 4, k = 3; v = 6, r = 2; \lambda_1 = 1, \lambda_2 = 0$.

(ii)

(1 3 4 6)
(1 2 5 6)
(2 3 4 5)

with $b = 3, k = 4; v = 6, r = 2; \lambda_1 = 1, \lambda_2 = 2$

In both these examples we have $n_1 = 4, n_2 = 1$;

$$p^1_{11} = 2, p^1_{12} = 1, p^1_{22} = 0, p^2_{11} = 4, p^2_{12} = 0, p^2_{22} = 0.$$

These values result from the association scheme and hence are the same for the two examples.

13.4 Intrablock Analysis

We shall consider the intrablock analysis for PBIB designs with two associate classes only. The adjusted treatment totals are denoted as before by Q_j ($j = 1, \ldots, v$). The reduced normal equations are

$$\mathbf{C\tau} = \mathbf{Q} \tag{13.2}$$

The analysis can be easily carried out once the C-matrix of the design is known.

The C-matrix of the design has all the elements of the principal diagonal equal to $r(k - 1)/k$. The off-diagonal elements are $-\lambda_1/k$ and $-\lambda_2/k$ occurring n_1 and n_2 times respectively in any row. In fact, the off-diagonal element $c_{jj'}$ is $-\lambda_1/k$ or $-\lambda_2/k$ according as treatments j and j' are first or second associates.

Let $S(j, 1)$, $S(j, 2)$ denote the sets of treatments which are first and second associates respectively of the jth treatment. If we denote the sums of the treatment parameters occurring in $S(j, 1)$ and $S(j, 2)$ by G_{j1} and G_{j2} respectively, the reduced normal equations can be written as

$$Q_j = \{r(k-1)/k\}\,\tau_j - (\lambda_1/k)\,G_{j1} - (\lambda_2/k)\,G_{j2} \tag{13.3}$$

or

$$kQ_j = r(k-1)\,\tau_j - \lambda_1 G_{j1} - \lambda_2 G_{j2} \qquad j = 1, \ldots, v. \tag{13.4}$$

Denoting the sum $\tau_1 + \ldots + \tau_v$ of the treatment parameters for the v treatments by $\tau.$ we have

$$G_{j1} + G_{j2} = \tau. - \tau_j \tag{13.5}$$

Using (13.5) we can write (13.4) as

$$kQ_j = \{r(k-1) + \lambda_2\}\,\tau_j - (\lambda_1 - \lambda_2)\,G_{j1} - \lambda_2\tau.,\qquad j = 1, \ldots, v \tag{13.6}$$

These equations have to be solved to obtain estimates of treatment contrasts and the adjusted treatment SS, $S^2_{V/B}$. We try to obtain a solution which satisfies the further condition $\tau. = 0$. Using this condition (13.6) becomes

$$kQ_j = \{r(k-1) + \lambda_2\}\,\tau_j - (\lambda_1 - \lambda_2)\,G_{j1},\ j = 1, \ldots, v \tag{13.7}$$

If we sum the equations (13.7) over all treatments belonging to $S(j, 1)$, we get, denoting the sum of the adjusted treatment totals of all first associates of the jth treatment by Q_{j1}

$$kQ_{j1} = \{r(k-1) + \lambda_2\}\,G_{j1} - (\lambda_1 - \lambda_2)\{n_1\tau_j + p^1_{11}G_{j1} + p^2_{11}G_{j2}\} \tag{13.8}$$

This result is obtained by noting that when we add the terms $G_{j'1}$ for all j' which are the first associates of j, then j occurs n_1 times, every first associate of j occurs p^1_{11} times, and every second associate of j occurs p^2_{11} times in the sum. Substituting for G_{j2} from (13.5) and keeping $\tau. = 0$ equation (13.8) becomes

$$kQ_{j1} = \{r(k-1) + \lambda_2 - (\lambda_1 - \lambda_2)(p^1_{11} - p^2_{11})\}\,G_{j1} - (\lambda_1 - \lambda_2)(n_1 - p^2_{11})\,\tau_j \tag{13.9}$$

But we have $p^2_{11} + p^2_{12} = n_1$, so that (13.9) becomes

$$kQ_{j1} = \{r(k-1) + \lambda_2 - (\lambda_1 - \lambda_2)(p_{11}^1 - p_{11}^2)\} G_{j1} \\ - (\lambda_1 - \lambda_2) p_{12}^2 \tau_j \quad (13.10)$$

Putting

$$A_{12} = r(k-1) + \lambda_2;\ B_{12} = \lambda_2 - \lambda_1;\ A_{22} = (\lambda_2 - \lambda_1) p_{12}^2$$

$$B_{22} = r(k-1) + \lambda_2 + (\lambda_2 - \lambda_1)(p_{11}^1 - p_{11}^2) \quad (13.11)$$

we can write (13.7) and (13.10) as

$$kQ_j = A_{12}\tau_j + B_{12}G_{j1} \quad (13.12)$$

$$kQ_{j1} = A_{22}\tau_j + B_{22}G_{j1} \quad (13.13)$$

From these we easily obtain for the reduced normal equations the solution

$$\hat{\tau}_j = \{B_{22}(kQ_j) - B_{12}(kQ_{j1})\}/\Delta_1 \quad (13.14)$$

where

$$\Delta_1 = A_{12}B_{22} - A_{22}B_{12} \quad (13.15)$$

Since $\Sigma Q_j = \Sigma Q_{j1} = 0$, we see that the $\hat{\tau}_j$ satisfy the condition $\tau. = 0$, and hence (13.14) is a solution of the reduced normal equations.

If we had used (13.5) to eliminate G_{j1} instead of G_{j2} in (13.4) and (13.8) we would get the solution as

$$\hat{\tau}_j = [B_{21}(kQ_j) - B_{11}(kQ_{j2})]/\Delta_2 \quad (13.16)$$

where $A_{11} = r(k-1) + \lambda_1;\ B_{11} = \lambda_1 - \lambda_2;\ A_{21} = (\lambda_1 - \lambda_2) p_{12}^1$

$$B_{21} = r(k-1) + \lambda_1 + (\lambda_1 - \lambda_2)(p_{22}^2 - p_{22}^1)$$

$$\Delta_2 = A_{11}B_{21} - A_{21}B_{11} \quad (13.17)$$

and Q_{j2} is the sum of the adjusted treatment totals of the second associates of j.

We use (13.14), if $n_1 < n_2$ and (13.16) if $n_2 < n_1$, so that the work of adding up the adjusted treatment totals to get Q_{j1} or Q_{j2} is reduced.

The analysis of variance can now be carried out. The unadjusted block sum of squares is

$$S_B^2 = (\Sigma B_i^2)/k - (Y^2/bk) \quad (13.18)$$

where B_i is the total of the yields in the ith block and Y the total of all the yields. The adjusted treatment sum of squares is

$$S_{V/B}^2 = \sum_{j=1}^{v} \hat{\tau}_j Q_j \quad (13.19)$$

The degrees of freedom for these two sums of squares are $b-1$ and $v-1$ respectively. The intrablock error and other sums of squares can be obtained in the same manner as was done for the BIB design.

13.5 Variance of Estimates

The estimate of the difference $\tau_j - \tau_{j'}$ of the treatments j and j' is

$$\hat{\tau}_j - \hat{\tau}_{j'} = [B_{22}(kQ_j - kQ_{j'}) - B_{12}(kQ_{j1} - kQ_{j'1})]/\Delta_1 \qquad (13.20)$$

To obtain the variance of the estimate $\hat{\tau}_j - \hat{\tau}_{j'}$ we have only to obtain the coefficients of Q_j and $Q_{j'}$ in the above expression (see Sec. 11.4).

Suppose j and j' are first associates. Then j enters in $Q_{j'1}$ and j' in Q_{j1}, so that the cofficient of Q_j is

$$k(B_{22} + B_{12})/\Delta_1 \qquad (13.21)$$

and that of $Q_{j'}$ is

$$[-k(B_{22} + B_{12})]/\Delta_1 \qquad (13.22)$$

Hence,

$$\text{Var}(\hat{\tau}_j - \hat{\tau}_{j'}) = \{2k(B_{22} + B_{12})/\Delta_1\}\sigma^2 \qquad (13.23)$$

if j and j' are first associates.

If j and j' are second associates, treatments j and j' do not occur in Q_{j1} and $Q_{j'1}$. Therefore the coefficient of Q_j is kB_{22}/Δ_1 and that of $Q_{j'}$ is $-kB_{22}/\Delta_1$. We get then

$$\text{Var}(\hat{\tau}_j - \hat{\tau}_{j'}) = (2k\,B_{22}/\Delta_1)\,\sigma^2 \qquad (13.24)$$

Thus we see that the variance of any estimate $\hat{\tau}_j - \hat{\tau}_{j'}$ depends on whether j and j' are first or second associates. The design is not balanced. However, the differences between treatments which are first associates are all estimated with the same variance, and same is the case for differences of treatments which are second associates. It is due to this fact that we call the design partially balanced.

13.6 Notes on Computation

(i) Calculate the block totals B_i, and the treatment totals V_j.

(ii) Obtain the raw SS, the general correction factor Y^2/bk, the total SS and S_B^2.

(iii) Write down in tabular form for each treatment (with rows corresponding to treatments) the quantities V_j, T_j, and kQ_j. In the next column list the first associates if $n_1 < n_2$, or the second associates if $n_2 < n_1$. In the column following put Q_{j1} or Q_{j2}. The last column is for the solutions $\hat{\tau}_j$. As a check, we note that the column sums for the kQ_j, kQ_{j1} or kQ_{j2}, and $\hat{\tau}_j$ columns should be zero.

(iv) Use this table to obtain $S^2_{V/B} = [\Sigma\, \hat{\tau}_j\, (kQ_j)]/k$, and complete the analysis of variance table.

13.6 Example

The data in Table 13.1 give the yield in pounds of seed cotton per plot in a uniformity trial. The yields are arranged in 15 blocks each having four plots. On this data 15 dummy treatments have been superimposed (treatment numbers are shown in brackets) according to a PBIB design.

The association scheme of the design used is given by

1	6	11
2	7	12
3	8	13
4	9	14
5	10	15

where treatments occurring in any row are first associates. Thus 6 and 11 are the first associates and the remaining treatments the second associates of 1; we have $n_1 = 2$, $n_2 = 12$. The other parameters of the design are

$$b = 15,\ k = 4;\ v = 15,\ r = 4;\ \lambda_1 = 0,\ \lambda_2 = 1$$

$$p^1_{11} = 1,\ p^1_{12} = 0,\ p^1_{22} = 12;\ p^2_{11} = 0,\ p^2_{12} = 2,\ p^2_{22} = 9$$

Using these values we find

$$A_{12} = 13,\ B_{12} = 1,\ A_{22} = 2$$
$$B_{22} = 14,\ \Delta_1 = 180$$

Hence the solutions $\hat{\tau}_j$ are given by

$$\hat{\tau}_j = \frac{14(kQ_j) - (kQ_{j1})}{180}$$

In Table 13.2 we obtain the values of kQ_j, kQ_{j1}, and $180\,\hat{\tau}_j$.

Table 13.1

Block					Block Totals
1	(15) 2.4	(9) 2.5	(1) 2.6	(3) 2.0	9.5
2	(5) 2.7	(7) 2.8	(8) 2.4	(1) 2.7	10.6
3	(10) 2.6	(1) 2.8	(14) 2.4	(2) 2.2	10.0
4	(15) 3.4	(11) 3.1	(2) 2.1	(3) 2.3	10.9
5	(6) 4.1	(15) 3.3	(4) 3.3	(7) 2.9	13.6
6	(12) 3.4	(4) 3.2	(3) 2.8	(1) 3.0	12.4
7	(12) 3.2	(14) 2.5	(15) 2.4	(8) 2.6	10.7
8	(6) 2.3	(3) 2.3	(14) 2.4	(5) 2.7	9.7
9	(5) 2.8	(4) 2.8	(2) 2.6	(13) 2.5	10.7
10	(10) 2.5	(12) 2.7	(13) 2.8	(6) 2.6	10.6
11	(9) 2.6	(7) 2.6	(10) 2.3	(3) 2.4	9.9
12	(8) 2.7	(6) 2.7	(2) 2.5	(9) 2.6	10.5
13	(5) 3.0	(9) 3.6	(11) 3.2	(12) 3.2	13.0
14	(7) 3.0	(13) 2.8	(14) 2.4	(11) 2.5	10.7
15	(10) 2.4	(4) 2.5	(8) 3.2	(11) 3.1	11.2
				Grand total	164.0

Source: Cochran, W.G. and G.M., Cox, *Experimental Designs*, (2nd ed), John Wiley & Sons, New York, 1957, p. 456.

Table 13.2

Treatment	V_j	T_j	$kQ_j = 4V_j - T_j$	1st Ass.	kQ_{j1}	$180\, \hat{\tau}_j$
1	11.1	42.5	1.9	6,11	4.2	22.4
2	9.4	42.1	−4.5	7,12	3.7	−66.7
3	9.8	42.9	−3.7	8,13	−0.5	−51.3
4	11.8	47.9	−0.7	9,14	0.0	−9.8
5	11.2	44.0	0.8	10,15	−1.2	12.4
6	11.7	44.4	2.4	1,11	3.7	29.9
7	11.3	44.8	0.4	2,12	−1.2	6.8
8	10.9	43.0	0.6	3,13	−4.8	13.2
9	11.3	42.9	2.3	4,14	−3.0	35.2
10	9.8	41.7	−2.5	5,15	2.1	−37.1
11	11.9	45.8	1.8	1,6	4.3	20.9
12	12.5	46.7	3.3	2,7	−4.1	50.3
13	10.1	41.5	−1.1	3,8	−3.1	−12.3
14	9.7	41.1	−2.3	4,9	1.6	−33.8
15	11.5	44.7	1.3	5.10	−1.7	19.9
Total	164.0	656.0	0.0		0.0	0.0

The adjusted treatment sum of squares is obtained from the kQ_j and 180 $\hat{\tau}_j$ columns as $\Sigma\,(kQ_j)\,(180\,\hat{\tau}_j)/720$ and is equal to 1.5641. The total SS and the unadjusted block SS are obtained in the usual way. The intra-block analysis of variance is shown in Table 13.3.

Table 13.3

Source	df	SS	MS	F
Blocks (unadj.)	14	4.9233		
Treatments (adj.)	14	1.5641	0.1117	1.29
Error	31	2.6859	0.0866	
Total	59	9.1733		

The variance of the estimate $\hat{\tau}_j - \hat{\tau}_{j'}$ is

$$\{2k(B_{22} + B_{12})/\Delta_1\}\,\sigma^2 = \{8(14 + 1)/180\}\,\sigma^2 = (2/3)\sigma^2$$

if j and j' are first associates, and is

$$(2kB_{22}/\Delta_1)\sigma^2 = \{8(14)/180\}\,\sigma^2 = (28/45)\,\sigma^2$$

if j and j' are second associates. The estimates of these variances are, replacing σ^2 by $s^2 = 0.0866$, 0.0577 and 0.0539. Thus we see that in this case the two variances are close together.

REFERENCES

These designs were introduced by Bose and Nair (1939); Nair (1952) gives some examples to illustrate their analysis. The concept of associate classes is explicitly introduced in Bose and Shimamoto (1952). Tables of designs with two associate classes are given in Bose, Clatworthy and Shrikhande (1954).

Combinatorial aspects of PBIB designs are discussed in Raghavarao (1971).

For PBIB designs with Latin square properties see Mesner (1967); for missing values see Zelen (1954 b).

14

The General Factorial Experiment

14.1 Introduction

In agricultural experiments differences in yield from different plots depend upon soil heterogeneity, the variety of crop sown, the type of manure used, etc. Soil heterogeneity is beyond the experimenter's control and an appropriate design has to be used to isolate its effect on the differences in plot yields. The type of crop sown or the manure used can, however, be selected by the experimenter. If different crop varieties are to be compared these could be assigned to different plots keeping the other factors (manure, irrigation, etc.) the same for all the plots. The conclusions resulting from the analysis of the observations from such an experiment would, however, be valid only for crops grown under similar conditions in respect of manuring, irrigation, etc. If different types of manures (or different doses of the same manure) are to be compared, a single crop could be sown on all the plots and the manurial treatment varied from plot to plot. However, in this case one can not be sure that the conclusions would be valid for another crop different from the one used in the experiment. It might well be that one variety of crop responds better than another to a particular type of manure whereas the situation may be quite different if another manure is used.

Given a number of crop varieties and manurial treatments we could compare the varieties in a series of experiments keeping the manurial treatment fixed in any particular experiment. For example, with two crop varieties a and b, and three manurial treatments A, B and C, we could use two randomised block designs each having, say, three blocks of three plots each. The treatments are assigned at random to the three plots in any block and the two crop varieties at random to the two randomised block designs. A possible arrangement would be as follows.

bB	*bA*	*bC*	*aA*	*aB*	*aC*
bC	*bB*	*bA*	*aC*	*aA*	*aB*
bB	*bC*	*bA*	*aB*	*aC*	*aA*

In such an arrangement, however, the differences among the manurial treatments can be estimated but not those among the crop varieties. The latter are entangled with the differences between blocks.

If we use, instead, three sets of three blocks each, with each block having two plots, randomising the varieties inside each block and assigning the treatments at random to the three sets we would have an arrangement as follows.

bB	*aB*	*aC*	*bC*	*aA*	*bA*
aB	*bB*	*bC*	*aC*	*bA*	*aA*
bB	*aB*	*aC*	*bC*	*bA*	*aA*

The differences between the crop varieties are now estimable but those between the manurial treatments are not. The first approach amounts to testing the manurial treatments keeping the crop variety fixed, and the second tests the crop varieties keeping the manurial treatment fixed.

The factorial experiment is designed to get over this difficulty. If we combine each crop with each manurial treatment, we have six composite treatments (or treatment combinations) *aA*, *aB*, ..., *bC*. Keeping the total number of observations equal to 18, as in the two examples above, we may use a randomised block design with three blocks of six plots each, assigning the six composite treatments at random inside each block. A possible arrangement now is as follows.

bA	*aC*	*aB*	*bB*	*aA*	*bC*
aA	*aC*	*bC*	*aB*	*bB*	*bA*
bB	*aB*	*bA*	*aC*	*aA*	*bC*

With the above arrangement we can estimate the differences between the crop varieties as well as those between the manurial treatments.

More precisely, let us assume, in all the three cases, that the expected yield from a plot is the sum of three components; a block effect β_i, a varietal effect ν_j, and a treatment effect τ_k. In the first example, differences between treatments are estimated by

$$(T_k - T_{k'})/6,$$

T_k denoting the total yield from the plots receiving the kth treatment. The differences $\nu_j - \nu_{j'}$ and $\beta_i - \beta_{i'}$ are not estimable.

In the second case differences between varieties are estimated by $(V_j - V_{j'})/9$ where V_j stands for the total yield from the plots sown with the jth variety. The differences $\tau_k - \tau_{k'}$, $\beta_i - \beta_{i'}$ are not estimable.

In the third case, differences $\beta_i - \beta_{i'}$ between the blocks are estimated by $(B_i - B_{i'})/6$, differences $\nu_j - \nu_{j'}$ between varieties by $(V_j - V_{j'})/6$, and differences $\tau_k - \tau_{k'}$ between treatments by $(T_k - T_{k'})/6$. Here B_i is the total of the yields from the ith block, V_j the total yield from the jth variety, and T_k the total yield from the kth treatment.

The advantage of using a factorial experiment (the third example above) is obvious. The only drawback is an increase in block size. We shall see later how this is overcome by the technique of confounding.

In the above analysis we had assumed that the expected yield from any plot was given by $\beta_i + \nu_j + \tau_k$. We now examine this assumption in detail. Consider four plots in the ith block, two of which receive the kth and the other two the k'th treatment, one plot in each pair being sown with one of the two crop varieties. If the above assumption is true the expected value of the difference in yield from the two plots of each pair would be $\nu_j - \nu_{j'}$. In other words, the expected increase in yield from one variety as compared to the other is the same no matter which manurial treatment is used. As mentioned earlier, it usually happens that one variety responds better in the presence of one treatment as compared to another. In fact it can happen that a particular manurial treatment increases the yield of one variety whereas its effect on another variety may be insignificant. If such be the case the assumption of additivity of the varietal effect ν_j and the treatment effect τ_k would be invalidated. Factorial experiments have the additional advantage that we can get a test for the additivity assumption from the experimental data of the factorial experiment. Such a test cannot be carried out in experiments of the type described by the first two examples.

14.2 Terminology

In a factorial experiment we study the effect on the yield of one or more factors. The term *factor* is used in a general sense to denote a feature of the experimental conditions which may be assigned at will to the different experimental plots. The different states of a factor are called the *levels* of that factor; the levels of the factor have to be selected before the experiment is carried out.

There are two main types of factors: qualitative and quantitative. A qualitative factor is one in which the different levels cannot be arranged in order of magnitude; varieties of crops, the sex of an animal, the different pieces of plant used in industrial production, etc., are examples of qualitative factors. A quantitative factor is one for which the different levels can be arranged in order of magnitude. Examples of quantitative factors are temperature and humidity in a chemical experiment, a fertiliser with levels representing different doses, irrigation where the levels denote the amount of water used, etc. So far as design and analysis are concerned, both types may be treated in the same way, but they admit of differences in interpretation.

Each experimental unit (or plot) in a factorial experiment receives a *composite treatment* (or *treatment combination*) comprising one level of each of the factors under investigation. If there are m factors with $s_1, \ldots, s_m$ levels respectively there are $s_1 \cdot s_2 \cdot \ldots \cdot s_m = s$ treatment combinations to be assigned to the different plots, and the experiment itself is called an $s_1 \times s_2 \times \ldots \times s_m$ *factorial experiment*. Thus in the example considered in the preceding section we had a 2×3 factorial experiment with two factors; crop variety at two levels, and manurial treatment at three levels, giving in all six different treatment combinations.

The s^m factorial experiment in which each of the m factors is at s levels is called a *symmetrical factorial experiment*. If the numbers of levels of the different factors are unequal, the experiment is termed an *asymmetrical factorial experiment*. The simplest examples of symmetrical factorial experiments are the 2^n and 3^n factorials where each factor is at two and three levels respectively.

14.3 Main Effects and Interactions

Consider an $s_1 \times \ldots \times s_m$ factorial experiment with m factors which we denote by $A_1, \ldots, A_m$. The ith factor A_i is at s_i levels which we may denote by 0, 1, ..., $s_i - 1$. Then there are $s = s_1 s_2 \ldots s_m$ treatment combinations; a particular treatment combination may be written as

$$(x_1, \ldots, x_m), \qquad x_i = 0, 1, \ldots, s_i - 1 \qquad (14.1)$$
$$i = 1, \ldots, m$$

where x_i stands for the level of the ith factor A_i entering in the treatment combination. Thus in a three factor experiment the symbol (0, 1, 3) stands for the treatment combination obtained by taking the level 0 of the first, level 1 of the second, and level 3 of the third factor.

Let us denote the treatment parameter corresponding to the treatment combination $(x_1, \ldots, x_m)$ by $\tau(x_1, \ldots, x_m)$. Then the expected yield from the plot receiving the treatment $(x_1, \ldots, x_m)$ will be the sum of $\tau(x_1, \ldots, x_m)$ and other terms representing the contribution to the expected yield of the known characteristics of the experimental material. For example, if the different treatment combinations are studied in a randomised block design, the expected yield from any plot will be the sum of the block effect and $\tau(x_1, \ldots, x_m)$.

There are s treatment parameters $\tau(x_1, \ldots, x_m)$ and a typical contrast among these may be written as

$$\sum_{x_1, \ldots, x_m} l(x_1, \ldots, x_m)\, \tau(x_1, \ldots, x_m) \qquad (14.2)$$

where $l(x_1, \ldots, x_m)$ are constants with

$$\sum_{x_1, \ldots, x_m} l(x_1, \ldots, x_m) = 0 \qquad (14.3)$$

For example in a 2^2 factorial the four treatment parameters may be written as $\tau(0, 0)$, $\tau(0, 1)$, $\tau(1, 0)$, $\tau(1, 1)$. A typical contrast among these four is

$$l(0, 0)\,\tau(0, 0) + l(0, 1)\,\tau(0, 1) + l(1, 0)\,\tau(1, 0) + l(1, 1)\,\tau(1, 1)$$

with

$$l(0, 0) + l(0, 1) + l(1, 0) + l(1, 1) = 0$$

There are $s - 1$ independent treatment contrasts in all, and these are sub-divided into subsets called *main effects* and *interactions*.

The contrast (14.2) is said to be a main effect contrast of the factor A_i if the coefficient $l(x_1, \ldots, x_m)$ of $\tau(x_1, \ldots, x_m)$ depends only on the level x_i of the factor A_i, i.e. if

$$l(x_1, \ldots, x_m) = \lambda(x_l)$$
$$\sum_{x_l} \lambda(x_l) = 0 \qquad (14.4)$$

For example, in the 2^2 factorial considered above the main effect contrast of A_1 may be written as

$$\tau(0, 0) + \tau(0, 1) - \tau(1, 0) - \tau(1, 1)$$

Here we have taken $\lambda(x_1) = 1$ for $x_1 = 0$ and $\lambda(x_1) = -1$ for $x_1 = 1$. We can also write it as $-\tau(0, 0) - \tau(0, 1) + \tau(1, 0) + \tau(1, 1)$. or in general as $a\{\tau(0, 0) + \tau(0, 1)\} - a\{\tau(1, 0) + \tau(1, 1)\}$.

The main effect is also called the 0*th order interaction* (or *single factor interaction*). The set of all main effect contrasts of A_l is called the *main effect* of A_l and denoted by the same symbol A_l as used for the factor. The coefficients $\lambda(x_l)$ can be chosen arbitrarily except that we must have $\sum_{x_l} \lambda(x_l) = 0$. Thus there are $s_l - 1$ linearly independent contrasts in the main effect A_l.

The contrast (14.2) will be said to belong to the $(k-1)$*th order interaction* (or *k-factor interaction*) between the factors $A_{i_1}, \ldots, A_{ik}$ if

$$l(x_1, \ldots, x_m) \text{ depends only on } x_{i_1}, \ldots, x_{ik}$$

and $$\sum_{x_{i_1}} l(x_1, \ldots, x_m) = \sum_{x_{i_2}} l(x_1, \ldots, x_m) = \sum_{x_{ik}} l(x_1, \ldots, x_m) \qquad (14.5)$$
$$= 0$$

For example, in the 2^2 factorial the 2-factor interaction between A_1 and A_2 may be written as $a\{\tau(0,0) - \tau(0,1)\} - a\{\tau(1,0) - \tau(1,1)\}$.

The set of all contrasts belonging to the k factor interaction between $A_{i_1}, \ldots, A_{ik}$ is denoted by $A_{i_1} \times A_{i_2} \times \ldots \times A_{ik}$ (or by $A_{i_1} A_{i_2} \ldots A_{ik}$). The contrast (14.2) belongs to the interaction $A_{i_1} \times \ldots \times A_{ik}$ if

$$l(x_1, \ldots, x_m) = \lambda(x_{i_1}, \ldots, x_{ik})$$
$$\sum_{x_{ij}} \lambda(x_{i_1}, \ldots, x_{ik}) = 0, j = 1, \ldots, k \qquad (14.6)$$

Fix the values of $x_{i_2}, \ldots, x_{ik}$. Then we can choose $s_{i_1} - 1$ of the $\lambda(x_{i_1}, \ldots x_{ik})$ arbitrarily for we must have

$$\sum_{x_{i_1}} \lambda(x_{i_1}, \ldots, x_{ik}) = 0$$

It follows therefore that there are

$$(s_{i_1} - 1)(s_{i_2} - 1)\ldots(s_{ik} - 1)$$

linearly independent contrasts belonging to the k factor interaction $A_{i_1} \times A_{i_2} \times \ldots \times A_{i_k}$.

Consider a constrast belonging to A_1 with

$$l(x_1, \ldots, x_m) = \lambda(x_1), \sum_{x_1} \lambda(x_1) = 0 \tag{14.7}$$

and another belonging to $A_1 A_2$, with

$$l(x_1, \ldots, x_m) = \pi(x_1, x_2),$$

$$\sum_{x_1} \pi(x_1, x_2) = \sum_{x_2} \pi(x_1, x_2) = 0 \tag{14.8}$$

Then the scalar product of the coefficient vectors of these two contrasts is

$$\sum_{x_1, \ldots, x_m} \lambda(x_1)\, \pi(x_1, x_2) = \frac{s}{s_1 s_2} \sum_{x_1, x_2} \lambda(x_1)\, \pi(x_1, x_2)$$

$$= \frac{s}{s_1 s_2} \sum_{x_1} \lambda(x_1) \left(\sum_{x_2} \pi(x_1, x_2)\right) = 0$$

Thus the two contrasts are orthogonal. In the same way it can be shown that,

(a) contrasts belonging to interactions of different orders are mutually orthogonal, and

(b) contrasts belonging to interactions of the same order differing at least in one factor are mutually orthogonal.

For example, consider a contrast belonging to $A_1 A_2$ with

$$\begin{aligned} &l(x_1, \ldots, x_m) = \phi(x_1, x_2) \\ &\sum_{x_1} \phi(x_1, x_2) = \sum_{x_2} \phi(x_1, x_2) = 0 \end{aligned} \tag{14.9}$$

and another belonging to $A_2 A_3$ with

$$l(x_1, \ldots, x_m) = \psi(x_2, x_3)$$

$$\sum_{x_2} \psi(x_2, x_3) = \sum_{x_3} \psi(x_2, x_3) = 0 \tag{14.10}$$

Then the scalar product of the coefficient vectors is

$$\sum_{x_1, \ldots x_m} \phi(x_1, x_2)\, \psi(x_2, x_3) = \frac{s}{s_1 s_2 s_3} \sum_{x_1, x_2, x_3} \phi(x_1, x_2)\, \psi(x_2, x_3)$$

$$= \frac{s}{s_1 s_2 s_3} \sum_{x_1, x_2} \phi(x_1, x_2) \left(\sum_{x_3} \psi(x_2, x_3)\right)$$

$$= 0$$

Thus the total number $s-1$ of independent contrasts between the treatment effects $\tau(x_1, \ldots, x_m)$ can be divided into the following mutually orthogonal sets:

(a) $s_1 - 1$ independent contrasts belonging to the main effect A_1, $s_2 - 1$ to $A_2, \ldots, s_m - 1$ to A_m,

(b) $(s_1 - 1)(s_2 - 1)$ independent contrasts belonging to the 1st-order interaction $A_1 A_2$, $(s_1 - 1)(s_3 - 1)$ to $A_1 A_3, \ldots$, $(s_{m-1} - 1)(s_m - 1)$...... to $A_{m-1} A_m$.

....

(k) $(s_1 - 1)(s_2 - 1)\ldots(s_m - 1)$ independent contrasts belonging to the $(m-1)$th order interaction $A_1 A_2 \ldots A_m$.

These exhaust all the $s-1$ independent contrasts between the treatment combinations since we have

$$\begin{aligned} s - 1 &= s_1 s_2 \ldots s_m - 1 \\ &= \{(s_1 - 1) + 1\}\{(s_2 - 1) + 1\}\ldots\{(s_m - 1) + 1\} - 1 \\ &= \sum_i (s_i - 1) + \sum_{i \neq j} (s_i - 1)(s_j - 1) + \ldots \\ &\qquad + (s_1 - 1)(s_2 - 1)\ldots(s_m - 1) \end{aligned}$$

The reason for defining the main effects and interactions in the above manner will be clear from the following discussion. Suppose we have

$$\tau(x_1, x_2, \ldots, x_m) = \alpha(x_1) + \beta(x_2, \ldots, x_m),$$

i.e. the expected yield from $(x_1, x_2, \ldots, x_m)$ is the sum of two quantities, one depending on the first factor A_1 and the second on the remaining factors. Consider now a contrast belonging to the main effect of A_1; it may be written as

$$\sum_{x_1, \ldots, x_m} \lambda(x_1)\{\alpha(x_1) + \beta(x_2, \ldots, x_m)\} = \frac{s}{s_1} \sum_{x_1} \lambda(x_1)\,\alpha(x_1) \tag{14.11}$$

since $\sum_{x_1} \lambda(x_1) = 0$. Each contrast belonging to the main effect of A_1 is thus a contrast between the effects of the different levels of A_1. Consider next, a contrast belonging to the interaction $A_1 A_2$. It may be written as

$$\begin{aligned} &\sum_{x_1, \ldots, x_m} \lambda(x_1, x_2)\{\alpha(x_1) + \beta(x_2, \ldots, x_m)\} \\ &= \frac{s}{s_1 s_2} \sum_{x_1, x_2} \lambda(x_1, x_2)\,\alpha(x_1) + \sum_{x_1, \ldots, x_m} \lambda(x_1, x_2)\,\beta(x_2, \ldots, x_m) \\ &= 0 \end{aligned}$$

since $\sum_{x_1} \lambda(x_1, x_2) = \sum_{x_2} \lambda(x_1, x_2) = 0$. Thus every contrast belonging to $A_1 A_2$ reduces identically to zero. More generally, every contrast belonging to $A_1 A_{i_1} A_{i_2} \ldots A_{ik}$ ($i_j \geqslant 2$) is identically equal to zero.

Let us suppose now that the m factors split up additively into two groups, i.e. (say)

$$\tau(x_1, \ldots, x_m) = \alpha(x_1, x_2, \ldots, x_r) + \beta(x_{r+1}, \ldots, x_m)$$

Consider any interaction involving factors from both the groups. Let us take, for example, the interaction $A_1 A_2 A_{r+1} A_{r+2} A_{r+3}$.

A contrast belonging to it may be written as

$$\sum_{x_1, \ldots, x_m} \lambda(x_1, x_2, x_{r+1}, x_{r+2}, x_{r+3}) \{\alpha(x_1, \ldots, x_r) + \beta(x_{r+1}, \ldots, x_m)\}$$

$$= \frac{s}{s_1 s_2 \ldots s_r s_{r+1} s_{r+2} s_{r+3}} \sum_{x_1, \ldots, x_r, x_{r+1}, x_{r+2}, x_{r+3}} \lambda(x_1, x_2, x_{r+1}, x_{r+2}, x_{r+3}) \alpha(x_1, \ldots, x_r)$$

$$+ \frac{s}{s_1 s_2 s_{r+1} \ldots s_m} \sum_{x_1, x_2, x_{r+1}, \ldots, x_m} \lambda(x_1, x_2, x_{r+1}, x_{r+2}, x_{r+3}) \times \beta(x_{r+1}, \ldots, x_m) = 0$$

since

$$\sum_{x_1} \lambda(x_1, x_2, x_{r+1}, x_{r+2}, x_{r+3}) = \sum_{x_2} = \sum_{x_{r+1}} = \sum_{x_{r+2}} = \sum_{x_{r+3}} = 0.$$

If all the factors are additive, i.e.

$$\tau(x_1, x_2, \ldots, x_m) = \phi_1(x_1) + \phi_2(x_2) + \ldots + \phi_m(x_m)$$

then all the interactions are identically zero and the contrasts belonging to the main effect of any factor A_i are simply contrasts among the effects $\phi_i(x_i)$ ($x_i = 0, \ldots, s_i - 1$) of that factor.

14.4 Analysis of Variance in Factorial Experiments

The different treatment combinations of a factorial experiment can be assigned to the experimental units according to any design selected for the experiment. The choice of the design will be determined by the nature of the heterogeneity present, or expected, in the experimental material. What we try in general (except in the case of confounding, to be discussed later) is to use a connected design, i.e. a design in which all the contrasts among the treatment parameters $\tau(x_1, \ldots, x_m)$ are estimable. The analysis is most convenient if the design used has the further property that the orthogonality between the main effect and interaction contrasts holds also for their estimates.

Then the sums of squares due to the estimates of the different main effects and interactions are additive and independently distributed.

Suppose the $s_1 \times s_2 \times \ldots \times s_m$ factorial experiment is carried out in r randomised blocks, each block consisting of $s = s_1 s_2 \ldots s_m$ plots. There will be r plots receiving the treatment combination $(x_1, x_2, \ldots, x_m)$ i.e., with A_1 at level x_1, A_2 at level $x_2, \ldots,$ and A_m at level x_m. We denote the total yield from these r plots by

$$T_{1,2,\ldots m}(x_1, x_2, \ldots, x_m) \tag{14.12}$$

Similarly, there will be $r \dfrac{s}{s_{i_1} s_{i_2} \ldots s_{i_k}}$ plots in which A_{i_1} occurs at level x_{i_1}, A_{i_2} at level $x_{i_k}, \ldots, A_{i_k}$ at the level x_{i_k} and the total yield from these plots will be denoted by

$$T_{i_1, i_2, \ldots, i_k}(x_{i_1}, x_{i_2}, \ldots, x_{i_k}) \tag{14.13}$$

In this case all contrasts among the $\tau(x_1, x_2, \ldots, x_m)$ are estimable and the estimate of the contrast

$$\sum_{x_1, \ldots, x_m} l(x_1, \ldots, x_m)\, \tau(x_1, \ldots, x_m), \quad \sum_{x_1, \ldots, x_m} l(x_1, \ldots, x_m) = 0$$

is

$$\sum_{x_1, \ldots, x_m} l(x_1, \ldots, x_m) \frac{T_{1,2,\ldots,m}(x_1, \ldots, x_m)}{r} \tag{14.14}$$

It is easily verified that the orthogonality relations between main effects and interactions hold also for the estimates.

Consider now the contrasts belonging to the main effect of A_1. The estimate of such a contrast is of the form

$$\sum_{x_1, \ldots, x_m} \lambda(x_1) \frac{T_{1,2,\ldots,m}(x_1, \ldots, x_m)}{r}$$

i.e.

$$\sum_{x_1} \lambda(x_1) \frac{T_1(x_1)}{r}, \quad \sum_{x_1} \lambda(x_1) = 0 \tag{14 15}$$

We now calculate the SS due to these estimates, called the SS of the main effect A_1. There are $s_1 - 1$ independent contrasts belonging to A_1, and they are all linear functions of the totals $T_1(x_1)$ $(x_1 = 0, 1, \ldots, s_1 - 1)$. The totals are mutually orthogonal, each is a total of $\frac{rs}{s_1}$ observations, and any contrast among these totals is orthogonal to the grand total T.

Hence, the SS for the main effect

$$A_1 = \frac{s_1}{rs} \sum_{x_1=0}^{s_1-1} \{T_1(x_1)\}^2 - (T^2/rs) \tag{14.16}$$

and has $s_1 - 1$ degrees of freedom (T stands for the total of all the observations.

The contrasts belonging to the interaction A_1A_2 are estimated by

$$\sum_{x_1, \ldots, x_m} \pi(x_1, x_2)\, T_{1, 2, \ldots, m}(x_1, x_2, \ldots, x_m)/r$$

i.e.

$$\sum_{x_1, x_2} \pi(x_1, x_2)\, T_{1, 2}(x_1, x_2)/r \text{ with } \sum_{x_1} \pi(x_1, x_2) = \sum_{x_2} = 0$$

There are $(s_1 - 1)(s_2 - 1)$ independent estimates of this type, they are all linear functions of the totals $T_{1, 2}(x_1, x_2)$ $(x_1 = 0, 1, \ldots, s_1 - 1;$ $x_2 = 0, 1, \ldots, s_2 - 1)$. The totals are mutually orthogonal and contrasts among them are orthogonal to the grand total T. Hence the SS due to the contrasts among the totals $T_{1, 2}(x_1, x_2)$ is

$$\frac{s_1 s_2}{rs} \sum_{x_1, x_2} \{T_{1, 2}(x_1, x_2)\}^2 - (T^2/rs)$$

with $s_1 s_2 - 1$ degrees of freedom. But some of the contrasts among the totals $T_{1, 2}(x_1, x_2)$ are the same as the estimates of the main effects of A_1 and A_2. Hence, keeping in view the orthogonality relations among the estimates, the SS for the interaction A_1A_2 is

$$\frac{s_1 s_2}{rs} \sum_{x_1, x_2} \{T_{1, 2}(x_1, x_2)\}^2 - (\text{SS for } A_1) - (\text{SS for } A_2) - (T^2/rs)$$

with $s_1 s_2 - 1 - (s_1 - 1) - (s_2 - 1) = (s_1 - 1)(s_2 - 1)$ degrees of freedom.

In the same manner the SS for the $(k - 1)$th order interaction berween $A_{i_1}, A_{i_2}, \ldots, A_{i_k}$ is given by

$$\frac{s_{i_1} s_{i_2}, \ldots, s_{i_k}}{rs} \sum_{x_{i_1}, \ldots, x_{i_k}} \{T_{i_1, \ldots, i_k}(x_{i_1}, \ldots, x_{i_k})\}^2 \tag{14.18}$$

$-$ (SS for all lower order interactions formed out of $A_{i_1}, A_{i_2}, \ldots, A_{i_k}$) $-$ (SS for the main effects of

$$A_{i_1}, \ldots, A_{i_k}) - (T^2/rs)$$

The SS for the main effects and interactions are thus successively obtained. The total of all these is the SS due to treatment contrasts in the randomised block design. If the block totals are denoted by B_i $(i = 1, \ldots, r)$ the SS due to block contrasts is

$$\frac{1}{s} \sum_i B_i^2 - (T^2/rs) \tag{14.19}$$

The error SS is obtained as usual by subtraction. The analysis of variance is usually presented as follows:

Table 14.1

Source	df	SS
Blocks	$r-1$	S_B^2
A_1	$s_1 - 1$	$S_{A_1}^2$
$\vdots$	$\vdots$	$\vdots$
A_m	$s_m - 1$	$\vdots$
A_1A_2	$(s_1-1)(s_2-1)$	$\vdots$
$\vdots$	$\vdots$	$\vdots$
$A_1A_2 \ldots A_m$	$(s_1-1)(s_2-1) \ldots (s_m-1)$	$S_{A_1 \ldots A_m}^2$
Error	$(r-1)(s-1)$	S_e^2
Total	$rs-1$	$\mathbf{y}'\mathbf{y} - T^2/(rs)$

The mean squares and *F*-ratios are then obtained as required.

We can also use an $s \times s$ Latin square. The main effect and interaction sums of squares are calculated as in the randomised block design except that the number of replicates r is now equal to s. The row, column and error SS are obtained in the usual manner.

We have mentioned earlier that a factorial experiment enables us to discover if the effects of various factors are *additive* or, in the contrary case, if there is *interaction* between the factors. Suppose we wish to test the hypothesis that all factors are additive, i.e.

$$\tau(x_1, \ldots, x_m) = \phi_1(x_1) + \ldots + \phi_m(x_m) \tag{14.20}$$

If the hypothesis is true, all the interaction contrasts are identically zero and hence their estimates are error functions (i.e. have zero expectations). The mean square due to these estimates thus provides another unbiased estimate of σ^2. This estimate and the estimate provided by the error mean square are independently distributed. The hypothesis of additivity can, therefore, be tested by means of the usual F-test using these two mean squares.

More generally, if we wish to test the hypothesis

$$\tau(x_1, \ldots, x_m) = \alpha(x_1, \ldots, x_r) + \beta(x_{r+1}, \ldots, x_m) \qquad (14.21)$$

then the mean square due to the estimates of all the interaction contrasts involving factors from both the groups $(A_1, \ldots, A_r)$ and $(A_{r+1}, \ldots, A_m)$ provides, if the hypothesis is true, an unbiased estimate of σ^2, and the hypothesis can be tested by the usual F-test.

If the experiment is carried out in a single block with s plots, the total SS carries $s-1$ d.f. All these are accounted for by the main effects and interactions and no degree of freedom is left for error. In such a situation, which occurs frequently in factorial experiments, the usual practice is to regard some higher order interactions as zero and to use the corresponding SS to get an estimate of σ^2. We saw earlier that if there are some additivity relations among the factors, then certain interactions are identically zero and the SS due to their estimates may be taken as the error SS. Thus in a $3 \times 4 \times 5$ factorial with three factor A, B, and C, laid out in a single block of 60 plots, the partitioning of the d.f. is as follows.

A	2
B	3
C	4
AB	6
BC	12
AC	8
ABC	24
Total	59

If we have reason to believe that factor A does not interact with B and C, we can pool the SS due to AB, AC and ABC to get an error SS with $6+8+24=38$ df. In the absence of any evidence of additivity of the factors we could use the SS due to the highest order interaction ABC to get an estimate of σ^2 with 24 df. In experiments with more than three factors and no additivity assumption we could pool together all interactions involving more than three factors to get the error SS.

14.5 Notes on Computation

(i) Obtain first the raw SS, the general correction factor (CF), and the total SS.

(ii) Form two-way tables for each pair of factors; the entry in the i-th row and jth column being the total of yields from all plots where one factor of the pair is at the ith level and the other at the jth level. Enter the row and column totals. If the two factors are A_1 and A_2 the entries in the table are the totals $T_{1,2}(x_1, x_2)$ and the row and column totals are the totals $T_1(x_1)$, $T_2(x_2)$ respectively. From these we obtain the SS due the main effects A_1, A_2 and the interaction A_1A_2.

(iii) For the three factor interactions calculate the totals of the type $T_{1,2,3}(x_1, x_2, x_3)$. If necessary, three-way tables for such sums may be prepared, there being one table for every triplet of factors. The totals are then used to get the SS due to the three factor interactions.

(iv) Where every treatment combination is replicated only once, the SS due to the highest order interaction is obtained by subtraction.

14.6 Example

In order to assess the uniformity of a chemical product with respect to its moisture content six samples were taken, one on each day of the week. Each sample was divided into twelve subsamples. The moisture content (expressed as percentage of total weight) was measured by four different technicians using three different methods of measurement, each technician working on three sub-samples, one for each method of measurement. The results of these measurements are given in Table 14.2.

The observations may be regarded as resulting from a $4 \times 3 \times 6$ factorial experiment with one observation on each treatment combination. The three factors are: Technician (T) with 4 levels, Measurement Method (M) with 3 levels, and Samples (S) with 6 levels.

For these 72 observations we have:

$$\begin{aligned} \text{Raw SS} &= 246053 \\ \text{Grand Total} &= 4205 \\ \text{CF} &= (4205)^2/72 \\ &= 245583.68 \\ \text{Total SS} &= 469.32 \end{aligned}$$

Table 14.2

	Technician											
	1			2			3			4		
	Measurement Method			Measurement Method			Measurement Method			Measurement Method		
Sample	1	2	3	1	2	3	1	2	3	1	2	3
1	59	61	61	57	60	58	55	58	62	54	56	59
2	57	58	60	57	58	58	61	60	57	60	56	58
3	55	57	59	55	55	56	54	52	58	53	55	55
4	60	57	58	56	57	57	54	58	55	61	59	58
5	61	61	60	59	58	59	61	57	60	62	60	60
6	63	59	60	62	63	61	64	62	59	59	60	61

We next obtain (Table 14.3) the three two-way tables taking two factors at a time.

Table 14.3

Level of factor M	Level of factor T 1	2	3	4	Sum
1	355	346	349	349	1399
2	353	351	347	346	1397
3	358	349	351	351	1409
Sum	1066	1046	1047	1046	4205

Level of factor S	Level of factor T 1	2	3	4	Sum
1	181	175	175	169	700
2	175	173	178	174	700
3	171	166	164	163	664
4	175	170	167	178	690
5	182	176	178	182	718
6	182	186	185	180	733
Sum	1066	1046	1047	1046	4205

Level of factor S	Level of factor M 1	2	3	Sum
1	225	235	240	700
2	235	232	233	700
3	217	219	228	664
4	231	231	228	690
5	243	236	239	718
6	248	244	241	733
Sum	1399	1397	1409	4205

Each entry in the main body of the two-way table of factors T and M is a sum of six observations corresponding to the six levels of S. Hence, we have

SS due to the main effect of

$$T = [\{(1066)^2 + (1046)^2 + (1047)^2 + (1046)^2\}/18] - \text{CF}$$
$$= 245599.83 - 245583.68 = 16.15$$

SS due to the main effect of

$$M = [\{1399)^2 + (1397)^2 + (1409)^2\}/24] - \text{CF}$$
$$= 245587.12 - 245583.68 = 3.44$$

SS due to the interaction

$$TM = [\{(355)^2 + (346)^2 + ... + (351)^2 + (351)^2\}/6] - SS(T)$$
$$- SS(M) - \text{CF}$$
$$= 245607.5 - 16.15 - 3.44 - 245583.68 = 4.23$$

The SS due to other main effects and two factor interactions are obtained in the same manner from the remaining two-way tables. For Example,

SS due to the main effect of

$$S = [\{(700)^2 + (700)^2 + ... + (733)^2\}/12] - \text{CF}$$
$$= 245817.41 - 245583.68 = 233.73$$

SS due to the interaction

$$TS = [\{(181)^2 + (175)^2 + ... + (185)^2 + (180)^2\}/3] - SS(T)$$
$$- \text{SS}(S) - \text{CF}$$
$$= 245899.67 - 16.15 - 233.73 - 245583.68 = 66.11$$

The SS due to the three factor interaction TMS can be obtained in this case by subtraction. The complete analysis of variance is given in Table 14.4.

In the absence of an independent estimate of error (i.e. of σ^2), the mean square corresponding to the three factor interaction TMS has been used as the error mean square to obtain the F-values. Except for the main effect of S for which the F value is highly significant, no other main effect or interaction is significant.

We may thus conclude that the different methods of measurement yield similar values and that there is no individual bias in measurement due to the technician. However, there is considerable variation in the moisture content of the chemical produced on different days. The product is not uniform in its moisture content and there is need to improve the working of the plant producing the chemical.

Table 14.4

Source	df	SS	MS	F
T	3	16.15	5.38	1.84
M	2	3.44	1.72	—
S	5	233.73	46.75	15.96
TM	6	4.23	0.71	—
MS	10	57.90	5.79	1.98
TS	15	66.11	4.41	1.51
TMS	30	87.76	2.93	—
Total	71	469.32		

Instead of looking at the six samples as six levels of a factor we can also regard them as blocks with 12 plots each, and the experiment as a 4×3 factorial carried out in a randomised block design. There are now only two factors T and M with 4 and 3 levels respectively. The SS due to the main effect of S now appears as the between blocks sum of squares of the randomised block design. The between treatments sum of squares is made up of the SS due to T, M and TM. The resulting analysis of variance is give in Table 14.5.

The error mean square now provides an independent estimate of σ^2, and has been used to get the F-values. Only the block sum of squares is significant and we arrive at the same conclusion as derived earlier.

Table 14.5

Source	df	SS	MS	F
Blocks	5	233.73	46.75	12.14
T	3	16.15	5.38	1.40
M	2	3.44	1.72	—
TM	6	4.23	0.71	—
Error	55	211.77	3.85	
Total	71	469.32		

14.7 Tukey's Test for Non-additivity in a Randomised Block

Suppose we have a randomised block design with q varieties in p blocks with pq observations in all. In our analysis the model was taken as

$$y_{ij} = \beta_i + \tau_j + e_{ij} \qquad (i = 1, \ldots, p; j = 1, \ldots, q)$$

In such a model we assume that the block and treatment effects are additive, or that there is no interaction between the blocks and treatments, a "good" block being good for all the varieties.

If we wish to to examine the presence of interaction, i.e. to test for non-additivity of the block and treatment effects, we could take the model as

$$y_{ij} = \mu_{ij} + e_{ij}$$

and analyse the experiment as a $p \times q$ factorial with one observation for each treatment combination. We have then $p - 1$ degrees of freedom for the main effect for blocks, $q - 1$ for the main effect for treatments, and $(p - 1)(q - 1)$ for the interaction between blocks and treatments. Thus there are no degrees of freedom left for error against which to test the interaction SS for non-additivity. Tukey has devised a special test to get over this difficulty,

We take the model as

$$y_{ij} = \beta_i + \tau_j + \theta(\beta_i - \bar{\beta})(\tau_j - \bar{\tau}) + e_{ij}$$

where $\bar{\beta} = (\sum_i \beta_i)/p$, $\bar{\tau} = (\sum_j \tau_j)/q$, and devise a test for the hypothesis $\theta = 0$. This model is non-linear in the parameters θ, β_i and τ_j. Thus the general least squares method is not applicable. However, let us assume for the moment that the β_i and τ_j are known parameters. Then θ is estimable, the normal equation being

$$\sum_{i,j} (\beta_i - \bar{\beta})(\tau_j - \bar{\tau})\, y_{ij} = \theta \sum_{i,j} (\beta_i - \bar{\beta})^2 (\tau_j - \bar{\tau})^2$$

The estimate of θ is therefore

$$\hat{\theta} = \sum_{i,j} (\beta_i - \bar{\beta})(\tau_j - \bar{\tau})\, y_{ij} \Big/ \sum_i (\beta_i - \bar{\beta})^2 \sum_j (\tau_j - \bar{\tau})^2$$

and the SS due to the estimate $\hat{\theta}$ of θ is

$$\Big\{\sum_{i,j} (\beta_i - \bar{\beta})(\tau_j - \bar{\tau})\, y_{ij}\Big\}^2 \Big/ \sum_i (\beta_i - \bar{\beta})^2 \sum_j (\tau_j - \bar{\tau})^2$$

We now assume that $\theta = 0$ (i.e. the block and treatment effects are additive). We have then a randomised block design and $\beta_i - \bar{\beta}$ $\tau_j - \bar{\tau}$ being contrasts are estimable, the estimates being $\bar{y}_{i\cdot} - \bar{y}_{\cdot\cdot}$, $\bar{y}_{\cdot j} - \bar{y}_{\cdot\cdot}$ respectively. We substitute these estimates for $\beta_i - \beta$ and $\tau_j - \bar{\tau}$ in the SS for $\hat{\theta}$. (This was the reason for taking the non-additivity term as $\theta(\beta_i - \bar{\beta})(\tau_j - \bar{\tau})$ instead of simply $\theta\beta_i\tau_j$, since β_i, τ_j are not estimable in the randomised block model). Thus the "SS for $\hat{\theta}$" becomes

$$S_\theta^2 = \Big\{\sum_{ij} y_{ij}(\bar{y}_{i\cdot} - \bar{y}_{\cdot\cdot})(\bar{y}_{\cdot j} - \bar{y}_{\cdot\cdot})\Big\}^2 \Big/ \sum_i (\bar{y}_{i\cdot} - \bar{y}_{\cdot\cdot})^2 \sum_j (\bar{y}_{\cdot j} - \bar{y}_{\cdot\cdot})^2$$

This is strictly not a sum of squares arising from linear forms in the observations, but the quotient of a sixth-degree polynomial by a fourth degree polynomial. But, as we shall now show, it has, under the hypothesis $\theta = 0$, a chi-square distribution with 1 d.f.

Assume $\theta = 0$, and write

$$\hat{\gamma}_{ij} = y_{ij} - \bar{y}_{i\cdot} - \bar{y}_{\cdot j} + \bar{y}_{\cdot\cdot}$$

Then (since we have now a randomised block model) the $\hat{\gamma}_{ij}$ are error functions, $E(\hat{\gamma}_{ij}) = 0$, and these are orthogonal to the estimating functions $\bar{y}_{i\cdot} - \bar{y}_{\cdot\cdot}$, $\bar{y}_{\cdot j} - \bar{y}_{\cdot\cdot}$. In the analysis of variance for the randomised block design let us denote by S_β^2 the SS due to blocks, S_τ^2 the SS due to treatments, and by S_e^2 the error SS. These

SS are mutually orthogonal and hence independently distributed.

The quantity S_θ^2 can also be written as

$$S_\theta^2 = \{ \sum_{i,j} \hat{\gamma}_{ij} (\bar{y}_{i\cdot} - \bar{y}_{\cdot\cdot}) (\bar{y}_{\cdot j} - \bar{y}_{\cdot\cdot}) \}^2 / \sum_i (\bar{y}_{i\cdot} - \bar{y}_{\cdot\cdot})^2 \sum (\bar{y}_{\cdot j} - \bar{y}_{\cdot\cdot})^2$$

Consider now the linear form

$$f = \sum_{ij} a_i b_j \hat{\gamma}_{ij}$$

with SS

$$S_f^2 = f^2 / \sum_i a_i^2 \sum b_j^2$$

where $\{a_i, b_j\}$ are known constants subject only to the conditions $\Sigma a_i = 0, \Sigma b_j = 0, \Sigma a_i^2 > 0, \Sigma b_j^2 > 0$. Then f is an error function. It follows therefore (always under the assumption $\theta = 0$) that S_f^2/σ^2 is χ^2 with 1 d.f. and that $(S_e^2 - S_f^2)/\sigma^2$ is χ^2 with $(p - 1) q - 1) - 1$ d.f. the two χ^2's being independent.

Under the assumption $\theta = 0$ the quantities

$$\bar{y}_{i\cdot} - \bar{y}_{\cdot\cdot},\ \bar{y}_{\cdot j} - \bar{y}_{\cdot\cdot},\ \hat{\gamma}_{ij}$$

are mutually orthogonal and hence independently distributed. Hence the conditional distribution of the $\{\hat{\gamma}_{ij}\}$, given the values of $\{\bar{y}_{i\cdot} - \bar{y}_{\cdot\cdot}\}$, $\{\bar{y}_{\cdot j} - \bar{y}_{\cdot\cdot}\}$ for all i and j, is identical with the unconditional distribution of the $\{\hat{\gamma}_{ij}\}$. Let us consider $\bar{y}_{i\cdot} - \bar{y}_{\cdot\cdot}$ and $\bar{y}_{\cdot j} - \bar{y}_{\cdot\cdot}$ as fixed and take in f

$$a_i = \bar{y}_{i\cdot} - \bar{y}_{\cdot\cdot},\ b_j = \bar{y}_{\cdot j} - \bar{y}_{\cdot\cdot}$$

The random variables $\bar{y}_{i\cdot} - \bar{y}_{\cdot\cdot}, \bar{y}_{\cdot j} - \bar{y}_{\cdot\cdot}$ cannot all be simultaneously zero, and hence $\Sigma (\bar{y}_{i\cdot} - \bar{y}_{\cdot\cdot})^2 > 0, \Sigma (\bar{y}_{\cdot j} - \bar{y}_{\cdot\cdot})^2 > 0$ with probability one. Therefore the joint conditional distribution under the assumption $\theta = 0$, of S_θ^2/σ^2 and $(S_e^2 - S_\theta^2)/\sigma^2$, given the values of $\bar{y}_{i\cdot} - \bar{y}_{\cdot\cdot}, \bar{y}_{\cdot j} - \bar{y}_{\cdot\cdot}$ is that of two independent chi-squares with 1 and $(p - 1) (q - 1) - 1$ d.f. respectively. Since this does not depend on $\bar{y}_{i\cdot} - \bar{y}_{\cdot\cdot}, \bar{y}_{\cdot j} - \bar{y}_{\cdot\cdot}$, the unconditional distribution is the same as the conditional.

We thus see finally that under the assumption $\theta = 0$ the quantities S_θ^2/σ^2, $(S_e^2 - S_\theta^2)/\sigma^2$ are independent chi-squares. Hence the hypothesis $\theta = 0$ can be tested by means of the ratio

$$S_\theta^2 \{(p - 1) (q - 1) - 1\} / (S_e^2 - S_\theta^2)$$

which has the F-distribution with 1 and $pq - p - q$ d.f.

If we denote the block and treatment totals by B_i, T_j, and the grand total by Y, the SS for blocks and treatments are first obtained as before. We have

$$S_\beta^2 = (\sum_i B_i^2)/q - (Y^2/pq),\ S_\tau^2 = (\sum_j T_j^2/p) - (Y^2/pq)$$

After some simplification, we find that

$$S_\theta^2 = \{\sum_{ij} y_{ij} B_i T_j - Y[S_\beta^2 + S_\tau^2 + (Y^2/pq)]\}^2/pq - S_\beta^2 - S_\tau^2$$

which appears to be the most convenient form for computation. We already know S_β^2, S_τ^2 and the correction factor Y^2/pq. It remains to obtain $\sum_{ij} y_{ij}B_iT_j$ which is easily computed as

$$\sum_{i=1}^{p} B_i \left\{ \sum_{j=1}^{q} y_{ij}\, T_j \right\}$$

from the table giving y_{ij}, B_i and T_j. The analysis of variance table is as follows.

Source	df	SS	MS	F
Blocks	$p-1$	S_β^2		
Treatment	$q-1$	S_τ^2		
Non-additivity	1	S_θ^2	s_θ^2	s_θ^2/s^2
Error	$pq-p-q$	By subtraction	s^2	
Total	$pq-1$	$\mathbf{y}'\mathbf{y}-(Y^2/pq)$		

The F-value obtained is compared against the value obtained from the tables for F with $(1, pq-p-q)$ d.f.

COMPLEMENTS AND PROBLEMS

1. Let the composite treatments of a 2×2 factorial experiment, with two factors A and B, be denoted by

$$\tau_{11}, \tau_{12}, \tau_{21}, \tau_{22}$$

where τ_{ij} stands for the composite treatment with A at level i and B at level j. Then the interactions A, B, and AB may be represented by the contrasts

$$(\tau_{11} + \tau_{12}) - (\tau_{21} + \tau_{22})$$

$$(\tau_{11} + \tau_{21}) - (\tau_{12} + \tau_{22})$$

and

$$(\tau_{11} - \tau_{21}) - (\tau_{12} - \tau_{22})$$

respectively. These three contrasts are mutually orthogonal. If this experiment is carried out in a completely randomised design with τ_{ij} replicated n_{ij} times, and the mean of the n_{ij} observations on τ_{ij} is $\bar{y}_{ij}$, the estimates of A, B, and AB are

$$(\bar{y}_{11} + \bar{y}_{12}) - (\bar{y}_{21} + \bar{y}_{22})$$

$$(\bar{y}_{11} + \bar{y}_{21}) - (\bar{y}_{12} + \bar{y}_{22})$$

and

$$(\bar{y}_{11} - \bar{y}_{21}) - (\bar{y}_{12} - \bar{y}_{22})$$

respectively. These estimates are, however, not orthogonal in general. For instance, if $n_{11} = n_{22}$ and $n_{12} = n_{21}$, the estimates of A and B are orthogonal to estimate of AB, but estimates of A and B are not orthogonal to each other. A sufficient condition for all of them to be mutually orthogonal is that all n_{ij}'s be equal.

2. (a) Let the six composite treatments of a 2×3 factorial experiment, with two factors A and B, be denoted by

$$\tau_{11}, \tau_{12}, \tau_{13}, \tau_{21}, \tau_{22}, \tau_{23}$$

where τ_{ij} stands for the composite treatment with A at level i and B at level j. Then A has one degree of freedom and may be represented by the contrast

$$(\tau_{11} + \tau_{12} + \tau_{13}) - (\tau_{21} + \tau_{22} + \tau_{23})$$

B has two degrees of freedom and may be represented by the two orthogonal contrasts

$$(\tau_{11} + \tau_{21}) - (\tau_{12} + \tau_{22})$$

and

$$(\tau_{11} + \tau_{21}) + (\tau_{12} + \tau_{22}) - 2\,(\tau_{13} + \tau_{23})$$

The interaction AB (two degrees of freedom) may be represented by the two orthogonal contrasts

$$(\tau_{11} - \tau_{21}) - (\tau_{12} - \tau_{22})$$

and

$$(\tau_{11} - \tau_{21}) + (\tau_{12} - \tau_{22}) - 2\,(\tau_{13} - \tau_{23})$$

It is easily verified that the contrasts defining A, B and AB are mutually orthogonal.

(b) If the above factorial experiment is carried out in a completely randomised design with τ_{ij} replicated n_{ij} times then the estimates of A, B and AB are not necessarily orthogonal. For instance, if $n_{11} = n_{21}$, $n_{12} = n_{22}$, $n_{13} = n_{23}$, the estimates of B are orthogonal to estimates of A and AB, but the estimate of A is not orthogonal to the estimates of AB. If all the n_{ij}'s are equal then A, B and AB estimates are mutually orthogonal.

REFERENCES

Factorial experiments were introduced by Yates (1935) wherein he also discusses the advantages resulting from their use. General mathematical theory of the symmetric factorial design is given in Bose (1947).

Kurkjian and Zelen (1962, 1963) introduced a special calculus for the analysis of factorial experiments; see also Zelen and Federer (1964, 1965) and Federer and Zelen (1966). Margolin (1967) gives a systematic method for the $2^m \times 3^n$ experiment. Robinson and Balaam (1967) consider the case where the errors are correlated. Bradu (1965) uses pseudo-factors to make the analysis simpler.

See Finney (1948), Williams (1952), Rayner (1953) for a discussion of main effects and interactions. Shah (1958) and Kshirsagar (1966) discuss the problem of balance in factorial experiments; Cornish (1938) discusses the use of incomplete block designs for factorial experiments.

Tukey's test is given in Tukey (1949); for extention to other designs see Tukey (1955), Abraham (1960).

15

The 2^n Factorial Experiment

15.1 Notation

The simplest class of factorial designs is that involving factors at two levels, that is, the 2^n class, n being the number of factors. One of the two levels of any factor is conveniently designated as the *lower* level and the other as the *higher* level.

The factors are usually denoted by capital letters, and the two levels of the factor by (1) and the corresponding small letter. The levels of a factor A are thus denoted by (1) and a. By convention, (1) refers to the lower level of A, while a refers to the higher level.

The treatment combinations can then be easily written in terms of the symbols used for the individual factors. Thus, with three factors A, B and C the treatment combination with all the factors at the higher level is written as abc or the product of the individual levels a, b and c. If A is at the lower level, and B, C at the higher level, the resulting treatment combination is bc, the product of the levels (1), b and c. Obviously (1) denotes the treatment combination where all the factors are at the lower level.

With the two factors A and B the, four treatment combinations are

(1) — A and B at lower level
a — A at higher, B at lower level
b — A at lower, B at higher level
ab — A and B at higher level

Note that the expression bc may stand for different treatment combinations depending upon the number of factors in the the experiment. If there are only two factors B and C, bc is the treatment combination with both factors at the higher level. If there are other factors besides B and C, then bc denotes the treatment combination with B and C at the higher level and all the other factors at the lower level.

15.2 Standard Order for Treatment Combinations

The complete list of the 2^n treatment combinations is most conveniently written in what is called the *standard ord r*. For one factor A, the standard order is (1), a. For two factors, we add two more, b and ab, derived by multiplying the first two by the additional letter b. For three factors we add c, ac, bc and abc, derived by multiplying the first four by the additional letter c, and so on. Thus the standard order for any number of factors can be obtained step by step from that for a smaller number of factors, without disturbing the latter. For example, the standard order for the 2^4 experiment with factors A, B, C and D is

$$(1)\ a,\ \ b,\ \ ab,\ \ c,\ \ ac,\ \ bc,\ \ abc,$$

$$d,\ ad,\ bd,\ abd,\ cd,\ acd,\ bcd,\ abcd$$

The first row is the standard order for three factors A, B and C.

Let us define a multiplication rule for treatment combinations, viz. multiply the two expressions algebraically and replace every term with an even exponent by unity. Then, we have

$$(ab)\cdot(cd) = abcd$$

$$(abc)\,(ab) = c$$

With this multiplication the 2^n treatment combinations form a group with n generators. For example the group of 8 treatment combinations in a 2^3 factorial can be generated by the sets (a, b, c), (a, bc, ac), etc.

Note that (1), a; (1), a, b, ab; (1), a, b, ab, c, ac, bc, abc, etc. are subgroups of the group defined earlier.

15.3 Main Effects and Interactions

The capital letters A, B, ... used to denote the factors also serve to represent the main effects and interactions. Thus A, B, ... will stand for the main effect contrasts for the factors A, B, ... respectively. The first order (two-factor) interactions are denoted by AB, AC, BC; second order (three-factor) interactions by ABC, ABD, ...; and so on. These can also be written in the same standard order

$$B, AB, C, AC, BC, ABC, \ldots$$

as used for the treatment combinations. Introducing an identity symbol I, the set of all main effects and interactions with I forms a group under the same multiplication rule as given earlier for the treatment combinations.

In a 2^n experiment each main effect and interaction has one degree of freedom. There are n main effects, ${}^nC_1$ first order interactions, ${}^nC_2$ second order interactions, and so on.

The contrasts belonging to the main effects and interactions are conveniently written as follows:

$$A = (a - 1)(b + 1)(c + 1) \ldots (z + 1)$$

$$B = (a + 1)(b - 1)(c + 1) \ldots (z + 1)$$

$$\ldots\ldots\ldots\ldots\ldots\ldots\ldots\ldots\ldots\ldots\ldots\ldots$$

$$AB = (a - 1)(b - 1)(c + 1) \ldots (z + 1)$$

$$\ldots\ldots\ldots\ldots\ldots\ldots\ldots\ldots\ldots\ldots\ldots\ldots$$

$$ABC = (a - 1)(b - 1)(c - 1)(d + 1) \ldots (z + 1)$$

$$\ldots\ldots\ldots\ldots\ldots\ldots\ldots\ldots\ldots\ldots\ldots\ldots$$

$$ABC..Z = (a - 1)(b - 1)(c - 1) \ldots (z - 1)$$

The expressions on the right, when simplified algebraically, give the contrasts in terms of the treatment combinations. Thus in a 2^2 factorial with two factors A and B, the contrast belonging to the main effect of A is

$$(a - 1)(b + 1) = -(1) + a - b + ab$$

and that belonging to the interaction AB is

$$(a - 1)(b - 1) = (1) - a - b + ab$$

Another method to obtain these contrasts is to form a table in which the columns correspond to the treatment combinations and the rows to the main effects and interactions (or the other way round). The entries in the table are + and − indicating, for each main effect and interaction contrast, the treatment combinations entering with with positive and negative signs. The signs are determined according to a simple rule called the *rule of odds and evens*. If the interaction has an even number of letters (AB, $ABCD$, ...), a treatment combination having an even number of letters common with the interaction enters with a positive sign, and one with an odd number of letters common carries a negative sign. If the interaction has an odd number of letters (A, ABC, ...) the rule is reversed. Once a few rows have been filled the others can be obtained more easily by the multiplication rule, viz the signs for $ABCD$ are obtained by multiplying the corresponding signs for A and BCD, or for AB and CD, etc. We give below such a table for a factorial with three factors A, B, and C.

Table 15.1

	(1)	*a*	*b*	*ab*	*c*	*ac*	*bc*	*abc*
A	−	+	−	+	−	+	−	+
B	−	−	+	+	−	−	+	+
AB	+	−	−	+	+	−	−	+
C	−	−	−	−	+	+	+	+
AC	+	−	+	−	−	+	−	+
BC	+	+	−	−	−	−	+	+
ABC	−	+	+	−	+	−	−	+

Note: The portion outlined in the top left corner gives the representation for a 2^2 factorial.

The treatment combination (1) is taken to have an even number (zero) of letters common with every interaction.

15.4 Calculation of Sums of Squares

Suppose the 2^n-factorial experiment is carried out in a randomised block with r replicates. We denote the total yield from the r-plots receiving a particular treatment combination by the same symbol within brackets. Thus (ab) denotes the total yield from the plots receiving the treatments combination ab.

The SS due to any main effect or interaction is then obtained simply by combining the total yields according to the + and — signs from Table 15.1, squaring the results, dividing by the number of observations making the total ($= r \cdot 2^n$).

Thus in a 2^3 factorial the SS due to the interaction AC is

$$(1/r2^3)\,\{(1)-(a)+(b)-(ab)-(c)+(ac)-(bc)+(abc)\}^2$$

The same method applies if the design used is a $2^n \times 2^n$ Latin square, r being replaced by 2^n.

15.5 Yates' Method of Computation

An alternative computing scheme to calculate the SS due to the main effects and interactions in a 2^n factorial has been devised by Yates.

The treatment combinations are first written in the standard order in the column at the beginning of the table, called the *treatment column*. The next column called the *yield column* gives the total yield for each treatment combination. If necessary a common term is subtracted from each yield total. Columns (1), (2), ..., (n) are then obtained successively. Column (1) is obtained from the yield column, the upper half is obtained by adding the yields in pairs, the second by taking the differences in pairs, the difference obtained by subtracting the first term of the pair from the second. Columns (2), (3), ..., (n) are obtained from the preceding ones in the same manner as used for getting (1) from the yield column. The square of the column (n) divided by the total number observations, gives the SS due to the interaction, the entry in the row for *abc* giving the SS due to ABC.

The entry in column (n) in the first row, corresponding to the treatment combination (1), gives the total yield which must agree with the total of the yield column, thus providing a check. A more complete checking procedure has also been given by Yates. The table is extended further by introducing the following additional rows.

W = first half odds (i.e. sum of the odd terms in the first half of the table)

X = first half evens

Y = second half odds

Z = second half evens

$X + W$

$Z + Y$

$X - W$

$Z - Y$

$X + W + Z + Y$

$(X - W) + (Z - Y)$

As each column of the table is filled a check is made according to the following rule:

(i) $X + W$ in any column $= (X + W) + (Z + Y)$ in the previous column,

(ii) $Z + Y$ in any column $= (X - W) + (Z - Y)$ in the previous column, and

(iii) first entry in column $(n) = X + W + Z + Y$ in the yield column.

Thus any error is detected before the next column is begun and a clue to its location is provided. If it is found that $X + W$ is not equal to $(X + W) + (Z + Y)$ in the previous column, the error occurs most likely in the first half of the table. Similarly if (ii) shows disagreement the error is most likely to be in the second half.

Table 15.2 illustrates Yates procedure for a 2^2 factorial.

It is clear that the computation gives the various SS and provides the checks as described.

15.6 Example

An experiment was carried out to study the effect of inorganic and organic fertilisers on the yield of potatoes. There were four treatment combinations: blood and superphosphate, sulphate of ammonia and superphosphate, sulphate of ammonia and steamed bone flour, blood and steamed bone flour. We may regard these as the treatment combinations of a 2^2 experiment with n_0, n_1, p_0, p_1 as follows:

n_0: Blood
n_1: sulphate of ammonia
p_0: Superphosphate,
p_1: steamed bone flour.

The experiment was laid out in a 4 × 4 Latin square with plots of 1/50 acre in area. The yield of potatoes (in pounds) along with

Table 15.2

Treatment Combination	Yield	(1)	(2)
(1)	(1)	(1) + (*a*)	(1) + (*a*) + (*b*) + (*ab*)
a	(*a*)	(*b*) + (*ab*)	−(1) + (*a*) − (*b*) + (*ab*)
b	(*b*)	(*a*) − (1)	−(1) − (*a*) + (*b*) + (*ab*)
ab	(*ab*)	(*ab*) − (*b*)	(1) − (*a*) − (*b*) + (*ab*)
Total	T		
W	(1)	(1) + (*a*)	(1) + (*a*) + (*b*) + (*ab*)
X	(*a*)	(*b*) + (*ab*)	−(1) + (*a*) − (*b*) + (*ab*)
Y	(*b*)	(*a*) − (1)	−(1) − (*a*) + (*b*) + (*ab*)
Z	(*ab*)	(*ab*) − (*b*)	(1) − (*a*) − (*b*) + (*ab*)
$X + W$	(1) + (*a*)	(1) + (*a*) + (*b*) + (*ab*)	2 (*a*) + 2 (*ab*)
$Z + Y$	(*b*) + (*ab*)	(*a*) − (1) + (*ab*) − (*b*)	−2 (*a*) + 2 (*ab*)
$X - W$	(*a*) − (1)	(*b*) + (*ab*) − (1) − (*a*)	
$Z - Y$	(*ab*) − (*b*)	(*ab*) −(*b*) − (*a*) + (1)	
$X + W + Z + Y$	(1) + (*a*) + (*b*) + (*ab*)	2 (*a*) + 2 (*ab*)	
$X - W + Z - Y$	(*a*) − (1) + (*ab*) − (b)	2 (*ab*) − 2 (*a*)	

the treatment combinations is shown in Table 15.3.

Table 15.3

Row	Column 1	Column 2	Column 3	Column 4
I	645 (n_0p_1)	667 (n_1p_1)	670 (n_1p_0)	787 (n_0p_0)
II	752 (n_0p_0)	637 (n_1p_0)	655 (n_1p_1)	576 (n_0p_1)
III	642 (n_1p_1)	627 (n_0p_1)	686 (n_0p_0)	575 (n_1p_0)
IV	621 (n_1p_0)	762 (n_0p_0)	596 (n_0p_1)	660 (n_1p_1)

Source: Annual Report, Rothamstead Experimental Station, Harpenden, England, 1931.

The first step is to total the yields for each treatment combination; the totals are

$$(n_0\,p_0) = 2987$$

$$(n_1\,p_0) = 2503$$

$$(n_0\,p_1) = 2444$$

$$(n_1\,p_1) = 2624$$

We now follow the procedure described in Sec. 15.5. The details are shown in Table 15.4. We have subtracted 2400 from each total.

Table 15.4

Treatment	Yield	(1)	(2)	SS
$n_0 p_0$	587	690	958	
$n_1 p_0$	103	268	−304	5776
$n_0 p_1$	44	−484	−422	11,130.25
$n_1 p_1$	224	180	664	27,556
Total	958			44,462 25
W	587	690	958	
X	103	268	−304	
Y	44	−484	−422	
Z	224	180	664	
$X+W$	690	958	654	
$Z+Y$	268	−304	242	
$X+W$	−484	−422		
$Z-Y$	180	664		
$X+W+Z+Y$	958	654		
$X-W+Z-Y$	−304	242		

Subtracting 600 from each observation in Table 15.3 we get the following results

$$\text{Raw SS} = (45)^2 + (152)^2 + \ldots + (-25)^2 + (60)^2 = 115272$$

$$\text{CF} = (958)^2/16 = 57360.25$$

$$\text{Total SS} = 115272 - 57360.25 = 57911.75$$

$$\text{Row SS} = \{[(369)^2 + \dots + (339)^2]/4\} - \text{CF} = 7285.25$$

$$\text{Column SS} = \{[(260)^2 + \dots + (198)^2]/4\} - \text{CF} = 1515.25$$

The analysis of variance is shown in Table 15.5.

Table 15.5

Source	df	SS	MS	F
Rows	3	7,285.25		
Columns	3	1,515.25		
Treatments	3	44,462.25	14,820.75	19.13
N	1	5,776.00		7.45
P	1	11,130.25		14.38
NP	1	27,556.00		36.85
Error	6	4,649.00	774.83	
Total	15	57,911.75		

The main effect of N is significant at the 5% level, the main effect of P at the 1% level, and the interaction NP is significant at a somewhat higher level. An examination of the total yields of the four treatment combinations reveals that, in the presence of superphosphate (p_0), there is a decrease in yield when ammonia (n_1) is used in place of blood (n_0), whereas in the presence of steamed bone flour (p_1) the situation is reversed. The best yield is obtained from the combination $n_0 p_0$.

16

The 3^n Factorial Experiment

16.1 Notation

In the 3^n factorial we have every factor at three levels. The factors are denoted by capital letters $A, B, C, \ldots$. The three levels of a factor A are denoted by (1), a_1, a_2; of B by (1), b_1, b_2, and so on. We may call (1) the zeroth level, a_1 the first and a_2 the second level of A. The three levels of A may also be written a_0, a_1, a_2 or simply as 0, 1, 2.

As in the 2^n factorial a treatment combination can be written in terms of the individual levels. The nine treatment combinations of a 3^2 factorial with factors A and B are

$$(1) \quad a_1 \quad a_2 \quad b_1 \quad a_1b_1 \quad a_2b_1 \quad b_2 \quad a_1b_2 \quad a_2b_2,$$

where a_2b_1 denotes the treatment combination with A at the second level and B at the first. Similarly, (1) denotes the combination with both factors at the zero-th level.

16.2 Standard Order for Treatment Combinations

The *standard order* for one factor A is

$$(1), \quad a_1, \quad a_2 \tag{16.1}$$

With two factors there are nine treatment combinations. If B is the second factor with levels (1), b_1, b_2 these nine combinations are formed by multiplying the sequence (16.1) by b_1 and b_2 respectively. Thus the standard order with two factors A and B is

$$(1), \quad a_1, \quad a_2, \quad b_1, \quad a_1b_1, \quad a_2b_1, \quad b_2 \quad a_1b_2, \quad a_2b_2 \tag{16.2}$$

If a third factor C is introduced we multiply the sequence (16.2) by c_1, c_2 to get the standard order as

$$
\begin{array}{ccccccccc}
(1) & a_1 & a_2 & b_1 & a_1b_1 & a_2b_1 & b_2 & a_1b_2 & a_2b_2 \\
c_1 & a_1c_1 & a_2c_1 & b_1c_1 & a_1b_1c_1 & a_2b_1c_1 & b_2c_1 & a_1b_2c_1 & a_2b_2c_1 \\
c_2 & a_1c_2 & a_2c_2 & b_1c_2 & a_1b_1c_2 & a_2b_1c_2 & b_2c_2 & a_1b_2c_2 & a_2b_2c_2
\end{array} \qquad (16.3)
$$

16.3 Main Effects and Interactions

In a 3^n experiment we have two degrees of freedom for the main effect of any factor A, $2^2 = 4$ degrees of freedom for every two-factor interaction AB, $2^3 = 8$ for every three-factor interaction ABC, and so on.

We can express this in terms of orthogonal contrasts with one degree of freedom each. Thus in a 3^2 experiment with two factors A and B, the two degrees of freedom for the main effect A can be represented by the two orthogonal contrasts

$$(a_2 - 1)(1 + b_1 + b_2) = a_2 + a_2b_1 + a_2b_2 - (1) - b_1 - b_2 \qquad (16.4)$$

$$(1 - 2a_1 + a_2)(1 + b_1 + b_2) = (1) + b_1 + b_2 - 2a_1 - 2a_1b_1 - 2a_1b_2 + a_2 + a_2b_1 + a_2b_2 \qquad (16.5)$$

The first of these denoted by A_1, is called the *linear component* of A, and the second, denoted by A_2, is called the *quadratic component* of A.

Similarly the four degrees of freedom for the interaction AB can be represented by

$$(a_2 - 1)(b_2 - 1) = A_1B_1 \qquad (16.6)$$

$$(a_2 - 1)(1 - 2b_1 + b_2) = A_1B_2 \qquad (16.7)$$

$$(1 - 2a_1 + a_2)(b_2 - 1) = A_2B_1 \qquad (16.8)$$

$$(1 - 2a_1 + a_2)(1 - 2b_1 + b_2) = A_2B_2 \qquad (16.9)$$

The contrast A_1B_1 is called the *linear* $\times$ *linear component* of the interaction AB, A_1B_2 the *linear* $\times$ *quadratic component*, etc.

We can thus write these components, each having one degree of freedom, in the same standard order as was followed for the treatment combinations.

16.4 Calculation of Sums of Squares

Suppose a 3^n experiment is carried out in a randomised block design with b blocks. The sums of squares for the main effects and interactions can be calculated following the general method given in Chapter 14. These will account for $3^n - 1$ degrees of freedom. The sum of squares for blocks will have $(b-1)$ degrees of freedom. Since the total SS has $b \cdot 3^n - 1$ degrees of freedom. We shall have $(b-1)(3^n-1)$ degrees of freedom for error.

Similarly, if the experiment is carried out in a $3^n \times 3^n$ Latin square the degrees of freedom will be as follows:

Rows	$3^n - 1$
Columns	$3^n - 1$
Main Effects and Interactions	$3^n - 1$
Error	$(3^n - 1)(3^n - 2)$
Total	$3^n \times 3^n - 1$

The sums of squares for the main effects and interactions can also be obtained in terms of single degrees of freedom described earlier. The method used is an extension of the Yates' method for the 2^n experiment.

The 3^n treatment combinations are first written in standard order in the first column. In the next column we write the total of the yields for the particular treatment combination. The entries in these two columns are divided into sets of three consecutive entries beginning with the first. Columns (1), (2)..., are then obtained successively. Column (1) is made up of three sets of entries derived from the yield column. The first set occupying the first third of column (1), being the sums of the consecutive sets of three of the yield column. The second set, occupying the middle third of column (1), is formed by subtracting the first observation from the third in each set of three in the yield column. The third set, occupying the last third of column (1), is formed by taking the sum of the first and the third minus twice the second for each set of three in the yield column.

Each of the columns (2), ..., (n), n columns being required for a 3^n experiment, is then obtained from the preceding column by the same procedure as was used to obtain column (1) from the yield column.

The first entry in column (n) is the grand total of all the yields. Every other entry in column (n), when squared and divided by an appropriate divisor gives the sum of squares, with one degree of freedom, for the interaction component corresponding to the treatment combination in that row; in the row corresponding to the treatment combination $a_1b_2c_1$ we get the sum of squares for the component $A_1B_2C_1$ of the $A \times B \times C$ interaction.

The divisor is given by the formula,

$$2^m 3^{n-p} r \tag{16.10}$$

where $m =$ number of factors in the interaction, $p =$ number of linear terms in the component, $n =$ number of factors in the experiment, and $r =$ number of replications of the treatment combination (same for each combination). Thus the divisor for the component $A_1B_2C_1$ will be $2^3 3^{n-2} r = 8r3^n/9$.

We add three more columns after column (n); the first shows the components of the main effects and interactions and may be called the effect column, the second shows the divisor, and the last the sum of squares. As each of these sums of squares is based on one degree of freedom the last column will also give the mean squares.

A check on the calculations is provided by forming for each column, beginning with the column of yields and ending with column (n), the following quantities:

$$S_1 = \text{sum of the 1st, 4th, 7th,... entries}$$

$$S_2 = \text{sum of the 2nd, 5th, 8th, ... entries}$$

$$S_3 = \text{sum of the 3rd, 6th, 9th, ... entries}$$

$$X = S_1 + S_2 + S_3$$

$$Y = S_1 - S_2 + 3S_3$$

Then X in any column must equal Y in the preceding column.

The above method is illustrated in the example which follows.

16.5 Example

In an agricultural field trial three varieties of wheat were grown using three different doses of a chemical fertiliser, and three different rates of irrigation. The trial was planned as a 3^3 factorial experiment and each of the 27 treatment combinations was replicated twice. The yields are shown in Table 16.1 where factor V stands for variety, C for chemical fertiliser, and N for irrigation. To reduce computational labour a constant amount has been deducted from each yield.

From Table 16.1 we obtain the following

$$\text{Grand Total} = 1389 + 3597 + 2105 = 7091$$

$$\text{Raw } SS = 1171199$$

$$CF = (7091)^2/54 = 931153.35$$

$$\text{Total } SS = 240045.7 \simeq 240046$$

$$\text{Replicate Totals: } 687 + 1759 + 1154 = 3600, \text{ and}$$

$$702 + 1838 + 951 = 3491.$$

$$\text{Replicate } SS = [\{(3600)^2 + (3491)^2\}/27] - CF$$

$$= 931373.37 - 931153.35$$

$$= 220.02 \simeq 220$$

We now apply the method of Sec. 16.4 to calculate the sums of squares of the 26 components of the main effects and interactions, each with a single degree of freedom. The calculations are shown in Table 16.2, where the values in the yield column have been obtained by taking the sum of the yields of the two replicates of each treatment combination. The last column gives the mean squares.

The sums of squares of all the main effects and interactions in Table 16.2 add up to 219008 and carry 26 degrees of freedom. Subtracting this value and the replicate sum of squares from the total sum of squares we get the error sum of squares equal to 20818. Subtracting one degree of freedom for replicates and 26 for main effects and interactions from the 53 degrees of freedom for the total sum of squares, there are 26 degrees of freedom for error. Hence, the error mean square is $800.69 \simeq 801$.

The 5% and 1% values of F with 1 and 26 degrees of freedom are 4.22 and 7.72 respectively. Using these values of F we may draw the following conclusions from our analysis.

Table 16.1

Levels of factor C	Levels of factor N									
	n_0			n_1			n_2			Sum
	Levels of factor V			Levels of factor V			Levels of factor V			
	v_0	v_1	v_2	v_0	v_1	v_2	v_0	v_1	v_2	
c_0	73	12	68	111	45	129	70	55	124	687
	84	11	114	147	51	106	74	46	69	702
Sum	157	23	182	258	96	235	144	101	193	1389
c_1	210	109	198	165	217	219	200	215	226	1759
	183	144	163	287	203	141	215	238	264	1838
Sum	393	253	361	452	420	360	415	453	490	3597
c_2	73	146	194	46	145	197	89	101	163	1154
	74	103	182	64	123	164	59	69	113	951
Sum	147	249	376	110	268	361	148	170	276	2105

Table 16.2

			Column				
Treatment	Yield	(1)	(2)	(3)	Effect	Divisor	SS(=MS)
(1)	157	697	2141	7091	(T)		
c_1	393	525	2560	716	(C_1)	36	14240
c_2	147	919	2390	−3700	(C_2)	108	126759
v_1	23	820	410	610	(V_1)	36	10336
c_1v_1	253	784	150	557	(C_1V_1)	24	12927
c_2v_1	249	956	156	757	(C_2V_1)	72	7959
v_2	182	707	−880	992	(V_2)	108	9112
c_1v_2	361	724	−1136	−685	(C_1V_2)	72	6517
c_2v_2	376	759	−1684	335	(C_2V_3)	216	520
n_1	258	−10	222	249	(N_1)	36	1722
c_1n_1	452	226	136	−254	(C_1N_1)	24	2688
c_2n_1	110	194	252	−804	(C_2N_1)	72	8978
v_1n_1	96	−148	204	30	(V_1N_1)	24	38
$c_1v_1n_1$	420	172	274	−125	$(C_1V_1N_1)$	16	977
$c_2v_1n_1$	268	126	79	−291	$(C_2V_1N_1)$	48	1764
v_2n_1	235	4	318	−348	(V_2N_1)	72	1682
$c_1v_2n_1$	360	69	412	217	$(C_1V_2N_1)$	48	981
$c_2v_2n_1$	361	83	27	399	$(C_2V_2N_1)$	144	1105
n_2	144	−482	566	−589	(N_2)	108	3212
c_1n_2	415	−234	208	266	(C_1N_2)	72	983
c_2n_2	148	−164	218	−292	(C_2N_2)	216	395
v_1n_2	101	−536	−268	202	(V_1N_2)	72	567
$c_1v_1n_2$	453	−476	−366	−265	$(C_1V_1N_2)$	48	1463
$c_2v_1n_2$	170	−124	−51	−479	$(C_2V_1N_2)$	144	1594
v_2n_2	193	−538	−178	368	(V_2N_2)	216	627
$c_1v_2n_2$	490	−635	292	413	$(C_1V_2N_2)$	144	1185
$c_2v_2n_2$	276	−511	221	−541	$(C_2V_2N_2)$	432	677
S_1	1389	514	2535	8605			
S_2	3597	1155	2530	840			
S_3	2105	2438	1608	−4616			
X	7091	4107	6673	4829			
Y	4107	6673	4829				

The linear and quadratic effects for C and V are all highly significant. Neither effect for N is significant; but the quadratic effect approaches significance at the 5% level. All interactions $C \times V$ involving a linear component are significant. The only component of the $C \times N$ interaction which is significant is C_2N_1. None of the $V \times N$ and $C \times V \times N$ interactions are significant.

It is therefore sufficient to consider the two-way tables of $C \times V$ and $C \times N$. These are given in Table 16.3.

We see from Table 16.3 that the best condition of factor C is at c_1 whatever the condition of the other factors. There is a sharp peak at this level and yield is very sensitive to changes in the level of C, a higher or lower value than c_1 producing a marked drop in yield. From the $C \times V$ table, it is seen that the least value of V is to be preferred, and from the $C \times N$ table the best results are obtained from the highest level of N. The interaction between N and C is of considerable interest because although on the average there is no appreciable difference between n_1 and n_2, for the condition c_1, a change to n_2 produces an increase in yield, while for both conditions of c_0 and c_2 an increase in N results in a drop in yield.

The error mean square is 801, from which we deduce that the standard error of each result of Table 16.3 is $\sqrt{(801/6)} = 11.5$.

16.6 Calculation of Sums of Squares (Alternative Method)

Instead of representing the main effects and interactions in terms of contrasts having one degree of freedom, we can also represent them in terms of pairs of contrasts, each pair having two degrees of freedom. Thus A will be represented by one pair, AB by two pairs, ABC by four pairs, etc.

Let the 3^n treatment combinations be denoted by

$$(x_1, ..., x_n), \quad x_i = 0, 1, 2, \tag{16.11}$$

where x_i denotes the level of the i-th factor in the combination. With two factors A and B, the nine treatment combinations are (x_1, x_2) with x_1 and x_2 taking the values 0, 1 and 2. Let us divide these into three sets according as $x_1 = 0$, 1 or 2, i.e. into the sets

$$\begin{matrix} (0, 0) & (0, 1) & (0, 2) \\ (1, 0) & (1, 1) & (1, 2) \\ (2, 0) & (2, 1) & (2, 2) \end{matrix}$$

Table 16.3

	CV				CN			
	v_0	v_1	v_2	Mean	n_0	n_1	n_2	Mean
c_0	93.2	36.6	101.7	77.2	60.3	98.2	73.0	77.2
c_1	210.0	187.7	201.8	199.8	167.8	205.3	226.3	199.8
c_2	67.5	114.5	168.8	116.9	128.7	123.3	99.0	116.9
Mean	123.6	112.9	153.4		118.9	142.2	132.8	

The entries are the mean of all yields, e.g. the c_0v_0 entry is $(157 + 258 + 144)/6 = 93.2$.

In terms of our earlier notation the three sets of treatment combinations are

$$\begin{matrix} a_0b_0 & a_0b_1 & a_0b_2 \\ a_1b_0 & a_1b_1 & a_1b_2 \\ a_2b_0 & a_2b_1 & a_2b_2 \end{matrix} \tag{16.12}$$

If we write down a contrast in terms of these three sets, with the treatments of any set having the same coefficient, e.g. the contrast

$$\alpha\,[a_0b_0 + a_0b_1 + a_0b_2] + \beta\,[a_1b_0 + a_1b_1 + a_1b_2] + \gamma\,[a_2b_0 + a_2b_1 + a_2b_2]$$

with $\alpha + \beta + \gamma = 0$, we see easily that the contrast belongs to the main effect A.

Thus the set of all contrasts obtained in this manner is the same as the set of contrasts belonging to the main effect A. Since we can form two linearly independent contrasts out of the three sets (16.12) we say that the three sets (16.12) represent two degrees of freedom of the main effect A. Similarly, if we divide the treatment combinations into three groups according as $x_2 = 0, 1$ or 2, the three groups so obtained will be seen to represent the two degrees of freedom of the main effect of B.

Consider next the division of the nine treatment combinations into three sets according as $x_1 + x_2$ when divided by 3 leaves the remainder 0, 1 or 2, i.e. according as $x_1 + x_2 = 0, 1$ or $2 \pmod 3$. The three groups now are

$$\begin{matrix} a_0b_0 & a_1b_2 & a_2b_1 \\ a_0b_1 & a_1b_0 & a_2b_2 \\ a_0b_2 & a_1b_1 & a_2b_0 \end{matrix} \tag{16.13}$$

A general contrast using these three sets,

$$\alpha\,[a_0b_0 + a_1b_2 + a_2b_1] + \beta\,[a_0b_1 + a_1b_0 + a_2b_2] + \gamma\,[a_0b_2 + a_1b_1 + a_2b_0]$$

with $\alpha + \beta + \gamma = 0$ is easily seen to belong to the $A \times B$ interaction. Thus the sets (16.13) may be said to represent two out of the four degrees of freedom of the $A \times B$ interaction. The other two degrees of freedom of the $A \times B$ interaction are represented by the sets

$$\begin{matrix} a_0b_0 & a_1b_1 & a_2b_2 \\ a_0b_2 & a_1b_0 & a_2b_1 \\ a_0b_1 & a_1b_2 & a_2b_0 \end{matrix} \tag{16.14}$$

obtained by taking $x_1 + 2x_2 = 0, 1,$ or $2 \pmod 3$)

With three factors A, B and C, the division into three sets of nine treatment combinations each, with each set having two degrees of freedom is as follows. If the treatment combinations are denoted by (x_1, x_2, x_3), the sets corresponding to the two degrees of freedom of the main effect A are obtained by putting $x_1 = 0$ or 1 or 2, those of B by putting $x_2 = 0$ or 1 or 2, and of C by putting $x_3 = 0$ or 1 or 2. The $A \times B$ interaction has four degrees of freedom of which two are represented by sets for which $x_1 + x_2 = 0$ or 1 or $2 \pmod 3$ and two by the sets for which $x_1 + 2x_2 = 0$ or 1 or $2 \pmod 3$. Similarly, for the $B \times C$ interaction, we shall use $x_2 + x_3$ and $x_2 + 2x_3$ in place of $x_1 + x_2$ and $x_1 + 2x_2$. The eight degrees of freedom of the $A \times B \times C$ interaction are divided into four pairs obtained successively by putting $x_1 + x_2 + x_3$, $x_1 + x_2 + 2x_3$, $x_1 + 2x_2 + x_3$ and $x_1 + 2x_2 + 2x_3$ equal to 0 or 1 or $2 \pmod 3$). The extension of this method to more than three factors is obvious. For example, with four factors the 16 degrees of freedom of the interaction $A \times B \times C \times D$ are divided into eight pairs according as

$$
\begin{aligned}
&x_1 + x_2 + x_3 + x_4 \\
&x_1 + x_2 + x_3 + 2x_4 \\
&x_1 + x_2 + 2x_3 + x_4 \\
&x_1 + x_2 + 2x_3 + 2x_4 \\
&x_1 + 2x_2 + x_3 + x_4 \\
&x_1 + 2x_2 + x_3 + 2x_4 \\
&x_1 + 2x_2 + 2x_3 + x_4 \\
&x_1 + 2x_2 + 2x_3 + 2x_4
\end{aligned} \tag{16.15}
$$

is 0, 1 or $2 \pmod 3$).

The two degrees of freedom of the main effect A, represented by the sets (16.12) will be denoted by A, those of the main effect B, by B. Of the four degrees of freedom of the $A \times B$ interaction, the two degrees of freedom, represented by the sets (16.13) will be denoted by AB, and the other two, represented by the sets (16.14) by AB^2. Similarly, in the four factor case the eight sets of pairs of degrees of freedom (16.15) will be denoted by $ABCD$, $ABCD^2$, ABC^2D, ABC^2D^2, AB^2CD, AB^2CD^2, AB^2C^2D and $AB^2C^2D^2$ respectively.

It may be noted that if we divide the composite treatments into three sets for which $2x_1 + x_2 + 2x_3 + 2x_4 = 0$, 1 or 2 (mod 3), we will get the same three sets for which we have $x_1 + 2x_2 + x_3 + x_4 = 0$, 1, or 2 (mod 3). Similarly, $x_1 + 2x_2 + x_3 + 2x_4$ gives the same sets as $2x_1 + x_2 + 2x_3 + x_4$. It is for this reason that we usually take the coefficient of the levels of the first factor as unity. In fact, if the expression $x_1 + 2x_2 + x_3 + x_4$ is used for the interaction AB^2CD, the expression $2x_1 + x_2 + 2x_3 + 2x_4$ would represent $A^2BC^2D^2$. But, according to the multiplication rule for interactions introduced in the next chapter (Sec. 17.7), AB^2CD and its square $(AB^2CD)^2 = A^2B^4C^2D^4 = A^2BC^2D^2$ represent, together, the two degrees of freedom for the $A \times B \times C \times D$ interaction component represented in this chapter by AB^2CD.

The division of main effect and interaction contrasts into sets with two degrees of freedom each, has the additional property that for any interaction the contrasts belonging to two different sets are mutually orthogonal. For example, in a 3^2 factorial the interaction $A \times B$ gives two sets AB and AB^2. A contrast belonging to AB has from (16.13) the general form

$$l\,(a_0b_0 + a_1b_2 + a_2b_1) + m\,(a_0b_1 + a_1b_0 + a_2b_2) + n\,(a_0b_2 + a_1b_1 + a_2b_0)$$

with $l + m + n = 0$, and a contrast belonging to AB^2 has from (16.14) the general form

$$r\,(a_0b_0 + a_1b_1 + a_2b_2) + s\,(a_0b_2 + a_1b_0 + a_2b_1) + t\,(a_0b_1 + a_1b_2 + a_2b_0)$$

with $r + s + t = 0$. The scalar product of the coefficient vectors of these two contrasts is given by

$$l.r + l.t + l.s + m.t + m.s + m.r + n.s + n.r + n.t$$
$$= (l + m + n)(r + s + t) = 0$$

which shows that these two contrasts are orthogonal. Thus, if the estimates of these contrasts are also orthogonal the $A \times B$ sum of squares is decomposed into two orthogonal components which are the sums of squares due to AB and AB^2.

The sum of squares corresponding to the two degrees of freedom belonging to any main effect is obtained by the general method described in Chapter 14. The sums of squares corresponding to each pair of degrees of freedom of any multiple factor interaction is obtained by the procedure described in the example of Sec 16.7.

16.7 Example

A 3^3 factorial experiment was carried out to study the effect of the following factors:

D: date of sowing
S: spacing of rows
N: sulphate of ammonia

on the sugar content of sugarbeets. The three levels of each of the above factors were

d_0: March 15	s_0: 10-inch	n_0: none
d_1: April 18	s_1: 15-inch	n_1: 0.3 cwt
d_2: May 16	s_2: 20-inch	n_2: 0.6 cwt

Table 16.4 gives the percent sugar obtained for each treatment combination. These figures have been converted into the coded values $x = 100\,(y - 16)$, where y is the percent sugar. The analysis of variance is more conveniently carried out in terms of the coded values.

Every treatment combination was replicated twice, using blocks of nine plots. For the present we shall treat the data as arising from 2 blocks with 27 plots each.

There are 54 observations giving 53 degrees of freedom in all. These are partitioned as follows:

Blocks	1
Main effects and interactions	26
Error	26

The block totals are 1701, 1263 and the grand total is 2964. Hence we have

$$\text{Raw SS}: = [(70)^2 + (50)^2 + \ldots + (68)^2 + (27)^2] = 205916$$

$$\text{CF} = (2964)^2/54 = 162690.66$$

$$\text{Total SS} = 43225.34$$

$$\text{Block SS} = [(1701)^2 + (1263)^2]/27 - \text{CF}$$
$$= 166243.33 - 162690.66 = 3552.67$$

Table 16.4

Treatment dsn	% Sugar	Code	Treatment dsn	% Sugar	Code
012	16.70	70	121	16.79	79
202	16.50	50	100	16.88	88
001	17.00	100	112	16.39	39
210	17.05	105	002	16.18	18
020	16.50	50	210	16.88	88
122	16.79	79	011	16.53	53
111	17.05	105	020	16.21	21
100	16.85	85	222	16.39	39
221	16.36	36	201	16.39	39
201	16.53	53	211	16.43	43
000	16.79	79	202	16.42	42
011	17.05	105	122	16.30	30
022	15.98	−2	012	16.10	10
102	16.79	79	220	17.08	108
220	16.56	56	021	16.36	36
212	16.59	59	000	16.47	47
110	16.62	62	101	16.33	33
121	16.44	44	110	16.44	44
222	16.21	21	102	16.44	44
200	16.59	59	212	16.42	42
101	16.47	47	200	16.27	27
211	16.96	96	111	16.04	4
010	16.30	30	022	16.56	56
112	16.13	13	010	16.33	33
120	16.65	65	120	17.05	105
002	16.85	85	221	16.68	68
021	16.70	70	001	16.27	27

Source: *Annual Report*, Rothamsted Experiment Station, Harpenden, England, 1935.

Next we sum the observations over the replicates and then form the two-way tables (Table 16.5).

Table 16.5

	n_0			n_1			n_2		
	s_0	s_1	s_2	s_0	s_1	s_2	s_0	s_1	s_2
d_0	126	63	71	127	158	106	103	80	54
d_1	173	106	170	80	100	123	123	56	109
d_2	86	193	164	92	139	104	192	101	60

The marginal totals give the SS due to the main effects, each having 2 degrees of freedom. Thus

$$\text{SS for } D = [(888)^2 + (1045)^2 + (1031)^2]/18 - \text{CF} = 838.78$$

The entries in the main body of a two-way table give the SS, with four degrees of freedom, due to the two-factor interaction. Thus

$$\text{SS for } D \times S = [(356)^2 + (376)^2 + \ldots + (328)^2]/6 - (\text{SS for } D)$$
$$- (\text{SS for } S) - \text{CF}$$
$$= 5233.11$$

To partition the sums of squares of the two factor interactions into pairs of degrees of freedom, as well as to do the same for the three factor interactions we proceed as follows. We first obtain three two-way tables for D and S, one table for each level of N. Thus the two-way table for D and S with level n_0 is

	s_0	s_1	s_2
d_0	126	63	71
d_1	173	106	170
d_2	86	193	164

Each entry here is the sum of the two yields resulting from the two replicates of any treatment combination. From these two-way tables we obtain Table 16.6.

Table 16.6

		s_0	s_1	s_2		
		n_0				
d_0		126	63	71		
d_1		173	106	170		
d_2		86	193	164		
DS_2	263	126	63	71	396	DS_0^2
DS_0	489	173	106	170	437	DS_1^2
DS_1	342				319	DS_2^2

		n_1				
d_0		127	158	106		
d_1		80	109	123		
d_2		92	139	104		
DS_2	307	127	158	106	340	DS_0^2
DS_0	389	80	109	123	325	DS_1^2
DS_1	342				373	DS_2^2

		n_2				
d_0		103	80	54		
d_1		123	52	109		
d_2		92	101	60		
DS_2	198	103	80	54	215	DS_0^2
DS_0	313	123	52	109	278	DS_1^2
DS_1	263				281	DS_2^2

		n_0	n_1	n_2		
DS_0		489	389	343		
DS_1		400	342	263		
DS_2		263	307	198		
DSN_2	918	489	389	313	1029	DSN_0^2
DSN_0	1059	400	342	263	1020	DSN_1^2
DSN_1	987				915	DSN_2^2

		n_0	n_1	n_2		
DS_0^2		396	340	215		
DS_1^2		437	325	278		
DS_2^2		319	373	281		
DS^2N_2	859	396	340	215	1002	$DS^2N_0^2$
DS^2N_0	1047	437	325	278	1025	$DS^2N_1^2$
DS^2N_1	1058				937	$DS^2N_2^2$

Each of the first three tables is the two-way table described earlier with the first and second row repeated after the third. The diagonal sums are obtained as indicated. In this manner we obtain the sum of the yields for plots for which $x_1 + 2x_2 = 0, 1, 2 \pmod 3$, by summing downward towards the right, and successive tables yield this sum for $x_3 = 0$, 1 and 2. Similarly, summing diagonally downward to the left yields sums corresponding to $x_1 + x_2 = 0, 1, 2 \pmod 3$. The last two tables are formed from the above sums in a straightforward way, and the diagonal summation process is applied. The object of the process is now apparent. It was desired to obtain sums, for example, for which $x_1 + x_2 + 2x_3 = 0, 1, 2 \pmod 3$. The solutions, for example, to the equation $x_1 + x_2 + 2x_3 = 1 \pmod 3$ are those for which $x_1 + x_2 = 0$, $2x_3 = 1 \pmod 3$; $x_1 + x_2 = 1$, $2x_3 = 0$; and $x_1 + x_2 = 2$, $2x_3 = 2 \pmod 3$.

The entries DS_2, DS_0, DS_1, when summed over the three levels of N give the totals 768, 1191 and 1005. From these we obtain the SS, with two degrees of freedom, corresponding to DS. We have

$$\text{SS for } DS = [(768)^2 + (1191)^2 + (1005)^2]/18 - \text{CF}$$
$$= 167685 - 162690.66 = 4994.34$$

Similarly, the entries DS_2^2, DS_0^2, DS_1^2, give the SS corresponding to DS^2;

$$\text{SS for } DS^2 = [(951)^2 + (1040)^2 + (973)^2]/18 - \text{CF}$$
$$= 162929.44 - 162690.66 = 238.78$$

The sum of squares for the $D \times S$ interaction, with four degrees of freedom, is then

$$4994.34 + 238.78 = 5233.12$$

From the results of the diagonal summation process applied to the last two tables, we obtain the four sums of squares corresponding to the four pairs into which the eight degrees of freedom of the $D \times S \times N$ interaction are partitioned. Thus

$$\text{SS for } DSN^2 = [(1029)^2 + (1020)^2 + (915)^2]/18 - \text{CF}$$
$$= 163137 - 162690.66 = 446.34$$

The other sums of squares of the $D \times S \times N$ interaction are obtained in a similar manner. The error sum of squares can be obtained by subtraction.

17

Confounding in Factorial Experiments

17.1 Introduction

In a factorial experiment the number of treatment combinations increases rapidly as the number of factors or levels increases. When the number of treatment combinations is large it may not be possible to get blocks of sufficiently large size to accommodate all the treatment combinations. In such situations we are compelled to use incomplete block designs. We can either use completely connected incomplete block designs (a BIBD for example) so that all the main effect and interaction contrasts can be estimated. Or, we may use unconnected designs so that not all these contrasts can be estimated. In the second method, contrasts which are not estimable are said to be *confounded with the differences between blocks*. Contrasts which are estimable are said to be *unconfounded with blocks*, or *free from block effects* or *composed entirely of within-block comparisons.*

For example, a 2^3 factorial needs blocks of size 8. Suppose we can use blocks of size 4 only. We can use a BIBD with $b=14$, $k=4$; $v=8$, $r=7$; $\lambda=3$ (such a BIBD exists). Then all the main effects and interactions are estimable. We have, for this BIBD,

$$E=\frac{\lambda v}{kr}=\frac{3.8}{4.7}=\frac{6}{7},\ rE=6$$

Thus the variance of the estimate of any main effect or interaction is $8\sigma^2/6$.

Consider now the unconnected design in which 7 of the 14 blocks get treatment combinations

$$(1)\quad ab\quad bc\quad ca$$

and the other 7 get treatment combinations

$$a \quad b \quad c \quad abc$$

Then the main effect

$$\begin{aligned} A &= (a-1)(b+1)(c+1) \\ &= (1) + a - b - c + ab + ac - bc + abc \\ &= (ab - (1)) + (ac - bc) + (a - b) + (abc - c) \end{aligned}$$

is estimable as its estimate will be composed of differences of yields within the blocks. In the same manner, we verify easily that all the main effects and interactions are estimable, except *ABC*. We have

$$\begin{aligned} ABC &= -(1) + a + b + c - ab - bc - ca + abc \\ &= (a + b + c + abc) - ((1) + ab + bc + ca) \end{aligned}$$

and this function being a difference of treatment parameters of treatments occurring in different blocks is not estimable. However, the variance of the estimates of unconfounded main effects and interactions is now $8\sigma^2/7$.

We thus see that at the cost of not being able to estimate *ABC* we have, with the same number of replicates as in the BIBD used earlier, better estimates of *A*, *B*, *C*, *AB*, *BC*, and *AC*. Since higher order interactions are difficult to interpret and are usually not large, it is much better to use confounding arrangements which provide better estimates of the interactions in which we are more interested.

17.2 Confounding Arrangement

The arrangement of the treatment combinations in different blocks, whereby some pre-determined interaction (or main effect) contrasts are confounded, is called a *confounding arrangement.* Thus, in the example above, where the interaction ABC was confounded, the confounding arrangement consisted of dividing the eight treatment combinations into two sets

$$(1) \quad ab \quad bc \quad ca \qquad (17.1)$$

and

$$a \quad b \quad c \quad abc \qquad (17.2)$$

with the treatments of each set being assigned to the same block, and each of these sets was replicated the same number of times in the experiment. We say that we have a confounding arrangement of a 2^3 factorial in two blocks.

Any confounding arrangement has to be such that:

(i) Only predetermined interactions are confounded, and,

(ii) The estimates of the interactions which are not confounded (i.e. are estimable) are orthogonal whenever the interactions are orthogonal.

The latter property makes the analysis of variance easy to carry out, as the different sums of squares of the analysis of variance for the unconfounded experiment are still mutually orthogonal.

We shall consider confounding arrangements for 2^n and 3^n factorial experiments only.

17.3 Defining Contrasts

The interactions which are confounded are called the *defining contrasts* of the confounding arrangement. In the example of Sec. 17.2 there is only one defining contrast, ABC, and the resulting confounding arrangement is

$$\begin{matrix} (1) & ab & bc & ca \\ a & b & c & abc \end{matrix} \tag{17.3}$$

This arrangment reduced the block size to 4, and was obtained by putting all the treatment combinations occurring in ABC with a positive sign in one block, and those with a negative sign in the other block. The SS due to the difference between the two block totals is the same as the SS due to ABC if it had not been confounded. That is why we say that ABC has been confounded between blocks.

In the 2^3 factorial each main effect or interaction contrast has four treatment combinations occurring with a positive sign and four with negative signs. For any such contrast to be estimable, each block of the confounding arrangement must contain an equal number of positive and negative treatment combinations from that contrast. A confounded contrast, on the other hand, will have treatment combinations with the same sign in each block of the confounding arrangement.

Suppose we now wish to confound another contrast, in addition to ABC already confounded, so as to get blocks of size 2.

If A is still to be estimable then each of the four blocks of the confounding arrangement should contain one treatment combination from the set

$$a \quad ab \quad ac \quad abc \tag{17.4}$$

occurring with a positive sign in A, and one from the set

$$(1) \quad b \quad c \quad bc \tag{17.5}$$

occurring with a negative sign in A. Thus if a block has (1) it cannot have b, c or bc. Similarly, for B to be estimable the block containing (1) cannot have a, c or ac. Hence if wish to have both A and B estimable the block containing (1) can be given either the treatment ab or the treatment abc. If we take ab with (1), we notice that C and AB are confounded along with ABC. If, on the other hand, we take abc with (1) we find that AB and BC are confounded along with AC. In both cases all the other interactions are estimable.

Thus we see that if we wish to confound two interactions, a third interaction is automatically confounded. This situation is quite general. The defining contrasts for a confounding arrangement cannot be chosen arbitrarily, if some defining contrasts are selected some others will also get confounded.

17.4 Confounding in a 2^n Factorial Experiment

For a 2^n factorial without confounding we need blocks of size 2^n. If any interaction (or main effect) is confounded the block size is reduced by half and become 2^{n-1}. If each treatment combination is to be replicated r times we use $2r$ blocks of size 2^{n-1} of which r blocks get one set of treatments and the remaining r blocks get the other set. The minimum number of blocks required is 2 when each treatment combination occurs once only. In either case (i.e. with or without replication) the arrangement is called a *confounding arrangement in two blocks.*

If we wish to reduce the block size further we have to confound more than one interaction. If any main effect or interaction is to be estimable the confounding arrangement must be such that each block contains an equal number of treatment combinations with positive and negative signs in that interaction. Thus the block size in any confounding arrangement of a 2^n factorial experiment must be of the form 2^{n-p}. If the block size is 2^{n-p}, the 2^n treatment combinations are divided into 2^p sets of 2^{n-p} combinations each, with each set being assigned to one block. We say we have a *confounding arrangement in 2^p blocks.* If the treatment combinations are replicated then each of the 2^p sets must be replicated the same number of times. Thus a confounding arrangement of a 2^5 experiment in blocks of size $2^2 = 2^{5-3}$ is an arrangement in eight blocks; if each treatment is replicated 3 times, we need 3.2^3, blocks of size 2^2.

Let 2^{n-p} be the block size so that we have a confounding arrangement in 2^p blocks. To obtain the actual arrangement, i.e., the division of the 2^n treatment combinations into 2^p sets of 2^{n-p} combinations, we have to choose the interactions that are to be confounded. These are called the defining contrasts of the confounding arrangement. The notion of independent interactions, and generalised interactions is useful in describing the defining contrasts.

Given any two interactions, their *generalised interaction* is obtained by multiplying the expressions in capital letters and ignoring all terms with an even exponent. For example, the generalised interaction of ABC and BCD is $AB^2C^2D = AD$; that of AB, BC and ABE is $A^2B^3CE = BCE$.

A set of main effects and interaction contrasts is called *independent* if no member of the set can be obtained as a generalised interaction of the other members of the set. Thus AB, BC, AD is an independent set, but AB, BC, CD, AD is not, since $(AB)(BC)(CD) = AB^2C^2D = AD$. The defining contrasts along with I form a sub-group of the group defined in Sec. 15.3.

We now define the concept of orthogonality of a treatment combination and a contrast. The treatment combination $a^pb^qc^r...$ will be said to be *orthogonal* to the interaction $A^xB^yC^z...$ if

$$px + qy + rz + ... \tag{17.6}$$

is divisible by 2. Since $p, q, r, ..., x, y, z, ...$ are either zero or one we may say that a treatment combination is orthogonal to an interaction if the two have an even number of letters in common. The treatment combination (1) is orthogonal to every interaction. If $a^pb^qc^r ...$ and $a^{p'}b^{q'}c^{r'} ...$ are both orthogonal to $A^xB^yC^z ...$ then their product $a^{p+p'}b^{q+q'}c^{r+r'} ...$ is also orthogonal to $A^xB^yC^z$. Similarly, if two interactions are orthogonal to a treatment combination, their generalised interaction is orthogonal to it. In other words the set of all treatment combinations orthogonal to a given set of interactions is a sub group of the group defined in Sec. 15.2.

Suppose we wish to have a confounding arrangement in 2^p blocks of a 2^n-factorial experiment with each block having 2^{n-p} plots. Then $2^p - 1$ interactions have to be confounded, i.e. the set of defining contrasts has $2^p - 1$ elements. If any two interactions are confounded their generalised interaction is also confounded. It follows that only p of $2^p - 1$ defining contrasts are independent and the rest are obtained as generalised interactions. Thus in a 2^5 factorial with factors *A*, *B*, *C*, *D* and *E*, confounded in eight blocks of size four each, there will be seven defining contrasts. Of these three will be independent contrasts and can be chosen arbitrarily. If we take *ACE*, *ABDE* and *CDE* as the independent contrasts, the seven defining contrasts are

$$AD, BE, ABDE, ABC, ACE, BCD, CDE$$

If instead, we begin with the independent contrasts *ABCD*, *ACDE*, *ABCDE*, the defining contrasts are

$$B, E, BE, ACD, ABCD, ACDE, ABCDE$$

In the second case we find that some main effects are also confounded. As a rule we try to confound, as far as possible, higher order interactions only, as these are difficult to interpret. But, as the example above shows, the independent interactions to be confounded should be chosen with care. In trying to confound only high order interactions we may confound some main effects as well.

Having selected the p independent defining contrasts we have now to divide the 2^n-treatment combinations into 2^p-groups of 2^{n-p} combinations each, each group going into one block. The group containing the combination (1) is called the *principal block* (or *key block*). The principal block is easy to obtain; it contains all the treatment combinations which are orthogonal to the p independent defining contrasts. Obviously any treatment combination in the principal block is orthogonal to each of the $2^p - 1$ defining contrasts. The most convient way of getting the principal block is to write the treatment combinations in the standard order and then to check each one of them for orthogonality. If two treatment combinations belong to the principal block, their product (under the multiplication rule defined earlier) also belongs to the principal block. Hence when a few combinations of the principal block have been determined many of the others can be obtained easily by the multiplication rule.

Thus if we take the independent defining contrasts for confounding a 2^5 factorial as

$$AD, BE, ABC$$

the principal block has (1), *acd*, *bce* and *abde* = (*acd*) (*bce*). If the independednt defining contrasts are, *ACD*, *ABCD*, *ABCDE* the principal block has (1), *ac*, *ad*, *cd* = (*ac*) (*ad*).

The other blocks of the confounding arrangement are obtained from the principal block by multiplying the combinations of the principal block by a combination not occurring in it or in any other block already obtained. We give below the eight blocks of the confounding arrangement of a 2^5 factorial for which the principal block is (1), *acd*, *bce*, *abde*. The other blocks have been obtained by multiplying successively by *a*, *b*, *ab*, *c*, *ac*, *bc*, *abc*.

(1)	*a*	*b*	*ab*	*c*	*ac*	*bc*	*abc*
acd	*cd*	*abcd*	*bcd*	*ad*	*d*	*abd*	*bd*
bce	*abce*	*ce*	*ace*	*be*	*abe*	*e*	*ae*
abde	*bde*	*ade*	*de*	*abcde*	*bcde*	*acde*	*cde*

In the language of the theory of groups the principal block is a sub-group of the group of treatment combinations and the other blocks are the cosets of this sub-group.

17.5 Analysis of Variance for a 2^n Confounded Factorial

Suppose we have only one observation per treatment combination. Then we carry out the analysis of variance as if there had been no confounding (Yate's method may be followed). The total sum of squares is then split up into the SS due to the main effect and that due to interactions. Confounded interactions are not utilised. Other main effects and interactions can be tested in the usual manner, using some of unconfounded higher order interactions as error.

If each treatment is replicated r times an independent estimate of error is obtained. Suppose we have confounded $2^p - 1$ interactions, the experiment being carried out in $r \cdot 2^p$ blocks of size 2^{n-p} each. The sums of squares for the main effects and interactions are obtained as if there was no confounding. Of these we retain only the unconfounded main effects and interactions making a total of $(2^n - 1) - (2^p - 1) = 2^n - 2^p$ degrees of freedom. We next obtain the SS due to the block totals with $r \cdot 2^p - 1$ degrees of freedom. The total of these two subtracted from the total SS with $r \cdot 2^n - 1$ degrees of freedom gives the error SS with $(r - 1)(2^n - 2^p)$ degrees of freedom.

If necessary, a check can be made by calculating the SS due to those differences between blocks which are estimable, i.e. differences between blocks with common treatments. The blocks divide into 2^p sets of r blocks each and differences between blocks in the same set are estimable. This gives us an SS with $(r - 1) 2^p$ degrees of freedom. To this we add the SS due to the confounded interactions (total $df = 2^p - 1$) obtained by ignoring confounding. This sum equals the SS due to the block totals and provides a check.

17.6 Example

Table 17.1 gives the yield in lbs per plot of asparagus which had been subjected to all combinations of the three fertilisers, nitrogen (*n*), phoshate (*p*) and potash (*k*), each at two levels. Eight blocks each of four plots were used. Four of the eight blocks had one plot each with dressings of *n*, *p*, *k* and *npk*. The other four blocks each had one plot with no fertiliser (1), and one each with *np*, *nk* and *pk*.

Table 17.1

Block	Plots				Block total
1	*npk* (12.0)	*k* (16.2)	*p* (14.6)	*n* (12.7)	55.5
2	*nk* (10.2)	*np* (12.8)	*pk* (13.8)	(1) (13.9)	50.7
3	(1) (14.7)	*np* (9.3)	*nk* (8.8)	*pk* (9.0)	41.8
4	*n* (10.8)	*p* (8.9)	*k* (8.3)	*npk* (10.3)	38.3
5	*pk* (13.0)	(1) (12.7)	*nk* (11.3)	*np* (10.3)	47.3
6	*k* (13.3)	*p* (15.2)	*n* (12.1)	*npk* (11.5)	52.1
7	*npk* (11.4)	*k* (10.4)	*p* (11.7)	*n* (9.3)	42.8
8	*np* (9.1)	(1) (10.5)	*nk* (8.2)	*pk* (13.5)	41.3
				Grand total	369.8

Source: Mather, K. *Statistical Analysis in Biology*, Methuen, London, 1946, p. 104.

The design involves confounding, as each block carries only four of the eight treatment combinations. We therefore have to determine first which main effect or interaction has been confounded. The blocks divide into two groups, one having (1), *np*, *nk*, *pk* and the other *n*, *p*, *k*, *npk*. Comparing this with

$$NPK = -(1) + n + p + k - np - pk - nk + npk$$

we see that *NPK* has been confounded. The SS due to other main effects and interactions can be obtained by Yates' method and the calculations are shown in Table 17.2.

The SS due to block totals is

$$\frac{1}{4}[(55.5)^2 + (50.7)^2 + \ldots + (41.3)^2] - \frac{(369.8)^2}{32}$$

$$= 4338.5250 - 4273.5013 = 65.0237, \quad df = 7$$

The total SS with 31 *df*, is 4418.9800 − 4273.5013 = 145.4787.

Block differences between block numbers 1, 4, 6, 7 are estimable, as also differences between the other four blocks. The SS due to the estimates of the differences between the blocks of the first set is

$$\tfrac{1}{4}[(55.5)^2 + (38.3)^2 + (52.1)^2 + (42.8)^2]$$

$$- (1/16)(55.5 + 38.3 + 52.1 + 42.8)^2$$

$$= 2273.3475 - 2225.4806 = 47.8669$$

Similarly the SS due to the estimates of the differences between block numbers 2, 3, 5, 8 is

$$\tfrac{1}{4}[(50.7)^2 + (41.8)^2 + (47.3)^2 + (41.3)^2]$$

$$- \left(\frac{1}{16}\right)(50.7 + 41.8 + 47.3 + 41.3)^2$$

$$= 2065.1775 - 2049.8256 = 15.3519$$

If we add to the sum of these two SS for block differences the SS due to the confounded interaction NPK we have

$$47.8669 + 15.3519 + 1.8050 = 65.0238$$

which agrees with the SS due to block totals, thus providing a check

Table 17.2

Treatment	Yield	(1)	(2)	(3)	SS
(1)	51.8	96.7	188.6	369.8	
n	44.9	91.9	181.2	–29.6	27.3800
p	50.4	86.7	–15.8	3.0	0.2813
np	41.5	94.5	–13.8	3.6	0.4050
k	28.2	–6.9	–4.8	–7.4	1.7112
nk	38.5	–8.9	7.8	2.0	0.1250
pk	49.3	–9.7	–2.0	12.6	4.9613
npk	45.2	–4.1	5.6	7.6	1.8050
Total	369.8				
W	102.2	183.4	172.8	372.8	
X	86.4	186.4	167.4	–26.0	
Y	97.5	–16.6	–6.8	5.2	
Z	83.7	–13.0	13.4	9.6	
$X+W$	188.6	369.8	340.2	346.8	
$Z+Y$	181.2	–29.6	6.6	14.8	
$X-W$	–15.8	3.0	–5.4		
$Z-Y$	–13.8	3.6	20.2		
$X+W+Z+Y$	369.8	340.2	346.8		
$X-W+Z-Y$	–29.6	6.6	14.8		

on the calculations. The complete analysis of variance is given in Table 17.3.

Table 17.3

Source	df	SS	MS	F
Blocks	7	65.0237	9.2891	3.67
N	1	27.3800	27.3800	10.81
P	1	0.2813		
K	1	1.7112		
NP	1	0.4050		
NK	1	0.1250		
PK	1	4.9613		
Error	18	45.5912	2.5328	
Total	31	145.4787		

The 5% and 1% values of $F_{7,18}$ are 2.58 and 3.85, and those of $F_{1,18}$ are 4.41 and 8.28. Thus the blocks are significantly different, none of the interactions is significant and the main effect of N is highly significant.

Nitrogen is the only fertilizer which affects the yield. It should be noted, however, that addition of nitrogen decreases, rather than increases, the amount of asparagus harvested (column 3 corresponding to n in Table 17.2 is negative showing that yield is lower when nitrogen is at the higher level i.e. present).

17.7 Confounding in a 3^n Factorial

In a 3^n factorial, without any main effect or interaction contrast confounded, each block has 3^n plots. If the experiment is to be carried out in smaller blocks all having the same number of plots, the number of plots in any block must be a power of three, i.e. 3^{n-p}. The simplest case is when $p = 1$; the block size is then 3^{n-1} and the 3^n treatment combinations are divided into three sets of 3^{n-1} combinations each, each of the three sets being assigned to a different block. It is obvious that in such an arrangement two contrasts, each having one degree of freedom, will be confounded between blocks. If the block size is to be reduced further we shall have blocks of size 3^{n-2} so that the 3^n treatment combinations are now divided into 3^2 sets of 3^{n-2} combinations each. Consequently, eight degrees of freedom represented by eight interaction contrasts will be confounded. In general, if the block size is 3^{n-p}, the 3^n treatment combinations are divided into 3^p groups, each group of 3^{n-p} treatment combinations being assigned to one block. The minimum number of blocks needed is 3^p so that each treatment combination is replicated at least once. If each treatment combination is replicated r times we need $r.3^p$ blocks, each having 3^{n-p} plots. In either case, with one or more replications, we say we have a *confounding arrangement of the* 3^n *factorial experiment in* 3^p *blocks*. With such an arrangement $3^p - 1$ main effect or interaction contrasts, each having one degree of freedom, are confounded.

In 3^n factorial experiments the number of contrasts confounded is an even number. The contrasts to be confounded are most conveniently obtained in pairs; if the confounding arrangement is in 3^p blocks, the $3^p - 1$ confounded contrasts are described in terms of $(3^p - 1)/2$ pairs, each pair having two degrees of freedom. The partitioning of the main effect and interaction contrasts into pairs of degrees of freedom described in Chapter 16 (Sec. 16.6) is most convenient for obtaining confounding arrangements for 3^n factorial experiments.

For example, if the three sets of treatment combinations (16.12) of a 3^2 factorial are assigned to three different blocks the main effect contrasts of A will not be estimable but will be confounded with the blocks. All other contrasts belonging to the main effect B and the interaction $A \times B$ will however be estimable. Similarly, if we put the three sets (16.13) in three different blocks, the contrasts of the main effects A and B will be estimable. Out of the contrasts belonging to the $A \times B$ interaction, the contrasts formed from the sets (16.14) will be estimable. For example, the contrast

$$\alpha\,[a_0b_0 + a_1b_1 + a_2b_2] + \beta\,[a_0b_2 + a_1b_0 + a_2b_1] + \gamma\,[a_0b_1 + a_1b_2 + a_2b_0]$$

with $\alpha + \beta + \gamma = 0$, can be written as

$$(\alpha\,a_0b_0 + \beta\,a_2b_1 + \gamma\,\alpha_1b_2) + (\alpha\,a_2b_2 + \beta\,a_1b_0 + \gamma\,a_0b_1) + (\alpha\,a_1b_1 + \beta\,a_0b_2 + \gamma\,a_2b_0)$$

and is estimable, as it is the sum of three estimable intrablock contrasts. The contrasts formed from the sets (16.13) are confounded with the blocks. Thus we may say that the confounding arrangement resulting from the sets (16.12) confounds the two degrees of freedom represented by A (in the notation of Sec 16.6), and that resulting from (16.13) confounds the two degrees of freedom represented by AB.

17.8 Confounding Arrangement for 3^n Factorial Experiments

The confounding arrangement of a 3^n factorial in 3^p blocks is most easily obtained by an orthogonality rule similar to the one utilised in case of 2^n factorial experiments.

We introduce a multiplication rule for the treatment combinations $a_pb_qc_r$ ... in much the same way as for 2^n factorial experiments, but in this case the index of any term is replaced by the remainder on division by 3, i.e. every index is reduced to 0, 1 or 2 (mod 3).

Thus

$$a_3 = a_0,\ a_4 = a_1,\ a_1 = a_1$$

$$(a_2b_1c_2d_1) \times (a_2c_1d_2) \times (a_1b_1) = a_5b_2c_3d_3 = a_2b_2$$

With the above multiplication rule the treatment combinations form a group, and $a_0b_0c_0$... is the identity element of the group.

We shall say that the treatment combination $a_pb_qc_r$... is orthogonal to the pair of contrasts $A^xB^yC^z$... (see (16.6) if

$$px + qy + rz + ...$$

is a multiple of 3, i.e. equal to 0 (mod 3). It is clear that the set of all $a_pb_qc_r$... orthogonal to a given pair $A^xB^yC^z$ is a sub-group of the group defined earlier.

Any sub-group with 3^{n-p} elements of the group of treatment combinations gives a confounding arrangement in 3^p blocks. The sub-group, which contains $a_0b_0c_0$ .... is called the *principal block* (or *key block*) of the confounding arrangement. The other 3^{p-1} blocks are the cosets of this sub-group, and are obtained by multiplying the treatment combinations of the key block by a combination not occurring in the key block or any other block already obtained. Thus, with two factors A and B, a sub-group with three treatment combinations may be taken as

$$a_0b_0 \quad a_0b_1 \quad a_0b_2$$

Then the three blocks of the confounding arrangement are

a_0b_0	a_0b_1	a_0b_2	(key block)
a_1b_0	a_1b_1	a_1b_2	(multiplying by a_1b_0)
a_2b_0	a_2b_1	a_2b_2	(multiplying by a_2b_0)

which are the same as the sets (16.12) representing the two degrees of freedom of the main effect A.

The interactions which are confounded in any arrangement obtained from a given key block are those which are orthogonal to all the combinations in the key block. The description of the confounded interactions is facilitated by representing each pair of Sec. 16.6 by two symbols; one is the symbol already used in Sec. 16.6 and the other is its square. Thus the pair A wlll be represented by A, A^2; the pair AB by AB, A^2B^2; the pair AB^2 by AB^2 and $(AB^2)(AB^2) = A^2B^4 = A^2B$, and so on. In this representation we have introduced a multiplication rule similar to the one introduced for the treatment combinations, e.g.

$$(A^2BC^2E) \times (A^2CE^2) = A^4BC^3E^3 = AB$$

With this multiplication rule the interactions along with the identity element I form a group. Thus, in the confounding arrangement represented by the key block

$$a_0b_0 \quad a_0b_1 \quad a_0b_2$$

the confounded interactions representing *2* degrees of freedom, may be written as A, A^2, and form, along with I, a sub-group of the group of interactions. In fact, each pair of Sec. 16.6 when written in terms of two symbols forms a sub-group along with I.

Instead of starting with a sub-group of the treatment combinations to serve as the key block, we may, instead, choose a sub-group of the interactions. These will be confounded if the key block consists of all the treatment combinations which are orthogonal to the given sub-group of interactions selected for confounding, i.e., orthogonal to the defining contrasts of the confounding arrangement. Thus the sub-groups (I, AB, A^2B^2) and (I, AB^2, A^2B) give the confounding arrangement in three blocks represented by the sets (16.13) and (16.14) respectively.

In general, a confounding arrangement of a 3^n factorial in 3^p blocks is obtained either by choosing a sub-group of 3^{n-p} treatment combinations to serve as the key block, or a sub-group of 3^{p-1} interactions and I from which the key block can be obtained. For example, to get a confounding arrangement of a 3^3 factorial with factors A, B and C in 9 blocks we have to choose a set of eight defining contrasts which with I form a sub-group. Out of the four pairs of contrasts two pairs may be choosen at will, e.g. AB, A^2B^2 and BC, B^2C^2. The sub-group must then contain these and all others derivable by multiplication. The required sub-group is

$$I, AB, A^2B^2; \quad BC, B^2C^2; \quad AC^2, A^2C; \quad AB^2C, A^2BC^2$$

The key block is then

$$a_0b_0c_0 \qquad a_1b_2c_1 \qquad a_2b_1c_2$$

from which all the other eight blocks can be obtained.

It is to be noted that the interactions $A^xB^yC^z$ introduced here are not the same as those defined earlier in Sec. 16.3. The four degrees of freedom of the $A \times B$ interaction represented by AB, A^2B^2, AB^2 and A^2B are equivalent to the four linearly independent contrasts A_1B_1, A_1B_2, A_2B_1 and A_2B_2 of Sec 16.3 in the sense that both generate the set of all contrasts of the $A \times B$ interaction. But the interactions AB, A^2B^2 are not equivalent to the interactions

$$A_1B_1,\ A_2B_2.$$

COMPLEMENTS AND PROBLEMS

1. In a 2^6 factorial experiment the key-block for confounding *ABCD*, *ABEF* and *CDEF* is

   (1), *ab, cd, abcd, ef, abef, cdef, abcdef,*

   ace, bce, ade, bde, acf, bcf, adf, bdf

2. If the key-block in a 2^5 factorial experiment is taken as

   (1), *bc, de, bcde, abd, acd, abe, ace*

   we get a confounding arrangement in four blocks in which the interactions

   ABC, ADE, BCDE

   are confounded.

3. Suppose a 2^3-factorial is confounded in two 4×4 Latin squares, confounding one interaction say ABC. Then the treatments (1), *ab*, *bc*, *ac* will occur in one Latin square, and the treatments *a*, *b*, *c*, *abc* in the other. The break-up of the degrees of freedom in the analysis of variance will then be as follows.

Between squares	1*
Between rows within squares	6
Between columns within squares	6
Unconfounded main effects and interactions	6
Error	12
Total	31

(*This corresponds to the confounded interaction *ABC*)

The row and column sums of squares are obtained separately for the two Latin squares and then added up. The sums of squares for the main effects and interactions may be obtained by using Yates' method. The *ABC* sum of squares so obtained should turn out to be equal to the between squares sum of squares obtained directly.

4. In a 3^3 factorial let the key-block be takne as

(000, 011, 022, 101, 112, 120, 202, 210, 221)

This will give a confounding arrangement in three blocks. The two interactions thus confounded belong to the $A \times B \times C$ interaction and are represented by ABC^2.

5. Suppose we wish to find a confounding arrangement in nine blocks for a 3^4-factorial experiment. Then eight degrees of freedom are confounded between blocks. We may take ABC^2 as one pair, and BCD^2 as the second pair. The two remaining pairs will then be represented by $(ABC^2)(BCD^2) = AB^2D^2$, and $(BCD^2)(AB^2D)^2 = ACD$. The key-block is given by

(0000, 0112, 0221, 1011, 1120, 1202, 2022, 2101, 2210)

6. *Partial confounding.* If the factorial experiment consists of more than one replicate we need not confound the *same* interactions in all the replicates (*total confounding*) but may confound a different set of interactions in each replicate (*partial confounding*). The sums of squares for the unconfounded interactions are obtained by using the observations from all the replicates. The sum of squares for a partially confounded interaction is obtained by ignoring the observations in the replicate in which it is confounded. Thus, with partial confounding we are able to make maximum use of replications to estimate all the interactions, although the partially confounded interactions are estimated from a smaller number of observations.

For example, in a confounding arrangement in two blocks in a 2^3-factorial experiment the key blocks for confounding the interactions AB, BC, and AC are respectively [(1), *c*, *ab*, *abc*], [(1), *a*, *bc*, *abc*], and [(1), *b*, *ac*, *abc*]. If we have three replicates we may arrange the experiment in six blocks as follows.

Rep I	Rep II	Rep III
(1), *c*, *ab*, *abc*	(1), *a*, *bc*, *abc*	(1), *b*, *ac*, *abc*
a, *b*, *ac*, *bc*	*b*, *c*, *ab*, *ac*	*a*, *c*, *ab*, *bc*

In this experiment AB is confounded in the first replicate, BC in the second, and AC in the third. The unconfounded interactions will be estimated from the yields in all the three replicates; the partially confounded interaction AB will be estimated from the yields in the second and third replicates, BC from the yields in the first and third replicates, and AC from the first and second replicates. The break-up of the degrees of freedom in the analysis of variance will be as follows.

Blocks	5
Main effects and interactions	7
Error	11
Total	23

Similarly, in a confounding arrangement in three blocks in a 3^3-factorial experiment the key-blocks for confounding the pairs ABC and ABC^2 are respectively (000, 012, 021, 102, 111, 120, 201, 210, 222) and (000, 011, 022, 101, 112, 120, 202, 210, 221). A complete replicate with three blocks can be used to partially confound ABC, and a second replicate with three blocks can be used to partially confound ABC^2. The estimates and the sums of squares for the unconfounded interactions will be obtained by using all the 54 observations, ABC will be obtained from the 27 observations in the second replicate, and ABC^2 from the 27 observations in the first replicate. The breakup of the degrees of freedom in the analysis of variance would be as follows.

Blocks	5
Main effects and interactions	26
Error	22
Total	53

If we have four replicates we may partially confound the four pairs of interactions represented by ABC, ABC^2, AB^2C, and AB^2C^2, confounding one of the four pairs in each replicate. The break-up of the degrees of freedom will then be.

Blocks	11
Main effects and interactions	26
Error	70
Total	107

REFERENCES

General principles of confounding are given in Yates (1933 c) and Yates (1935).

For confounding in 2^n designs see Yates (1935, 1937), Fisher (1942); confounding plans are given in Cochran and Cox (1957), Kitagawa and Mitome (1953).

The general theory and methods of confounding in symmetrical factorial experiments will be found in Nair (1938), Bose and Kishen (1940), Bose (1947). For the asymmetrical case see Nair and Rao (1948), White and Hultquist (1965), Das and Rao (1967).

Muller (1966) uses pseudo-factors to study the problem of balance in confounding.

18

Analysis of Covariance

18.1 Introduction

In the designs considered so far we made observations on a single variable, called yield; the variation in the yields from different experimental units arose, on the one hand, from non-uniformity of the experimental material, and, on the other, from the differential response of yield to treatments. The use of a particular design i.e., the grouping of the experimental units into blocks, or rows and columns, and the associated technique of analysis of variance, eliminated the variation resulting from non-uniformity of the experimental material, making the experiment more precise for the comparison of treatments.

In many experiments we have, for each experimental unit, observations on one or more supplementary variables in addition to the yield. These are usually called *concomitant observations* or *concomitant variables.* If the concomitant variables are unrelated to treatments and influence the yield, the variation in yield caused by them should be eliminated before comparing treatments. For example, in a manurial trial, in which fertilisers have to be applied after establishment of the plants, the variation in the number of plants per plot before application of fertilisers will also contribute to the variation in the observed yields. The comparison of the effect of different fertilisers on yield should, therefore, be made after eliminating the variation due to the differences in the number of plants in each plot. Similarly, in a greenhouse experiment, it may be thought that variation in light due to differing distances of the experimental units from the side of the greenhouse may have produced some variation in yield, and that such variation is in no way related to treatments. We would then like to compare treatments only after the variation resulting from the distance factor has been eliminated. In animal feeding experiments designed to compare the effect of different diets on growth, the initial weight of the animal is expected to affect the increase in weight recorded at the end of the experimental period, and therefore comparison of diets should be made after the variation in weight increase resulting from the differences in initial weights of the animals has been eliminated.

Whereas the variation resulting from the heterogeneity of the experimental material is eliminated by the use of an appropriate design and that of the analysis of variance, a different technique of analysis is needed to eliminate the variation resulting from the influence on yield of the concomitant variables. We call it *analysis of covariance*. It is an adaptation of the methods of regression analysis to experimental designs and consists essentially of fitting a regression of yield on the concomitant variables. The yields are than "adjusted" by removing the portion explained by regression, and the comparison of treatments carried out by using the adjusted yields. At the same time, the variation due to non-uniformity of the experimental material is eliminated by the use of a suitable design. Analysis of covariance is only a technique of analysis, and can, like the analysis of variance, be used with any design.

Suppose the fertiliser experiment mentioned earlier had been carried out in a randomised block. If y_{ij} denotes the yield from the plot receiving jth treatment in the i-th block, and x_{ij} the number of plants in that plot, we take the expected value of the yield as

$$E(y_{ij}) = \beta_i + \tau_j + \gamma x_{ij}$$

assuming linear regression of yield on the number of plants. Here β_i represents the effect of the i-th block, τ_j of the jth treatment, and γ is a new parameter called the *regression parameter*. As in regression analysis the x_{ij} are assumed to be given non-random quantities.

The data in this experiment can be analysed by applying the general theory. Instead of a randomised block with $b + v$ parameters, we have now $b + v + 1$ parameters, and we can obtain the sums of squares for testing hypotheses about β_i, τ_i and γ as also the sum of squares for error, by using the general theory. We shall, however, show in the present chapter how the analysis in such cases can be conveniently obtained by modifying the usual analysis of variance which would have been done without taking the concomitant variables into consideration.

18.2 The Analysis of Covariance Model

Let $y_1, \ldots, y_n$ be the yields and $x_{11}, \ldots, x_{1n}; \ldots; x_{m1}, \ldots, x_{mn}$; the corresponding values of the m concomitant variables $x_1, \ldots, x_m$. Then we have

$$E(y_i) = a_{i1}\beta_1 + \ldots + a_{ip}\beta_p, \quad i = 1, \ldots, n \tag{18.1}$$

without the concomitant variables, and

$$E(y_i) = a_{i1}\beta_1 + \ldots + a_{ip}\beta_p + x_{1i}\gamma_i + \ldots + x_{mi}\gamma_m, \quad i = 1, \ldots, n \tag{18.2}$$

with the concomitant variables. We assume in both cases the y_i to be independent variables with a common variance σ^2.

Writing $\mathbf{y}' = (y_1, \ldots, y_n)$, $\mathbf{A} = [a_{ij}]$, $\boldsymbol{\beta}' = (\beta_1, \ldots, \beta_p)$, $\boldsymbol{\gamma}' = (\gamma_1 \ldots, \gamma_m)$, $\mathbf{x}_1' = (x_{11}, \ldots, x_{1n}), \ldots, \mathbf{x}_m' = (x_{m1}, \ldots, x_{mn})$, and $\mathbf{X} = [\mathbf{x}_1 \ldots \mathbf{x}_m]$, we may write the equations for the model without the concomitant variables as

$$\mathbf{y} = \mathbf{A}\boldsymbol{\beta} + \mathbf{e} \tag{18.3}$$

and with the concomitant variables as

$$\mathbf{y} = \mathbf{A}\boldsymbol{\beta} + \mathbf{X}\boldsymbol{\gamma} + \mathbf{e} = [\mathbf{A} \vdots \mathbf{X}]\begin{bmatrix} \boldsymbol{\beta} \\ \cdots \\ \boldsymbol{\gamma} \end{bmatrix} + \mathbf{e} \tag{18.4}$$

For the purpose of this chapter alone, we shall call the first the old model, and the second the new model. We shall call $\mathbf{A}$ the design matrix, $\boldsymbol{\beta}$ the design parameters, and $\boldsymbol{\gamma}$ the regression parameters. The parameters $\boldsymbol{\beta}$ include the parameters representing the effect of the experimental material and those representing the effect of the treatments. The main problem of analysis is to derive tests of hypotheses regarding the parameters $\boldsymbol{\beta}$, using the additional information provided by the concomitant variables.

18.3 Normal Equations and Estimates

The normal equations of the new model are

$$[\mathbf{A} \vdots \mathbf{X}]' [\mathbf{A} \vdots \mathbf{X}]\begin{bmatrix} \boldsymbol{\beta} \\ \cdots \\ \boldsymbol{\gamma} \end{bmatrix} = [\mathbf{A} \vdots \mathbf{X}]'\mathbf{y} \tag{18.5}$$

or,

$$\mathbf{A'A\beta} + \mathbf{A'X\gamma} = \mathbf{A'y}$$
$$\mathbf{X'A\beta} + \mathbf{X'X\gamma} = \mathbf{X'y} \qquad (18.6)$$

In most cases the regression parameters γ will be estimable. The conditions which ensure estimability of γ are given by the following theorem.

Theorem 18.1: The parameters $\gamma_1, \ldots, \gamma_m$ are estimable, if and only if,

(i) rank $\mathbf{X} = m$,

(ii) rank $[\mathbf{A} \vdots \mathbf{X}] = \text{rank } \mathbf{A} + m$

(In other words, the vectors $\mathbf{x}_i$ are linearly independent and cannot be expressed as linear combinations of the column vectors of the matrix A).

Proof: Let $\mathbf{u}_j$, $\mathbf{v}_j$ be the projections of $\mathbf{x}_j$ on the estimation and error space respectively of the old model, i.e, on $C(\mathbf{A})$ and its orthogonal complement. Writing

$$\mathbf{U} = [\mathbf{u}_1 \ldots \mathbf{u}_m], \mathbf{V} = [\mathbf{v}_1 \ldots \mathbf{v}_m] \qquad (18.7)$$

we have

$$\mathbf{X} = \mathbf{U} + \mathbf{V}, \mathbf{A'V} = \mathbf{O}, \mathbf{A'X} = \mathbf{A'U}$$
$$\mathbf{X'X} = \mathbf{U'U} + \mathbf{V'V}, C([\mathbf{AX}]) = C([\mathbf{AV}]) \qquad (18.8)$$
$$\text{rank } [\mathbf{A} \vdots \mathbf{X}] = \text{rank } [\mathbf{A} \vdots \mathbf{V}] = \text{rank } \mathbf{A} + \text{rank } \mathbf{V}$$

The conditions (i) and (ii) of the theorem are equivalent to the condition, rank $\mathbf{V} = m$.

Suppose rank $\mathbf{V} = m$. Then $\mathbf{V'V}$ is non-singular and the inverse $(\mathbf{V'V})^{-1}$ exists. The normal equations can be written as

$$\mathbf{A'A\beta} + \mathbf{A'U\gamma} = \mathbf{A'y}$$
$$\mathbf{U'A\beta} + (\mathbf{U'U} + \mathbf{V'V})\,\gamma = \mathbf{U'y} + \mathbf{V'y} \qquad (18.9)$$

Since $\mathbf{u}_j \in C(\mathbf{A})$, we may write $\mathbf{U} = \mathbf{AB}$ for some matrix $\mathbf{B}$, and the normal equations become

$$\mathbf{A'A\beta} + \mathbf{A'AB\gamma} = \mathbf{A'y}$$
$$\mathbf{B'A'A\beta} + (\mathbf{B'A'AB} + \mathbf{V'V})\,\gamma = \mathbf{B'A'y} + \mathbf{V'y} \qquad (18.10)$$

Multiplying the first equation by $\mathbf{B}'$ and subtracting from the second we get

$$\mathbf{V'V\gamma} = \mathbf{V'y} \qquad (18.11)$$

Since $\mathbf{V'V}$ is non-singular we conclude that $\boldsymbol{\gamma}$ is estimable and its best estimate is

$$\hat{\gamma} = (\mathbf{V'V})^{-1}\,\mathbf{V'y} \tag{18.12}$$

Conversely, suppose $\gamma_1, \ldots, \gamma_m$ are estimable.
Let $\mathbf{n}_1'\mathbf{y}, \ldots, \mathbf{n}_m'\mathbf{y}$ be the best estimates. Clearly, $\mathbf{n}_1, \ldots, \mathbf{n}_m$ are linearly independent.

If $\mathbf{N} = [\mathbf{n}_1 \ldots \mathbf{n}_m]$, then rank $\mathbf{N} = m$, and

$$E(\mathbf{N'y}) = \mathbf{N'}(\mathbf{A}\boldsymbol{\beta} + \mathbf{X}\boldsymbol{\gamma}) = \boldsymbol{\gamma}$$

for all $\boldsymbol{\beta}, \boldsymbol{\gamma}$. Hence,

$$\mathbf{N'A} = \mathbf{O} \qquad \mathbf{N'X} = \mathbf{I} \tag{18.13}$$

But, if $\mathbf{N'A} = \mathbf{O}$, we also have $\mathbf{N'U} = \mathbf{O}$. Thus

$$\mathbf{N'X} = \mathbf{N'}(\mathbf{U} + \mathbf{V}) = \mathbf{N'V} = \mathbf{I} \tag{18.14}$$

Now, rank $\mathbf{N'} = m$, rank $\mathbf{N'V}$ = rank $\mathbf{I} = m$. But rank $\mathbf{V} \leqslant m$, and if rank $\mathbf{V} < m$. we have rank $\mathbf{N'V} < m$. Hence, rank $\mathbf{V} = m$.

Corollary:

(i) $$E(\hat{\boldsymbol{\gamma}}) = \boldsymbol{\gamma}$$

(ii) $$D(\hat{\boldsymbol{\gamma}}) = (\mathbf{V'V})^{-1}\sigma^2 \tag{18.15}$$

We now consider the estimation of linear functions $\boldsymbol{\lambda}'\boldsymbol{\beta}$ of the design parameters $\boldsymbol{\beta}$.

Theorem 18.2: If $\boldsymbol{\lambda}'\boldsymbol{\beta}$ is estimable in the old model and its best estimate is $\mathbf{l'y}$ then it is estimable in the new model and its best estimate is $\mathbf{l}'(\mathbf{y} - \mathbf{X}\hat{\boldsymbol{\gamma}})$.

Proof: We know from the old model that $\mathbf{l} \in C(\mathbf{A})$ and $\mathbf{l'A} = \boldsymbol{\lambda}'$. In the new model

$$\begin{aligned} E[\mathbf{l}'(\mathbf{y} - \mathbf{X}\hat{\boldsymbol{\gamma}})] &= E(\mathbf{l'y}) - E(\mathbf{l'X}\hat{\boldsymbol{\gamma}}) \\ &= \mathbf{l}'(\mathbf{A}\boldsymbol{\beta} + \mathbf{X}\boldsymbol{\gamma}) - \mathbf{l'X}\boldsymbol{\gamma} \\ &= \mathbf{l'A}\boldsymbol{\beta} = \boldsymbol{\lambda}'\boldsymbol{\beta} \end{aligned} \tag{18.16}$$

Hence, $\boldsymbol{\lambda}'\boldsymbol{\beta}$ is estimable in the new model. Further,

$$\begin{aligned} \mathbf{l}'(\mathbf{y} - \mathbf{X}\hat{\boldsymbol{\gamma}}) &= \mathbf{l'y} - \mathbf{l'X}(\mathbf{V'V})^{-1}\mathbf{V'y} \\ &= [\mathbf{l} - \mathbf{V}(\mathbf{V'V})^{-1}\mathbf{X'l}]'\mathbf{y} \end{aligned} \tag{18.17}$$

Now, $\mathbf{l} \in C(\mathbf{A})$, and $\mathbf{V}(\mathbf{V'V})^{-1}\mathbf{X'l} = \mathbf{Vc} \in C(\mathbf{V})$. Thus the coefficient vector of the estimate $\mathbf{l}'(\mathbf{y} - \mathbf{X}\hat{\boldsymbol{\gamma}})$ belongs to $C([\mathbf{A} \,\vdots\, \mathbf{V}])$, i.e. to the estimation space $C([\mathbf{A} \; \vdots \; \mathbf{X}])$ of the new model, showing that $\mathbf{l}'(\mathbf{y}-\mathbf{X}\hat{\boldsymbol{\gamma}})$ is the best estimate of $\boldsymbol{\lambda}'\boldsymbol{\beta}$ in the new model.

We can also show that if $\boldsymbol{\lambda}'\boldsymbol{\beta}$ is estimable in the new model, it is estimable in the old model.

Corollary: If $\mathbf{l}'\mathbf{y}$ is the best estimate of $\boldsymbol{\lambda}'\boldsymbol{\beta}$ in the old model, then $\mathbf{l}'\mathbf{y}$ is orthogonal to each of the estimates $\hat{\gamma}_1, \ldots, \hat{\gamma}_m$ of the regression parameters $\gamma_1, \ldots, \gamma_m$.

Proof: We have $\mathbf{l} \in C(\mathbf{A})$, and hence $\mathbf{l} \perp C(\mathbf{V})$, i.e. $\mathbf{l}'\mathbf{V} = \mathbf{0}'$

Now,

$$\hat{\boldsymbol{\gamma}} = (\mathbf{V}'\mathbf{V})^{-1}\mathbf{V}'\mathbf{y} = [\mathbf{V}(\mathbf{V}'\mathbf{V})^{-1}]'\mathbf{y}$$

and

$$\mathbf{l}'\mathbf{V}(\mathbf{V}'\mathbf{V})^{-1} = \mathbf{0}'$$

which completes the proof.

The vector

$$\mathbf{y}-\mathbf{X}\hat{\boldsymbol{\gamma}} = \mathbf{y} - \hat{\gamma}_1\mathbf{x}_1 - \ldots - \hat{\gamma}_m\mathbf{x}_m \tag{18.18}$$

is called the *vector of adjusted yields*, the yields having been adjusted for regression on the concomitant variables.

18.4 Sums of Squares for Estimates and Error

The sum of squares for estimates is, by definition, equal to the square of the projection of y on the estimation space, i.e. on $C([\mathbf{A} \; \vdots \; \mathbf{X}]) = C([\mathbf{A} \; \vdots \; \mathbf{V}]) = C(\mathbf{A}) + C(\mathbf{V})$. Since $C(\mathbf{A})$ is orthogonal to $C(\mathbf{V})$, the square of the projection of $\mathbf{y}$ on $C([\mathbf{A} \; \vdots \; \mathbf{V}])$ is given by

Square of projection on $C(\mathbf{A})$ + Square of projection on $C(\mathbf{V})$

Hence, the SS for estimates in the new model is given by

SS (Est) in the old model + Square of projection of y on $C(\mathbf{V})$

Now, the projection of y on $C(\mathbf{V})$ is $\mathbf{V}\boldsymbol{\delta}$ where $\boldsymbol{\delta}$ is any solution of (Theorem 2.7)

$$\mathbf{V}'\mathbf{V}\boldsymbol{\delta} = \mathbf{V}'\mathbf{y} \tag{18.19}$$

i.e.

$$\boldsymbol{\delta} = (\mathbf{V}'\mathbf{V})^{-1}\mathbf{V}'\mathbf{y} \tag{18.20}$$

Hence, the square of the projection of y on $C(\mathbf{V})$ is

$$\boldsymbol{\delta}'\mathbf{V}'\mathbf{V}\boldsymbol{\delta} = \mathbf{y}'\mathbf{V}(\mathbf{V}'\mathbf{V})^{-1}\mathbf{V}'\mathbf{y} = \hat{\boldsymbol{\gamma}}'\mathbf{V}'\mathbf{y} \tag{18.21}$$

This also shows that $\hat{\boldsymbol{\gamma}}'\mathbf{V}'\mathbf{y}$ is the SS due to the estimates $\hat{\boldsymbol{\gamma}}$ of the regression coefficients.

Thus, the SS for estimates in the new model is given by

$$\text{SS (Est) in old model} + \hat{\boldsymbol{\gamma}}'\mathbf{V}'\mathbf{y} \tag{18.22}$$

Consequently, the error sum of squares in the new model is

$$\text{SS (Err) in old model} - \hat{\boldsymbol{\gamma}}'\mathbf{V}'\mathbf{y} \tag{18.23}$$

18.5 Tests of Hypotheses Regarding $\boldsymbol{\beta}$

Suppose the hypothesis to be tested is

$$\boldsymbol{\lambda}_1'\boldsymbol{\beta} = \ldots = \boldsymbol{\lambda}_k'\boldsymbol{\beta} = 0$$

where we assume $\boldsymbol{\lambda}_1, \ldots, \boldsymbol{\lambda}_k$ to be linearly independent. Let $\mathbf{l}_i'\mathbf{y}$ be the best estimate of $\boldsymbol{\lambda}_i'\boldsymbol{\beta}$ ($i = 1, \ldots, k$) in the old model. We may assume without loss of generality that the vectors $\mathbf{l}_i$ which belong to $C(\mathbf{A})$, form an orthonormal set.

The set of estimates $\mathbf{l}_i'\mathbf{y}, \ldots, \mathbf{l}_k'\mathbf{y}$ may be written as $\mathbf{L}'\mathbf{y}$ where $\mathbf{L}$ is the $n \times k$ matrix $[\mathbf{l}_1 \ldots \mathbf{l}_k]$. The estimates of $\boldsymbol{\lambda}_i'\boldsymbol{\beta}$ in the new model are $\mathbf{l}_i'(\mathbf{y} - \mathbf{X}\hat{\boldsymbol{\gamma}})$, i.e.

$$\mathbf{l}_i'\mathbf{y} - \mathbf{l}_i'\mathbf{X}\mathbf{N}'\mathbf{y},\ i = 1, \ldots, k \tag{18.24}$$

where $\mathbf{N}$ is the matrix obtained earlier from the estimates $\mathbf{n}_j'\mathbf{y}$ of γ_j, i.e. $\mathbf{N}' = (\mathbf{V}'\mathbf{V}^{-1})\,\mathbf{V}'$. The set of these estimates may be written as $\mathbf{L}'\mathbf{y} - \mathbf{L}'\mathbf{X}\mathbf{N}'\mathbf{y}$ that is

$$(\mathbf{L} - \mathbf{N}\mathbf{X}\mathbf{L})'\mathbf{y} \tag{18.25}$$

These k linearly independent linear forms lie in the space W spanned by the $k + m$ linear forms $\mathbf{l}_1'\mathbf{y}, \ldots, \mathbf{l}_k'\mathbf{y},\ n_1'\mathbf{y}, \ldots, \mathbf{n}_m'\mathbf{y}$. Since the $\mathbf{l}_i$'s are orthonormal, and each $\mathbf{l}_i$ is orthogonal to each $\mathbf{n}_j$, the SS due to these $k + m$ linear forms with $k + m$ degrees of freedom, is the sum of the SS due to the set $\{\mathbf{l}_1'\mathbf{y}, \ldots, \mathbf{l}_k'\mathbf{y}\}$ and the SS due to the set $\{\mathbf{n}_1'\mathbf{y}, \ldots, \mathbf{n}_m'\mathbf{y}\}$, i.e.

$$\sum_i (\mathbf{l}_i'\mathbf{y})^2 + \mathbf{y}'\mathbf{V}(\mathbf{V}'\mathbf{V})^{-1}\mathbf{V}'\mathbf{y} = \mathbf{y}'\mathbf{L}\mathbf{L}'\mathbf{y} + \mathbf{y}'\mathbf{V}(\mathbf{V}'\mathbf{V})^{-1}\mathbf{V}'\mathbf{y} \tag{18.26}$$

where we have $\mathbf{L}'\mathbf{L} = \mathbf{I}$. Thus, to obtain the SS for testing the hypothesis in the new model we have simply to subtract from (18.26), the SS due to m suitably chosen independent linear forms in W which are orthogonal to the linear forms (18.25).

A general set of m linear forms in W can be written as

$$\mathbf{PL'y} + \mathbf{QN'y} \tag{18.27}$$

where $\mathbf{P}$ is an $m \times k$ matrix, and $\mathbf{Q}$ an $m \times m$ matrix. For this set of linear forms to be orthogonal to the linear forms (18.25) we must have

$$(\mathbf{PL'} + \mathbf{QN'})\,(\mathbf{L} - \mathbf{NX'L}) = \mathbf{O} \tag{18.28}$$

i.e.,

$$\mathbf{PL'L} + \mathbf{QN'L} - \mathbf{PL'NX'L} - \mathbf{QN'NX'L} = \mathbf{O} \tag{18.29}$$

Since we have $\mathbf{N'A} = \mathbf{O}$, and hence $\mathbf{N'L} = \mathbf{O} = \mathbf{L'N}$,

and also $\mathbf{L'L} = \mathbf{I}$, we have to choose $\mathbf{P}$ and $\mathbf{Q}$ such that

$$\mathbf{P} - \mathbf{QN'NX'L} = \mathbf{O},$$

i.e.

$$\mathbf{P} - \mathbf{Q}\,(\mathbf{V'V})^{-1}\,\mathbf{U'L} = \mathbf{O} \tag{18.30}$$

since $\mathbf{N'N} = (\mathbf{V'V})^{-1}$, and $\mathbf{X'L} = (\mathbf{U'} + \mathbf{V'})\,\mathbf{L} = \mathbf{U'L}$. Thus, if we take

$$\mathbf{Q} = \mathbf{V'V},\ \mathbf{P} = \mathbf{U'L} \tag{18.31}$$

the linear forms (18.27) will be orthogonal to the linear forms (18.25). With the above values of $\mathbf{P}$ and $\mathbf{Q}$ the desired m linear forms are

$$\mathbf{U'LL'y} + (\mathbf{V'V})\mathbf{N'y} \tag{18.32}$$

i.e.

$$(\mathbf{LL'U} + \mathbf{V})'\mathbf{y} \tag{18.33}$$

since,

$$(\mathbf{V'V})\mathbf{N'} = (\mathbf{V'V})\,(\mathbf{V'V})^{-1}\,\mathbf{V'} = \mathbf{V'} \tag{18.34}$$

The m columns of $\mathbf{LL'U} = \mathbf{L}(\mathbf{L'U})$ being linear combinations of the columns of $\mathbf{L}$, i.e. linear combinations of the columns of $\mathbf{A}$, are orthogonal to the m columns of $\mathbf{V}$. This, along with the fact that the columns of $\mathbf{V}$ are linearly independent (rank $\mathbf{V} = m$) shows that the columns of $\mathbf{LL'U} + \mathbf{V}$ are also linearly independent. Thus the m linear forms (18.33) are linearly independent. The SS due to these m linear forms has m degrees of freedom, and is given by

$$\begin{aligned}&\mathbf{y'}\,(\mathbf{LL'U}+\mathbf{V})\,[(\mathbf{U'LL'}+\mathbf{V'})\,(\mathbf{LL'U}+\mathbf{V})]^{-1}\,(\mathbf{U'LL'}+\mathbf{V'})\mathbf{y}\\ &\quad= \mathbf{y'}\,(\mathbf{LL'U}+\mathbf{V})\,(\mathbf{U'LL'U}+\mathbf{V'V})^{-1}\,(\mathbf{U'LL'}+\mathbf{V'})\mathbf{y}\end{aligned} \tag{18.35}$$

This SS may be expressed as

$$\hat{\mathbf{z}}'(\mathbf{U}'\mathbf{L}\mathbf{L}' + \mathbf{V}')\,\mathbf{y} \tag{18.36}$$

where $\hat{\mathbf{z}}$ is the solution of the equations

$$(\mathbf{U}'\mathbf{L}\mathbf{L}'\mathbf{U} + \mathbf{V}'\mathbf{V})\,\mathbf{z} = \mathbf{U}'\mathbf{L}\mathbf{L}'\mathbf{y} + \mathbf{V}'\mathbf{y} \tag{18.37}$$

From (18.26) and (18.36) we see that the SS for testing the hypothesis, with k degrees of freedom, is

$$\mathbf{y}'\mathbf{L}\mathbf{L}'\mathbf{y} + \mathbf{y}'\mathbf{V}(\mathbf{V}'\mathbf{V})^{-1}\mathbf{V}'\mathbf{y} - \hat{\mathbf{z}}'(\mathbf{U}'\mathbf{L}\mathbf{L}' + \mathbf{V}')\,\mathbf{y} \tag{18.38}$$

18.6 Analysis of Covariance

The analysis of variance, if concomitant variables are ignored, consists principally in partitioning the raw sum of squares $\mathbf{y}'\mathbf{y}$ into two orthogonal components, the sum of squares for estimates and error. If $\mathbf{y}_s$ and $\mathbf{y}_e$ denote the projections of $\mathbf{y}$ on the estimation and error space respectively, then the sum of squares for estimates S_{yy} and that for error E_{yy} are given by

$$S_{yy} = \mathbf{y}_s'\mathbf{y}_s \quad E_{yy} = \mathbf{y}_e'\mathbf{y}_e \tag{18.39}$$

and we have

$$S_{yy} + E_{yy} = \mathbf{y}'\mathbf{y} \tag{18.40}$$

Since $\mathbf{u}_j$, $\mathbf{v}_j$ $(j = 1, \ldots, m)$ are the projections of $\mathbf{x}_j$ on the estimation and error space respectively in the old model, we may define

$$S_{jj} = \mathbf{u}_j'\mathbf{u}_j \qquad S_{jk} = \mathbf{u}_j'\mathbf{u}_k \qquad S_{jy} = \mathbf{u}_j'\mathbf{y}_s \tag{18.41}$$

and

$$E_{jj} = \mathbf{v}_j'\mathbf{v}_j \qquad E_{jk} = \mathbf{v}_j'\mathbf{v}_k \qquad E_{jy} = \mathbf{v}_j'\mathbf{y}_e \tag{18.42}$$

Then, we shall have

$$S_{jj} + E_{jj} = \mathbf{x}_j'\mathbf{x}_j$$

$$S_{jk} + E_{jk} = \mathbf{x}_j'\mathbf{x}_k \qquad S_{jy} + E_{jy} = \mathbf{x}_j'\mathbf{y} \tag{18.43}$$

The quantities (18.41) will be called the estimate components, and (18.42) the error components of the sums of squares and sums of products of the concomitant variables, and of the sums of products of the concomitant variables with the yield variable.

Once the method of obtaining S_{yy} and E_{yy} is known, the same method can be used to partition the raw sums of squares $\mathbf{x}_j'\mathbf{x}_j$, the raw sums of products $\mathbf{x}_j'\mathbf{x}_k$ and the raw sums of products $\mathbf{x}_j'\mathbf{y}$ into the estimate and error components (18.41) and (18.42). In fact, if any sum of squares for the yields in the old model is represented by $\sum_i (\mathbf{l}_i'\mathbf{y})^2$, the corresponding sums of squares and products arising from the concomitant variables will be

$$\sum_i (\mathbf{l}_i'\mathbf{x}_j)^2, \qquad j = 1, \ldots, m$$

$$\sum_i (\mathbf{l}_i'\mathbf{x}_j)(\mathbf{l}_i'\mathbf{x}_k), \qquad j \neq k = 1, \ldots, m$$

$$\sum_i (\mathbf{l}_i'\mathbf{x}_j)(\mathbf{l}_i'\mathbf{y}), \qquad j = 1, \ldots, m$$

The name analysis of covariance is given to this technique of partitioning the sums of squares and products arising out of the use of the concomitant variables.

Having partitioned the sums of squares and products as above, we use the error components (18.42) to obtain the estimates $\hat{\boldsymbol{\gamma}}$ of the regression parameters $\boldsymbol{\gamma}$. The quantities E_{jj} and E_{jk} are the elements of the matrix $\mathbf{V'V}$, and the vector $\mathbf{V'y}$ is given by

$$(\mathbf{V'y})' = (E_{1y}, \ldots, E_{my})$$

Thus, we can set up the equations

$$\mathbf{V'V}\boldsymbol{\gamma} = \mathbf{V'y} \tag{18.44}$$

using the error components (18.42). Solving these equations we get $\hat{\boldsymbol{\gamma}}$, as well as the sum of squares $\hat{\boldsymbol{\gamma}}'\mathbf{V'y}$. Then, the *adjusted error sum of squares*, i.e. the sum of squares for error in the new model, is given by

$$E_{yy} - \hat{\boldsymbol{\gamma}}'\mathbf{V'y} \tag{18.45}$$

Consider next the sum of squares for the hypothesis

$$\boldsymbol{\lambda}_1'\boldsymbol{\beta} = \ldots = \boldsymbol{\lambda}_k'\boldsymbol{\beta} = 0 \tag{18.46}$$

obtained in Sec. 18.5, viz.

$$\mathbf{y'LL'y} + \mathbf{y'V(V'V)^{-1}V'y} - \hat{\mathbf{z}}'(\mathbf{U'LL'} + \mathbf{V'})\,\mathbf{y}$$

$$= \mathbf{y'LL'y} + \hat{\boldsymbol{\gamma}}'\mathbf{V'y} - \hat{\mathbf{z}}'(\mathbf{U'LL'} + \mathbf{V'})\,\mathbf{y} \tag{18.47}$$

The term $\mathbf{y'LL'y}$ is the hypothesis sum of squares in the old model. Let us denote it by H_{yy}. Let the corresponding sums of squares and products for the concomitant variables be

$$H_{jj} = \mathbf{x}_j'\mathbf{LL}'\,\mathbf{x}_j = \mathbf{u}_j'\,\mathbf{LL}'\mathbf{u}_j \qquad (\because \quad \mathbf{L}'\mathbf{v}_j = 0)$$

$$H_{jk} = \mathbf{x}_j'\,\mathbf{LL}'\mathbf{x}_k = \mathbf{u}_j'\,\mathbf{LL}'\,\mathbf{u}_k \tag{18.48}$$

$$H_{jy} = \mathbf{x}_j'\,\mathbf{LL}'\mathbf{y} = \mathbf{u}_j'\,\mathbf{LL}'\mathbf{y}$$

The different H's are easily obtained once the method of obtaining H_{yy} is known.

To get the hypothesis sum of squares in the new model we have first to solve the equations

$$(\mathbf{U}'\mathbf{LL}'\mathbf{U} + \mathbf{V}'\mathbf{V})\,\mathbf{z} = \mathbf{U}'\mathbf{LL}'\mathbf{y} + \mathbf{V}'\mathbf{y} \tag{18.49}$$

to obtain $\hat{\mathbf{z}}$. These equations are easily written, once the hypothesis components (18.48) and the error components (18.42) are known. The elements of the $m \times m$ matrix $\mathbf{U}'\mathbf{LL}'\mathbf{U}+\mathbf{V}'\mathbf{V}$ are obtained by adding the hypothesis components (18.48) to the corresponding error components (18.42). The elements of the vector $\mathbf{ULL}'\mathbf{y} + \mathbf{V}'\mathbf{y}$ are obtained in the same manner.

Note that the procedure to obtain the term $\hat{\mathbf{z}}'\,(\mathbf{U}'\mathbf{LL}' + \mathbf{V}')\,\mathbf{y}$ is exactly similar to that followed earlier to get the term $\hat{\boldsymbol{\gamma}}'\mathbf{V}'\mathbf{y}$. The only difference is that each error component used in the latter case is replaced by the sum of the error and the corresponding hypothesis component in the former.

The hypothesis sum of squares in the new model may be written as

$$H_{yy} + \hat{\boldsymbol{\gamma}}'\mathbf{V}'\mathbf{y} - \hat{\mathbf{z}}'\,(\mathbf{U}'\mathbf{LL}' + \mathbf{V}')\,\mathbf{y}$$

$$= \{H_{yy} + E_{yy} - \hat{\mathbf{z}}'\,(\mathbf{U}'\mathbf{LL}' + \mathbf{V}')\,\mathbf{y}\} - \{E_{yy} - \hat{\boldsymbol{\gamma}}'\mathbf{V}'\mathbf{y}) \tag{18.50}$$

The second term is the adjusted error sum of squares, or the error SS in the new model. The first term is obtained by "adjusting" $H_{yy} + E_{yy}$, the sum of the hypothesis and error sums of squares in the old model, by the same method as was used to adjust E_{yy}. Thus we may summarise the above in the form of the following rule:

Hypothesis SS = (Adjusted hypothesis+error SS

— Adjusted error SS)

18.7 Example

Consider an experiment with two concomitant variables, x_1 and x_2, carried out in a balanced incomplete block design $(b, k; v, r; \lambda)$. The analysis of covariance, based on the inter block analysis of variance, will be done as follows.

Let y_{ij}, x_{ij1}, x_{ij2} denote the yield and concomitant observations from the plot receiving the jth treatment in the ith block, and Y, X_1, X_2 their grand totals. Then the total sums of squares and products are

$$T_{yy} = \sum_{i,j} y_{ij}^2 - (Y^2/bk)$$

$$T_{11} = \sum_{i,j} x_{ij1}^2 - (X_1^2/bk)$$

$$T_{22} = \sum_{i,j} x_{ij2}^2 - (X_2^2/bk)$$

$$T_{12} = \sum_{i,j} x_{ij1} x_{ij2} - (X_1 X_2/bk)$$

$$T_{1y} = \sum_{i,j} x_{ij1}\, y_{ij} - (X_1 Y/bk)$$

$$T_{2y} = \sum_{i,j} x_{ij2}\, y_{ij} - (X_2 Y/bk)$$

Let B_{iy}, B_{i1}, B_{i2} ($i = 1, \ldots, b$) denote the totals of the yields and concomitant observations in the ith block. From these we obtain the unadjusted block sums of squares and products

$$B_{yy} = (1/k) \sum B_{iy}^2 - (Y^2/bk)$$

$$B_{11} = (1/k) \sum B_{i1}^2 - (X_1^2/bk)$$

$$B_{22} = (1/k) \sum B_{i2}^2 - (X_2^2/bk)$$

$$B_{12} = (1/k) \sum B_{i1} B_{i2} - (X_1 X_2/bk)$$

$$B_{1y} = (1/k) \sum B_{i1}\, B_{iy} - (X_1 Y/bk)$$

$$B_{2y} = (1/k) \sum B_{i2}\, B_{iy} - (X_2 Y/bk)$$

To get the adjusted treatment sums of squares and products we use the adusted treatment totals Q_{jy}, Q_{j1}, Q_{j2} ($j = 1, \ldots, v$) for the yield and concomitant observations. The adjusted treatment sums of squares and products are

$$V_{yy} = \sum (kQ_{jy})^2/(k\lambda v)$$

$$V_{11} = \sum (kQ_{j1})^2/(k\lambda v)$$

....

$$V_{2y} = \sum (kQ_{j2})(kQ_{jy})/(k\lambda v)$$

The quantities kQ_{jy}, kQ_{j1}, kQ_{j2} are obtained first in tabular form as described in Chapter 12.

The error components of the sums of squares and products are then obtained by subtraction. The results are expressed in the form of the following analysis of covariance table (Table 18.1).

Table 18.1

Source	df	SS and SP					
		y^2	x_1^2	x_1x_2	x_2^2	x_1y	x_2y
Blocks (unadj.)	$b-1$	B_{yy}	B_{11}	B_{12}	B_{22}	B_{1y}	B_{2y}
Treatments (adj.)	$v-1$	V_{yy}	V_{11}	V_{12}	V_{22}	V_{1y}	V_{2y}
Error	$bk-v-b+1$	E_{yy}	E_{11}	E_{12}	E_{22}	E_{1y}	E_{2y}
Total	$bk-1$	T_{yy}	T_{11}	T_{12}	T_{22}	T_{1y}	T_{2y}

We now set up the equations for estimating the regression parameter γ:

$$\begin{bmatrix} E_{11} & E_{12} \\ E_{12} & E_{22} \end{bmatrix} \gamma = \begin{bmatrix} E_{1y} \\ E_{2y} \end{bmatrix}$$

If the solution is

$$\hat{\gamma} = \begin{bmatrix} E^{11} & E^{12} \\ E^{12} & E^{12} \end{bmatrix} \begin{bmatrix} E_{1y} \\ E_{2y} \end{bmatrix} \tag{18.51}$$

the adjusted error SS is

$$S_e^2 = E_{yy} - \hat{\gamma}_1 E_{1y} - \hat{\gamma}_2 E_{2y} \tag{18.52}$$

and has $(bk-b-v+1)-2 = bk-b-v-1$ degrees of freedom. The mean square error $s^2 = S_e^2/(bk-b-v-1)$ estimates σ^2.

The estimate of a contrast $\Sigma\, t_j\, \tau_j$ ($\Sigma\, t_j = 0$) in the absence of the concomitant variables was

$$(k/\lambda v)\ \Sigma\, t_j Q_{jy} \tag{18.53}$$

Hence the estimate now is

$$(k/\lambda v) \sum_j t_j (Q_{iy} - \hat{\gamma}_1 Q_{j1} - \hat{\gamma}_2 Q_{j2}) \tag{18.54}$$

To test the hypothesis

$$\tau_1 = \tau_2 = \ldots = \tau_v \tag{18.55}$$

we add the adjusted treatment and error sums of squares and products to set up the following equation.

$$\begin{bmatrix} V_{11} + E_{11} & V_{12} + E_{12} \\ V_{12} + E_{12} & V_{22} + E_{22} \end{bmatrix} \mathbf{z} = \begin{bmatrix} V_{1y} + E_{1y} \\ V_{2y} + E_{2y} \end{bmatrix}$$

If the solution is $\hat{\mathbf{z}}$, then adjusted hypothesis plus error SS is

$$V_{yy} + E_{yy} - \hat{z}_1 (V_{1y} + E_{1y}) - \hat{z}_2 (V_{2y} + E_{2y})$$

Subtracting from it the adjusted error SS

$$E_{yy} - \hat{\gamma}_1 E_{1y} - \hat{\gamma}_2 E_{2y}$$

we get the hypothesis sum of squares S_H^2 with $v - 1$ degrees of freedom. The test for the hypothesis can now be carried out using the ratio

$$\frac{S_H^2/(v-1)}{s^2} \tag{18.56}$$

which, under the hypothesis, has the F-distribution with $(v - 1, bk - b - v - 1)$ degrees of freedom.

To test the hypothesis

$$\gamma_1 = \gamma_2 = 0 \tag{18.57}$$

we use the SS due to the estimates $\hat{\gamma}_1, \hat{\gamma}_2$ of the regression parameters, viz. $\hat{\gamma}_1 E_{1y} + \hat{\gamma}_2 E_{2y}$ with two degrees of freedom. The test is done by using the ratio

$$\frac{\hat{\gamma}_1 E_{1y} + \hat{\gamma}_2 E_{2y}}{2s^2} \tag{18.58}$$

which, under the hypothesis, has the F-distribution with $(2, bk - v - b - 1)$ degrees of freedom. This test will indicate if there is regression of yield on the concomitant variables. If the regression is not significant, that is the hypothesis $\gamma_1 = \gamma_2 = 0$ is accepted, we may conclude that the concomitant variables do not affect the yield.

If we wish to test if any one concomitant variable, say x_1, influences the yield, we may set up the hypothesis $\gamma_1 = 0$. This will be tested by the usual *t*-test. The estimate of γ_1 is $\hat{\gamma}_1$ and $\text{Var}(\hat{\gamma}_1) = E^{11}\sigma^2$. Hence we use the ratio

$$\hat{\gamma}_1/s\sqrt{E^{11}} \tag{18.59}$$

which, under the hypothesis, has the *t*-distribution with $bk - v - b - 1$ degrees of freedom.

18.8 Example

In a feeding trial twenty cows were all fed on the same ration for a control period of three weeks, and then for a further period of seventeen weeks on four different treatments. The cows were divided into five groups or blocks, numbered *A, B, C, D* and *E*. The division was made partly on the basis of age and partly on that of the date of calving. Treatments and shed positions were allotted to cows of the same block at random. Unfortunately the difficulty of obtaining cows calving at the appropriate period led to the inclusion in block *A* of cows which ran dry before the experiment was finished, so that for the full period only blocks *B, C, D* and *E* were available. In addition, one cow in block *E* developed a hard quarter, and the observations made on it are somewhat erratic. However, for the purpose of the present analysis we shall ignore this complication. The average weekly milk yields in pounds for the control period (x) and the experimental period proper (y) are given in Table 18.2.

Table 18.2

Diet	Block							
	B		*C*		*D*		*E*	
	x	y	x	y	x	y	x	y
1. Control	279	208	245	212	197	157	250	208
2. Dried grass	223	172	224	164	179	156	232	166
3. Molassed silage	269	183	256	204	191	136	311	192
4. A.I.V. Fodder	342	247	208	150	210	159	274	210

Source: Bartlett, M.S. *Journal of the Royal Statistical Society*, Supplement, Vol. 4, 1937, p. 137.

We shall use y as the main variable and x as the concomitant variable. The computations are carried in the same manner as for the randomised block design.

	Block totals			Treatment totals	
	x	y		x	y
B	1113	810	(1)	971	785
C	933	730	(2)	858	658
D	777	608	(3)	1027	715
E	1067	776	(4)	1034	766
Total	3890	2924	Total	3890	2924

	y^2	xy	x^2
Raw SS (SP)	547848	728172	975448
CF	$(2924)^2/16$	$(3890)\,(2924)/16$	$(3890)^2/16$
	$= 534361$	$= 710897.5$	$= 945756.25$
Total SS (SP)	13487	17274.5	29691.75

$$\text{SS}(y) \text{ for blocks} = (1/4)\,[(810)^2 + (730)^2 + (608)^2 + (776)^2] - 534361 = 5849.0$$

$$\text{SP}(xy) \text{ for blocks} = (1/4)\,[(1113)(810) + (933)(730) + (777)(608) + (1067)(776)] - 710897.5 = 9859.5$$

$$\text{SS } (x) \text{ for blocks} = (1/4)[(1113)^2 + (933)^2 + (777)^2 + (1067)^2] - 945756.25$$
$$= 17112.75$$

The remaining sums of squares and products (treatments and error) are obtained similarly. The analysis of covariance is given in Table 18.3.

Table 18.3

Source	df	SS and SP			Adjusted SS	
		y^2	xy	x^2	df	SS
Blocks	3	5849.0	9859.5	17112.75		
Treatments	3	2431.5	2389.5	4966.25		
Error	9	5206.5	5025.5	7612.75	8	1888.9538
Total	15	13487.0	17274.5	29691.75		
Treatment+ Error	12	7638.0	7415.0	12579.00	11	3267.0464
		Adjusted treatment SS			3	1378.0926

SS due to regression (i.e. due to the estimate $\hat{\gamma}$ of γ)

$$= 5206.5 - 1888.9538 = 3317.5462$$

To test if the regression is significant we calculate the *F*-ratio

$$\frac{3317.5462}{(1888.9538)/8} = 14.0503$$

The value obtained is highly significant, the 1% value of F(1, 8) being 11.26. Thus we find that the use of the mean milk yield during the control period as a concomitant variable was of importance. The estimate of the regression γ is

$$\hat{\gamma} = \frac{5025.5}{7612.75} = 0.6601$$

The use of the concomitant variable has reduced the error MS from $\frac{5206.5}{9} = 578.5$ to $\frac{1888.9538}{8} = 236.1192$, thus resulting in a gain in precision of $\frac{578.5}{236.1192} - 1 = 1.45$ or 145%.

The differences between the treatments are tested by using the F-ratio

$$\frac{\text{Adjusted treatment MS}}{\text{Adjusted error MS}} = \frac{1378.0926}{3} \times \frac{8}{1888.9538} = 1.945$$

The value obtained is not significant since the 5% value of F (3, 8) is 4.07.

COMPLEMENTS AND PROBLEMS

1. *Completely randomised design with one concomitant variable.* Taking the model as

$$y_{ij} = \tau_i + \gamma\, x_{ij} + e_{ij},\ i = 1, \ldots, m$$

$$j = 1, \ldots, n_i,\ \Sigma n_i = n$$

the estimate sum of squares is

$$\Sigma\, n_i\, \bar{y}_{i.}^2 + [\Sigma\Sigma\, y_{ij}\,(x_{ij} - \bar{x}_{i\cdot})]^2/\Sigma\Sigma\,(x_{ij} - \bar{x}_{i\cdot})^2 = \Sigma\, n_i\, \bar{y}_{i.}^2 + E_{xy}^2/E_{xx}$$

with $m + 1$ degrees of freedom, and the error sum of squares is

$$\Sigma\Sigma\, y_{ij}^2 - \Sigma\, n_i\, \bar{y}_{i.}^2 - E_{xy}^2/E_{xx} = E_{yy} - E_{xy}^2/E_{xx},$$

with $n - m - 1$ degrees of freedom.

Consider now the hypothesis

$$\tau_1 = \tau_2 = \ldots\ldots = \tau_m = \tau \text{ (say)}$$

Under the hypothesis the model becomes

$$y_{ij} = \tau + \gamma\, x_{ij} + e_{ij}$$

for which the estimate sum of squares is

$$n\, \bar{y}_{..}^2 + [\Sigma\Sigma\, x_{ij}\, y_{ij} - n\, \bar{x}_{..}\, \bar{y}_{..}]^2/[\Sigma\Sigma\, x_{ij}^2 - n\, \bar{x}_{..}^2] = n\, \bar{y}_{..}^2 + T_{xy}^2/T_{xx}$$

with 2 degrees of freedom, and the error sum of squares is

$$\Sigma\Sigma\, y_{ij}^2 - n\, \bar{y}_{..}^2 - T_{xy}^2/T_{xx} = T_{yy} - T_{xy}^2/T_{xx}$$

with $n - 2$ degrees of freedom.

Thus the hypothesis sum of squares is

$$(T_{yy} - T_{xy}^2/T_{xx}) - (E_{yy} - E_{xy}^2/E_{xx})$$

= (adjusted hypothesis + error SS) — (adjusted error SS)

with $m - 1 = (n - 2) - (n - m - 1)$ degrees of freedom.

2. *Randomised block design with one concomitant variable.* The model is

$$y_{ij} = \beta_i + \tau_j + \gamma\, x_{ij} + e_{ij},\ i = 1, \ldots, b, j = 1, \ldots, v$$

The normal equations are

$$v\,\beta_i + \tau. + \gamma\, x_{i\cdot} = y_{i\cdot}$$

$$\beta. + b\,\tau_j + \gamma\, x._j = y._j$$

$$\Sigma\, \beta_i\, x_{i\cdot} + \Sigma\, \tau_j\, x._j + \gamma\, \Sigma\Sigma\, x_{ij}^2 = \Sigma\Sigma\, x_{ij}\, y_{ij}$$

of which a solution is

$$\hat{\beta}_i = \bar{y}_{i\cdot} - \hat{\gamma}\, \bar{x}_{i\cdot} - (\bar{y}_{\cdot\cdot} - \hat{\gamma}\, \bar{x}_{\cdot\cdot})/2$$

$$\hat{\tau}_j = \bar{y}._j - \hat{\gamma}\, \bar{x}._j - (\bar{y}_{\cdot\cdot} - \hat{\gamma}\, \bar{x}_{\cdot\cdot})/2$$

$$\hat{\gamma} = [\Sigma\Sigma\, x_{ij}\, y_{ij} - \Sigma\, x_{i\cdot}\, \bar{y}_{i\cdot} - \Sigma\, x._j\, \bar{y}._j + \Sigma\, x_{\cdot\cdot}\, \bar{y}_{\cdot\cdot}]/$$

$$[\Sigma\Sigma\, x_{ij}^2 - \Sigma\, x_{i\cdot}\, \bar{x}_{i\cdot} - \Sigma\, x._j\, \bar{x}._j + \Sigma\, x_{\cdot\cdot}\, \bar{x}_{\cdot\cdot}]$$

$$= E_{xy}/E_{xx}$$

From which the estimate and error sums of squares are easily obtained as

$$S_\beta^2 + S_\tau^2 + y_{\cdot\cdot}^2/bv + E_{xy}^2/E_{xx}$$

and

$$S_e^2 - E_{xy}^2/E_{xx} = E_{yy} - E_{xy}^2/E_{xx}$$

with $b + v$ and $bv - b - v$ degrees of freedom respectively.

(Here S_β^2, S_τ^2, $S_e^2 = E_{yy}$ refer to the sums of squares of the randomised block design without the concomitant variable)
Under the hypothesis

$$\tau_1 = \tau_2 = \ldots = \tau_v = \tau \text{ (say)}$$

the model becomes

$$y_{ij} = \beta_i + \tau + \gamma\, x_{ij} + e_{ij}$$

for which the normal equations are

$$v\beta_i + v\tau + \gamma x_{i\cdot} = y_{i\cdot}$$

$$v\beta_{\cdot} + bv\tau + \gamma x_{\cdot\cdot} = y_{\cdot\cdot}$$

$$\Sigma \beta_i x_{i\cdot} + \tau x_{\cdot\cdot} + \gamma \Sigma\Sigma x_{ij}^2 = \Sigma\Sigma x_{ij} y_{ij}$$

A solution of these equations may be taken as

$$\beta_i^* = \bar{y}_{i\cdot} - \gamma^* \bar{x}_{i\cdot}$$

$$\tau^* = 0$$

$$\gamma^* = [\Sigma\Sigma\ x_{ij} y_{ij} - \Sigma\ \bar{x}_{i\cdot}\ y_{i\cdot}]/[\Sigma\Sigma\ x_{ij}^2 - \Sigma\ \bar{x}_{i\cdot}\ x_{i\cdot}]$$

which gives the estimate sum of squares as

$$\Sigma\ y_{i\cdot}\ \bar{y}_{i\cdot} - \gamma^*\ \Sigma\ x_{i\cdot}\ \bar{y}_{i\cdot} + \gamma^*\ \Sigma\Sigma\ x_{ij}\ y_{ij}$$

with $b+1$ degrees of freedom, and the error sum of squares as

$$\Sigma\Sigma\ y_{ij}^2 - \Sigma\ y_{i\cdot}\ \bar{y}_{i} - \gamma^*\ [\Sigma\Sigma\ x_{ij} y_{ij} - \Sigma\ x_{i\cdot}\ \bar{y}_{i\cdot}]$$

with $bv - b - 1$ degrees of freedom.

The sum of the squares for testing the given hypothesis, with $(bv - b - 1) - (bv - b - v) = v - 1$ degrees of freedom, is obtained by subtracting the error sum of squares of the original model from the error sum of squares under the hypothesis model, and is easily seen to be the same as the general result obtained in this chapter.

REFERENCES

Good review articles are Delury (1948), Cochran (1957), and Finney (1957). An example with many concomitant variables is discussed in Day and Fisher (1957). See Mallios (1967) for the case where independent variables are uncontrolled. Bartlett (1936) discusses the choice of covariates and interpretation of analysis of covariance.

Rao (1946) and Coons (1957) derive the analysis of covariance results by other methods. Analysis for incomplete data is given in Nair (1940), Wilkinson (1957).

Atiqullah (1964) considers the problem of robustness of the tests. Nonparametric tests are discussed by Quade (1967) and Puri and Sen (1969).

Hazel (1946) and Federer (1957) give the analysis of covariance for unbalanced experiments, Zelen (1957) for incomplete block designs with recovery of interblock information, and Pasternack (1960) for PBIB designs.

Quenouille (1948) uses analysis of covariance for making non-orthogonal comparisions, Federer and Schlotfeld (1954) for controlling fertility gradients.

Smith (1957) discusses the relation of analysis of covariance with regression.

Bibliography

Abraham, J.K. (1960). On an alternative method of computing Tukeys' statistic for the latin square model. *Biometrics*, **16**, 686–691.

Agrawal, H. (1963). A note on the omission of variables in multiple regression analysis. *J. Indian Statist. Assoc.*, **1**, 226–229.

Agrawal, H. (1966). Some methods of construction of designs for two-way elimination of heterogeneity–I. *J. Amer. Statist. Assoc.*, **61**, 1151–1171.

Aitken, A.C. (1933). On fitting polynomials to data with weighted and correlated errors. *Proc. Roy. Soc. Edin.*, **54**, 12–16.

Aitken, A.C. (1935). On least squares and linear combination of observations. *Proc. Roy. Soc. Edin.*, **55**, 42–48.

Allan, F.E. and Wishart, J. (1930). A method of estimating yield of a missing plot in field experimental work, *J. Agr. Sci.*, **20**, 399–406.

Anderson, R.L. (1946). Missing-plot techniques. *Biometrics*, **2**, 41–47.

Anderson, R.L. and Houseman, E.E. (1942). Tables of orthogonal polynomial values extended to $N = 104$. *Iowa Agr. Exp. Sta. Res. Bull.*, **297**, Ames, Iowa.

Anscombe, F.J. (1948). The validity of comparative experiments, *J. Roy. Statist. Soc.*, Ser. A, **61**, 181–211.

Anscombe, F.J. and Tukey, J.W. (1963). The examination and analysis of residuals. *Technometrics*, **5**, 141–160.

Atiqullah, M. (1964). The robustness of the covariance analysis of a one-way classification. *Biometrika*, **51**, 365–372.

Atiqullah, M. (1969). On a restricted least squares estimator. *J. Amer Statist. Assoc.*, **64**, 964–968.

Bartlett, M.S (1934). The vector representation of a sample. *Proc. Camb. Phil. Soc.*, **30**, 327–340.

Bartlett, M.S. (1936): "A note on the analysis of covariance," *J. Agr. Sci.*, **26**, 488–491.

Bartlett, M.S. (1947). The use of transformations. *Biometrics*, **3**, 39–52.

Bartlett, M.S. and Kendall, D.G. (1946). The statistical analysis of variance-heterogeneity and the logarithmic transformation. *J. Roy. Statist. Soc.*, Ser. B, **8**, 128–138.

Baten, W.D. (1952). Variances of differences between two means when there are two missing values in randomised block designs. *Biometrics*, **8**, 42–50.

Beall, G. (1942). The transformation of data from entomological field experiments so that the analysis of variance becomes applicable. *Biometrika*, **32**, 243–262

Bennet, C.A. and Franklin, N.L. (1954). *Statistical Analysis in Chemistry and the Chemical Industry*. John Wiley, New York.

Birge, R.T. (1947). Least squares fitting of data by means of polynomials, *Rev. Mod. Phys.*, **19**, 298–347.

Bose, R.C. (1944). The fundamental theorems of linear estimation. *Proc. 31st. Indian Sci. Cong.*, pp. 2–3 (abstract).

Bose, R.C. (1947). Mathematical theory of the symmetrical factorial design. *Sankhyā*, **8**, 107–166.

Bose, R.C. (1949). *Least Squares Aspects of Analysis of Variance*. Inst. Stat. Mimeo, series 9, Chapell Hill, N.C.

Bose, R.C., Clatworthy, W.H. and Shrikhande, S.S. (1954). *Tables of Partially Balanced Designs with Two Associate Classes* Inst. Stat. Univ. N.C., reprint series 50, Raleigh.

Bose, R.C. and Kishen, K. (1940). On the problem of confounding in the general symmetrical factorial design. *Sankhyā*, **5**, 21–36.

Bose, R.C. and Nair, K.R. (1939). Partially balanced incomplete block designs. *Sankhyā*, **4**, 337–372.

Bose, R.C. and Shimamoto, T. (1952). Classification and analysis of partially balanced designs with two associate classes. *J. Amer. Statist. Assoc.*, **47**, 151–184.

Bose, S.S. and Mahalanobis, P.C. (1938). On estimating individual yields in the case of mixed-up yields of two or more plots in field experiments. *Sankhyā*, **4**, 103–120.

Box, G.E.P. (1954). Some theorems on quadratic forms applied in the study of analysis of variance problems: I and II. *Ann. Math. Statist.*, **25**, 290–302 and 484–498.

Box, G.E.P. and Cox, D.R. (1964). An analysis of transformations. *J. Roy. Statist, Soc.*, Ser. B, **26**, 211–252.

Box, G.E.P. and Watson, G.S. (1962). Robustness to non-normality of regression tests. *Biometrika*, **49**, 93–106.

Bradu, D. (1965). Main-effect analysis of the general non-orthogonal layout with any number of factors *Ann. Math. Statist.*, **36**, 88–97.

Chakrabarti, M.C. (1962). *Mathematics of Design and Analysis of Experiments*. Asia Publishing House, Bombay.

Clark, V. (1965). Choice of levels in polynomial regression with one or more variables. *Technometrics*, **7**, 325–333.

Cochran, W.G. (1938). Recent work on the analysis of variance, *J. Roy. Stat. Soc.*, **101**, 434–449.

Cochran, W.G. (1947). Some consequences when the assumptions for the analysis of variance are not satisfied. *Biometrics*, **3**, 22–38.

Cochran, W.G. (1957). Analysis of covariance: Its nature and uses. *Biometrics*, **13**, 261–281.

Cochran, W.G. and Cox, G.M. (1957). *Experimental Designs*, 2nd Edition, John Wiley, New York.

Coons, I. (1957). The analysis of covariance as a missing plot technique. *Biometrics*, **13**, 387–405.

Cornish, E.A. (1938). Factorial treatments in incomplete randomised blocks. *J. Austr. Inst. Agr. Sci.*, **4**, 199–203.

Cornish, E.A. (1940). The estimation of missing values in incomplete randomised block experiments. *Ann. Eug.*, **10**, 112–118.

Cox, D.R. (1952). Some recent work on systematic experimental designs. *J. Roy. Statist. Soc.*, Ser. B, **14**, 211–219.

Cox, D.R. (1958). *Planning of Experiments*. John Wiley, New York.

Cunningham, E.P. and Henderson, C.R. (1966). Analytical techniques for incomplete block experiments. *Biometrics*, **22**, 829–842.

Curtiss, J.H. (1943). On transformations used in the analysis of variance. *Ann. Math. Statist.*, **14**, 107–122.

Das, M.N. and Kulkarni, G.A. (1966). Incomplete block designs for bioassays. *Biometrics*, **22**, 706–729.

Das, M.N. and Rao. P.S. (1967). Construction and analysis of some new series of confounded asymmetrical factorial designs. *Biometrics*, **23**, 813–822.

David, F.N. and Johnson, N.L. (1951). The effect of non-normality on the power function of the *F*-test in the analysis of variance. *Biometrika*, **38**, 43–57.

David, F.N. and Neyman, J. (1938). Extension of the Markoff theorem on least squares. *Statist. Res. Mem.*, **2**, 105–116.

Davies, P. (1969). The choice of variables in the design of experiments for linear regression. *Biometrika*, **56**, 55–63.

Davies, O.L. (1956). *The Design and Analysis of Industrial Experiments*, 2nd. Edition, Oliver and Boyd, London.

Day, B. and Fisher, R.A. (1937). The comparision of variability in populations having unequal means. An example of the analysis of covariance with multiple dependent and independent variates, *Ann. Eug.* **7**, 333–348.

DeLury, D.B. (1946). The analysis of latin squares when some observations are missing. *J. Amer. Statist. Assoc.*, **41**, 370–389.

DeLury, D.B. (1948). The analysis of covariance. *Biometrics*, **4**, 153–170.

Draper, N.R. and Cox, D.R. (1969). On distributions and their transformation to normality. *J. Roy. Statist. Soc.*, Ser. B, **31**, 472–476.

Draper, N.R. and Smith, H. (1966). *Applied Regression Analysis*, John Wiley, New York.

Durbin, J. and Watson, G.S. (1950). Testing for serial correlation in least squares regression, I. *Biometrika*, **37**, 409–428.

Durbin, J. and Watson, G.S. (1951). Testing for serial correlation in least squares regression, II. *Biometrika*, **38**, 159–178.

Dunlop, G. (1933). Methods of experimentation in animal nutrition. *J. Agr. Sci.*, **23**, 580-614.

Eisenhart, C. (1947). The assumptions underlying the analysis of variance. *Biometrics*, **3**, 1–21.

Federer, W.T. (1955). *Experimental Designs*. Macmillan, New York.

Federer, W.T. (1957). Variance and covariance analysis for unbalanced classifications. *Biometrics*, **13**, 333–362.

Federer, W.T. and Schlottfeldt, C.S. (1954). The use of covariance to control gradients in experiments. *Biometrics*, **10**, 282–290.

Federer, W.T. and Zelen, M. (1966). Analysis of multi-factor classifications with unequal numbers of observations. *Biometrics*, **22**, 525–552.

Finney, D.J. (1946). Standard errors of yields adjusted for regression on an independent measurement. *Biometrics*, **2**, 53–55.

Finney, D.J. (1948). Main effects and interactions. *J. Amer. Statist. Assoc.*, **43**, 566–571.

Finney, D.J. (1955). *Experimental Design and its Statistical Basis*. Cambridge Univ. Press, London.

Finney, D.J. (1957) Stratification, balance and covariance. *Biometrics*, **13**, 373–386.

Fisher, R.A. (1926). The arrangement of field experiments. *J. Ministry Agr.*, **33**, 503–513; included in *Contributions to Mathematical Statistics* by R.A. Fisher, John Wiley, New York, 1950.

Fisher, R.A. (1940). An examination of the different possible solutions of a problem in incomplete blocks. *Ann. Eug.*, **10**, 52–75.

Fisher, R.A. (1942). The theory of confounding in factorial experiments in relation to the theory of groups. *Ann. Eug*, **11**, 341–353.

Fisher, R.A. (1951). *The Design of Experiments*. 6th Edition, Oliver and Boyd, London.

Fisher, R.A. and Yates, F. (1953). *Statistical Tables for Biological, Agricultural and Medical Research*. 4th Edition, Oliver and Boyd, London.

Fix, E. (1949). Tables of non-central chi-square. *Univ. Calif. Publ. Statist.*, **1**, 15–1 .

Fox, M. (1956). Charts of the power of the *F*-test. *Ann. Math. Statist.*, **27**, 484-497.

Freeman, G.H. (1959). The use of the same experimental material for more than one set of treatments. *Appl. Statist.*, **8**, 13–20.

Freeman, G.H. (1964). The addition of further treatments to latin square designs. *Biometrics*, **20**, 713–729.

Freeman, M.F. and Tukey, J.W. (1950). Transformations related to the angular and the square root transformations. *Ann. Math. Statist.*, **21**, 607–611.

Grant, D.A. (1948). The latin square principle in the design and analysis of psychological experiments. *Psychol. Bull.*, **45**, 427–442.

Graybill, F.A. (1961). *An Introduction to Linear Statistical Models*—I. McGraw-Hill, New York.

Grundy, P.M. (1951). A general technique for the analysis of experiments with incorrectly treated plots. *J. Roy. Statist. Soc.*, Ser. B, **13**, 272–283.

Hazel, L.N. (1946). The covariance analysis of multiple classification tables with unequal subclass numbers. *Biometrics*, **2**, 21–25.

Herzberg, A.M. and Cox, D.R. (1969). Recent work on the design of experiments: A bibliography and a review. *J. Roy. Statist. Soc.*, Ser. A, **132**, 29–67.

Hill, B.M. (1969). Foundations for the theory of least squares. *J. Roy. Statist. Soc.*, Ser. B, **31**, 89–97.

Hocking, R.R. and Leslie, R.N. (1967). Selection of the best subset in regression analysis. *Technometrics*, **9**, 531–540.

Hooper, J. and Zellner, A. (1961). The error of forecast for multivariate regression models. *Econometrica*, **29**, 544–555.

Hsu, P.L. (1938). On the best unbiased quadratic estimate of the variance. *Statist. Res. Mem.*, **2**, 91–104.

Hsu, P.L. (1941). Analysis of variance from the power function standpoint. *Biometrika*, **32**, 62–69.

Huang, D.S. (1970). *Regression and Econometric Models.* John Wiley, New York.

Irwin, J.O. (1931). Mathematical theorems involved in the analysis of variance. *J. Roy. Statist. Soc.*, **94**, 284–300.

Ishii, G. and Kudō, H. (1963). Tolerance region for missing variables in linear statistical models. *J. Math. Osaka City Univ.*, **14**, 117–130.

John, J.A. (1965). A note on the analysis of incomplete block experiments. *Biometrika*, **52**, 633–636.

Johnson, N.L. and Welch, B.L. (1940). Applications of the non-central *t*-distribution. *Biometrika*, **31**, 362–389.

Johnson, N.L. and Pearson, E.S. (1965). Tables of percentage points of the non-central chi-square. *Biometrika*, **56**, 255–272.

Kempthorne, O. (1947). A simple approach to confounding and fractional replication in factorial experiments. *Biometrika*, **34**, 255–272.

Kempthorne, O. (1952). *Design and Analysis of Experiments.* John Wiley, New York.

Kempthorne, O. (1953). A class of experimental designs using blocks of two plots. *Ann. Math. Statist.*, **24**, 876–884.

Kempthorne, O. and Barclay, W.D. (1953). The partition of error in randomised blocks. *J. Amer. Statist. Assoc.*, **48**, 610–614.

Kitagawa, T. and Mitome, M. (1953). *Tables for the Design of Factorial Experiments.* Baifukan Co. Ltd., Tokyo.

Kogan, L.S. (1953). Variance designs in psychological research. *Psychol. Bull.*, **50**, 1–40.

Kolodziejczyk, S. (1935). On an important class of statistical hypotheses. *Biometrika*, **27**, 161-190.

Kruskal, J.B. (1965). Analysis of factorial experiments by estimating monotone transformations of the data. *J. Roy. Statist. Soc.*, Ser. B, **27**, 251–263.

Kruskal, W. (1968). When are Gauss-Markov and least squares estimators identical? A coordinate-free approach. *Ann. Math. Statist.*, **39**, 70–75.

Kshirsagar, A.M. (1966). Balanced factorial designs. *J. Roy. Statist. Soc.*, Ser. B, **28**, 559–567.

Kurkjian, B. and Zelen, M. (1962). A calculus for factorial arrangements. *Ann. Math. Statist.*, **33**, 600–619.

Kurkjian, B. and Zelen, M. (1963). Applications of the calculus for factorial arrangements. I—Block and direct product designs. *Biometrika*, **50**, 63–73.

Lehmer, E. (1944). Inverse tables of probabilities of errors of the second kind. *Ann. Math. Statist.*, **15**, 388–398.

Lindley, D.V. (1968). The choice of variables in multiple regression. *J. Roy. Statist. Soc.*, Ser. B, **30**, 31–66.

Mallios, W.S. (1967). A structural regression approach to covariance analysis when the covariable is uncontrolled. *J. Amer. Statist. Assoc.*, **62**, 1037–1049.

Margolin, B.H. (1967). Systematic methods for analysing 2^m3^n factorial experiments with applications. *Technometrics*, **9**, 245–259.

McElroy, F.W. (1967). A necessary and sufficient condition that ordinary least squares estimators be best linear unbiased. *J. Amer. Statist. Assoc.*, **62**, 1302–1304.

Mesner, D.M. (1967). A new family of partially balanced incomplete block designs with some latin square design properties. *Ann. Math. Statist.*, **38**, 571–581.

Muller, E.R. (1966), Balanced confounding of factorial experiments. *Biometrika*, **53**, 507–524.

Nair, K.R. (1938). On a method of getting confounded arrangements in the general symmetrical type of experiment. *Sankhyā*, **4**, 121–138.

Nair, K.R. (1940). The application of the technique of analysis of covariance to field experiments with several missing or mixed-up plots. *Sankhyā*, **4**, 581–588.

Nair, K.R. (1944). The recovery of inter-block information in incomplete block designs. *Sankhyā*, **6**, 383–390.

Nair, K.R. (1952). Analysis of partially balanced incomplete block designs illustrated on the simple square and rectangular lattices. *Biometrics*, **8**, 122–155.

Nair, K.R. and Rao, C.R. (1948). Confounding in asymmetrical factorial experiments. *J. Roy. Statist. Soc.*, Ser. B. **10**, 109–131.

Pasternack, B.S. (1960): Analysis of covariance for a 3×4 triple rectangular lattice design (3 associate P.B.I.B.), *Biometrics*, **16**, 117–118.

Patnaik, P.B. (1949). The non-central chi-square and *F*-distributions and their approximations. *Biometrika*, **36**, 202–232.

Pearce, S.C. (1948). Randomised blocks with interchanged and substituted plots. *J. Roy. Statist. Soc.*, Supp. **10**, 252–256.

Pearce, S.C. (1960). Supplemental balance. *Biometrika*, **47**, 263–271.

Pearson, E.S. and Hartley, H.O. (1951). Charts of the power function of the analysis of variance tests, derived from the non-central *F*-distribution. *Biometrika*, **38**, 112–130.

Pearson, E.S. and Hartley, H.O. (1954). *Biometrika Tables for Statisticians*, Vol. 1, Cambridge Univ. Press, Cambridge.

Pitman, E.J.G. (1937). Significance tests which may be applied to samples from any populations: III the analysis of variance test. *Biometrika*, **29**, 322–335.

Plackett, R.L. (1949). A historical note on the method of least squares. *Biometrika*, **36**, 458–460.

Plackett, R.L. (1960). *Principles of Regression Analysis.* Oxford Univ. Press, London.

Plackett, R.L. (1960). Models in the analysis of variance. *J. Roy Statist. Soc.*, Ser. B, **22**, 195–217.

Pothoff, R.F. (1962). Three-factor additive designs more general than the Graeco-Latin square. *Technometrics*, **4**, 361–366.

Press, S.J. (1966). Linear combinations of non-central chi-square variates. *Ann. Math. Statist.* **37**, 480–487.

Puri, M.L. and Sen, P.K. (1969). Analysis of covariance based on general rank scores. *Ann. Math. Statist.*, **40**, 610–618.

Putter, J. (1967). Orthonormal bases of error spaces and their use for investigating the normality and variances of residuals. *J. Amer. Statist. Assoc.* **62**, 1022–1036.

Quade, D. (1967). Rank analysis of covariance. *J, Amer. Statist. Assoc.*, **62** 1187–1200.

Quenouille, M.H. (1948). The analysis of covariance and non-orthogonal comparisons. *Biometrics*, **4**, 240–246.

Quenouille, M.H. (1953). *Design and Analysis of Experiments.* Griffin, London.

Raghavarao, D. (1971). *Constructions and Combinatorial Problems in Design of Experiments.* John Wiley, New York.

Rahman, N.A. (1968). *A Course in Theoretical Statistics.* Griffin, London.

Rao, C.R. (1946). On the linear combination of observations and the general theory of least squares. *Sankhyā*, **7**, 237–256.

Rao, C.R. (1947). General methods of analysis for incomplete block designs. *J. Amer. Statist. Assoc.*, **42**, 541–561.

Rao, C.R. (1965). *Linear Statistical Inference and Its Applications.* John Wiley, New York.

Rao, C.R. (1971). Unified theory of linear estimation. *Sankhyā*, Ser. A, **33**, 371–394.

Rao, C.R and Mitra, S.K. (1969). Conditions for optimality and validity of simple least squares theory. *Ann. Math. Statist.*, **40**, 1617–1624.

Rao, C.R. and Mitra, S.K. (1971). *Generalised Inverse of Matrices and Its Applications.* John Wiley, New York.

Rayner, A.A. (1953). Quality × quantity interaction. *Biometrics*, **9**, 387–411.

Resnikoff, G.J. and Lieberman, G.J. (1957). *Tables of the Non-central t-distribution.* Stanford Univ. Press, Stanford, California.

Robinson, J. (1966). Balanced incomplete block designs with double grouping of blocks into replications. *Biometrics*, **22**, 368–373.

Robinson, J. and Balaam, L.N. (1967). Variance heterogeneity and error correlation in factorial experiments. *Austral. J. Statist.*, **9**, 126–130.

Robson, D.S. (1959). A simple method for constructing orthogonal polynomials when the independent variable is unequally spaced. *Biometrics*, **15**, 187–191.

Robson, D.S. and Atkinson, G.F. (1960). Individual degrees of freedom for testing homogeneity of regression coefficients in a one-way analysis of variance. *Biometrics*, **16**, 593-605.

Roy, J. and Shah, K.R. (1962). Recovery of inter block information, *Sankhyā*, Ser. A, **24**, 269–280.

Scheffé, H. (1959). *The Analysis of Variance.* John Wiley, New York.

Seber, G.A.F. (1964). The linear hypothesis and idempotent matrices. *J. Roy. Statist. Soc.*, Ser. B, **26**, 261–266.

Seshadri, V. (1966). Comparison of combined estimators in balanced incomplete blocks. *Ann. Math. Statist.*, **37**, 1832–1835.

Shah, B.V. (1958). On balancing in factorial experiments. *Ann. Math. Statist.*, **29**, 766–779.

Shah, B.V. (1959). A generalisation of partially balanced incomplete block designs. *Ann. Math. Statist.*, **30**, 1041–1050.

Shah, K.R. (1964). Use of inter-block information to obtain uniformly better estimators. *Ann. Math. Statist.*, **35**, 1064–1078.

Shrikhande, S.S. (1951). Designs for two-way elimination of heterogeneity. *Ann. Math. Statist.*, **22**, 235–247.

Smith, H.F. (1957). Interpretation of adjusted treatment means and regressions in analysis of covariance. *Biometrics*, **13**, 282–308.

Sprott, D.A. (1956). A note on combined interblock and intra-block estimation in incomplete block designs. *Ann. Math. Statist.*, **27**, 633–641.

Stein, C. (1966). An approach to the recovery of interblock information in balanced incomplete block designs, *Research Papers in Statistics* (Festschrift J. Neyman), pp. 351–366, John Wiley, London.

Tang, P.C. (1938). The power function of the analysis of variance tests with tables and illustrations of their use. *Statist. Res. Mem.*, **2**, 126–149.

Theil, H. (1965). The analysis of disturbances in regression analysis. *J. Amer. Statist. Assoc.*, **60**, 1067–1079.

Theil, H. (1968). A simplification of the BLUS procedure for analysing regression disturbances. *J. Amer. Statist. Assoc.*, **63**, 242–251.

Theil, H. and Nagar, A.L. (1961). Testing the independence of regression disturbances. *J. Amer. Statist. Assoc.*, **56**, 793–806.

Tiao, G.C. and Draper, N.R. (1968). Bayesian analysis of linear models with two random components with special reference to the balanced incomplete block design. *Biometrika*, **55**, 101–117.

Tocher, K.D. (1952). The design and analysis of block experiments. *J. Roy. Statist. Soc.*, Ser. B, **14**, 45–100.

Tukey, J.W. (1949). One degree of freedom for non-additivity, *Biometrics*, **5**, 232–242.

Tukey, J.W. (1955). Querry 113, *Biometrics*, **11**, 111–113.

Tukey J.W. (1957). The comparative anatomy of transformations. *Ann. Math. Statist.*, **28**, 602–632.

Wald, A. (1942). On the power function of the analysis of variance test. *Ann. Math. Statist.*, **13**, 434–439.

Watson, G.S. (1967). Linear least squares regression. *Ann. Math. Statist.*, **38**, 1679–1699.

Welch, B.L. (1937). On the *z*-test in randomised blocks and latin squares. *Biometrika*, **29**, 21–52.

White, D. and Hultquist, R.A. (1965). Construction of confounding plans for mixed factorial designs. *Ann. Math. Statist.*, **36**, 1256–1271.

Wilkinson, G.N. (1957). The analysis of covariance with incomplete data. *Biometrics*, **13**, 363–372.

Williams, E.J. (1952). The interpretation of interactions in factorial experiments. *Biometrika*, **39**, 65–81.

Williams, E.J. (1959). *Regression Analysis*. John Wiley, New York.

Wishart, J. (1936). Tests of significance in analysis of covariance. *J. Roy. Statist. Soc.*, Suppl. **3**, 79–82.

Yates, F. (1933 a). The analysis of replicated experiments when the field results are incomplete. *Emp. J. Exp. Agr.*, **1**, 129–142. Reprinted in Yates (1970).

Yates, F. (1933 b). The formation of latin squares for use in field experiments. *Emp. J. Exp. Agr.*, **1**, 235–244. Reprinted in Yates (1970).

Yates, F. (1933 c). The principles of orthogonality and confounding in replicated experiments. *J. Agr. Sci.*, **23**, 108–145. Reprinted in Yates (1970).

Yates, F. (1944). The analysis of multiple classifications with unequal numbers in the different classes. *J. Amer. Statist. Assoc.*, **29**, 51–66.

Yates, F. (1935). Complex experiments. *J. Roy. Statist. Soc.*, Suppl. **2**, 181–247. Reprinted in Yates (1970).

Yates, F. (1936), Incomplete randomised blocks *Ann. Eug.*, **7**. 121–140. Reprinted in Yates (1970).

Yates, F. (1937). *The Design and Analysis of Factorial Experiments.* Imperial Bureau of Soil Sciences, Tech. Comm. No. 35.

Yates, F. (1939). The comparative advantages of systematic and randomised arrangements in the design of agricultural and biological experiments. *Biometrika*, **30**, 440–466.

Yates, F. (1940). The recovery of inter-block information in balanced incomplete block design. *Ann. Eug.*, **10**, 317–325. Reprinted in Yates (1970).

Yates, F. (1965). A fresh look at the basic principles of the design and analysis of experiments. *Proc. Fifth Berk. Symp.*, vol. IV, 777–790. Reprinted in Yates (1970).

Yates, F. (1970). *Experimental Design: Selected Papers of Frank Yates.* Griffin, London.

Yates, F. and Hale, R.W. (1939). The analysis of latin squares when two or more rows, columns, or treatments are missing. *Suppl. J. Roy. Statist., Soc.*, 6, 67–79.

Zelen, M. (1954 a). A note on partially balanced designs. *Ann. Math. Statist.*, **25**, 599–602.

Zelen, M. (1954 b). Analysis for some partially balanced incomplete block designs having a missing block. *Biometrics*, **10**, 273–281.

Zelen, M. (1957). The analysis of covariance for incomplete block designs. *Biometrics*, **13**, 309–332.

Zelen, M. (1958). The use of group divisible designs for confounded asymmetrical factorial arrangements. *Ann. Math. Statist.*, **29**, 22–40.

Zelen, M. and Federer, W. T. (1964). Applications of the calculus for factorial arrangements II: Designs with two-way elimination of heterogeneity. *Ann. Math. Statist.*, **35**, 658–672.

Zelen, M. and Federer, W.T. (1965). Applications of the calculus for factorial arrangements III: Analysis of factorials with unequal numbers of observations. *Sankhyā*, Ser. A, **27**. 383–400.

Index

Additional observations, effect of 50.
Additive factors 194.
Additivity 82.
Additivity of varietal and treatment effects 185.
Adjusted: block sum of squares 136; block total 136, 151; error sum of squares 261; hypothesis plus error sum of squares 262; treatment sum of squares 136; treatment total 130, 143.
Analysis of covariance 252, 260;—model 254.
Analysis of variance 82;—in BIBD 146;—in completely randomised design 89; —in factorial experiments 191ff;—in general incomplete block design 134;—in Latin Square design 120;—in PBIBD 177;—in randomised block design 102;—in 2^n confounded factorials 240.
Associates 173.
Association scheme 172; rectangular—173; triangular—173.
Asymmetrical factorial experiment 186.
Augmented matrix 14.

Balance 139.
Balanced design 140.
Balanced incomplete block design (BIBD) 142ff; symmetrical—143.
Basis 4;—of space of linear forms 7; orthonormal—5.
Best estimates 38; variances and covariances of—43.
Best linear unbiased estimates 38.
Between groups sum of squares 91.
Block 80, 99;—contrast 101;—mean 101;—total 101; key—239, 247; principal —239, 246.

Calculation of sum of squares 57;—in 3^n factorials 218;—in 2^n factorials 210.
Chi-square distribution 31; non-central—31.
C-matrix 130.
Coefficient: matrix 14, 34; vector 6.

Column: space 12; vector 1.
Complement of BIBD 167.
Completely randomised design 88ff.
Composite treatment 186.
Concomitant observation 252.
Concomitant variable 252; completely randomised design with one—269; randomised block design with one—270.
Confounding arrangement 235, 236;—for 3^n factorial experiments 246ff.;—in 3^p blocks 245;—in two blocks 237;—in 2^p blocks 237.
Confounding in factorial experiments 234ff;—in 3^n factorials 244;—in 2^n factorials 237.
Connected: components 139; design 139.
Connectedness 139.
Consistent equation 14.
Contrast 7.
Coordinates of a vector 5.
Crude sum of squares 62.

Defining contrasts 236.
Degrees of freedom 55; error—55; estimate—55.
Derived design of a BIBD 167.
Design 80;—of experiment 79.
Determinant of a matrix 11.
Diagonal elements of a matrix 9.
Diagonal matrix 9.
Dimension of a vector space 5.
Disjoint subspaces 4.
Dispersion matrix 25.
Distribution of sums of squares 59.
Divisor (in Yates' method) 220.
Dot notation 86.

Efficiency: of BIBD 147; of Latin Square design 125; of randomised block design 109.
Equivalent hypothesis 66.
Error: degrees of freedom 55; function 37; mean square 64; space 37; sum of squares 56, 62; vector 34.
Estimable: linear hypothesis 66; parametric function 36.
Estimate: degrees of freedom 55; sum of squares 56, 62.
Estimates in general incomplete block design 132ff.
Estimation space 37.
Expectation vector 25.
Experimental: material 79; unit 79.

Factor 186.
Factorial experiments 183ff.
F-distribution 32; non-central—32.
Field trial 79.
Fisher's "*t*" 70.
F-test, generalised 66.

Generalised: *F*-test 66; interaction 238; *t*-test 68.
Graeco-Latin square 117; —design 117.

Higher level 207.
Homogeneous equation 15.
Hypothesis 65; —mean square 67; —of rank k 65; —sum of squares 67; equivalent—66; estimable linear—66; linear—65.

Identity matrix 10.
Image 18; —space of a linear transformation 18.
Incidence matrix 170.
Incomplete block design 128ff.
Inconsistent equation 14.
Independent contrasts 238.
Inner product 2.
Interaction 187, 188; —between factors 194; —in 3^n factorials 218; —in 2^n factorials 208; k-factor—188; $(k—1)$th order—188; single factor—188; zero order—188.
Interblock analysis 147; distributions in—153.
Interblock estimate 156, 158.
Interblock information, recovery of 160.
Intrablock analysis 143, 147.
Intrablock error 136; —mean square 153.
Intrablock estimate 156, 158.
Inverse of a matrix 13.

Key block 239, 247.

Latin Square 115; – design 115.
Least significant difference 104.
Least squares estimate 46, 64.
Least squares, theory of 45.
Length of a vector 2.
Level 186; higher—207; lower—207.
Linear combination (of vectors) 3; —of linear forms 7.
Linear: component 218; estimate 35; estimation 34.
Linear forms 6; basis for space of—7; linear combination of—7; linear independence of—6; multiplication by scalars of—6; orthogonal basis of space of—7; orthogonal complement of space of—7; orthogonality of—6; orthogonality of subspaces of—6; sum of—6.

Linear function 6; —of random variables 26.
Linear × linear component 218.
Linear model 34; —with correlated observations 47.
Linear parametric function 35.
Linear × quadratic component—218.
Linear regression 72; multivariate—72.
Linear subspace 3.
Linear transformation 18; matrix as a—19; matrix of a—20; nullity of a—19; rank of a—19.
Local control 81.
Lower triangular matrix 9.

Main effect 187, 188; —contrasts 187; —in 3^n factorials **218**; —in 2^n factorials 208.
Matrix 8; —as a linear transformation 19; —of a linear transformation 20; —of a quadratic form 13; —of full rank 12; —sum 9; inverse of a—13; lower triangular—9; non-singular—12; null—9; orthogonal—20; partitioned—10; singular—12; skew-symmetric—9; square—9; symmetric—9; transpose of a —9, upper triangular—9.
Mean square: due to error 64; due to set of functions 64.
Method of paired comparisons 52.
Missing plot technique 84; —in Latin Square design 126; —in randomised block design 111.
Model equations 34.
Multivariate linear regression 72.
Multivariate normal distribution 27; singular—28.
Mutually orthogonal Latin Squares 117.
Mutually orthogonal sets of vectors 4.

Non-central: chi-square distribution 31; *F*-distribution 32; *t*-distribution 32.
Non-centrality parameter of non-central distributions: chi-square 31, *F* 32, *t* 32.
Non-singular: matrix 12; multivariate normal distribution 28; transformation 19.
Normal equations 40ff.; reduced—130.
Norm of a vector 2.
Nullity of a linear transformation 19.
Null: matrix 9; space of a linear transformation 18; vector 1.

Observation vector 34.
One-to-one transformation 18.
Orthogonal complement 4; —of space of linear forms 7.
Orthogonal: matrix 20; projection 21; sums of squares 57; transformation 20; vectors 3.
Orthogonality: of a vector to a set 4; of interaction contrasts 189, 190; of linear forms 6; of subspaces of linear forms 6; of treatment contrasts and treatment combinations 238, 246.
Orthonormal basis 5.

Parameter of the first kind 174; —second kind 174.
Parameter: space 65; vector 34.
Partial confounding 249ff.
Partially balanced incomplete block (PBIB) design 172, 174.
Partitioned matrix 10.
Patnaik's approximation 31.
Plot 79.
Polynomial regression 74.
Pooled estimate 157, 158.
Positive: definite 13; semi-definite 13.
Principal block 239, 246.
Product: of a scalar and a matrix 9; of a vector and a scalar 1; of two matrices 9.
Projection 21.

Quadratic: component 218; form 13.

Randomisation 83; —in Latin Square design 117; —in randomised block design 100.
Randomised block design 99ff; —with controls 112; —with one concomitant variable 270.
Random vector 25.
Rank: of a matrix 12; of a transformation 19.
Raw sum of squares 62.
Recovery of inter-block information 160.
Rectangular association scheme 173.
Reduced normal equations 130.
Regression parameters in analysis of covariance 253, 254.
Reparametrisation 36.
Replication 80, 81.
Representation: of main effects and interactions in 2^n-factorials 209; of treatment combinations 208, 217.
Residual: in BIBD 149; of a BIBD 167; sum of squares 46; vector 46.
Resolvable BIBD 168.
Row: space 12; vector 1.
Rule of odds and evens 209.

Scalar product 2.
Simultaneous linear equations 14.
Single factor interaction 188.
Singular: matrix 12; multivariate normal distribution 28.
Skew-symmetric matrix 9.
Span of a vector space 3.
Square matrix 9.
Standard order for treatment contrasts 208, 217.
Student's "*t*" 69.

Submatrices 10.
Subspace 3; —of linear forms 6; disjoint—4; linear—3.
Sum: of linear forms 6; of matrices 9; of vectors 1.
Sum of squares 55; —for blocks eliminating treatments 136; —for blocks ignoring treatments 135; —for interactions 193; —for main effects 193; —for treatments eliminating blocks 136; —for treatments ignoring blocks 136; crude—62; error—56; estimate—56; raw—62; total—62.
Symmetrical factorial experiment 186.
Symmetric: BIBD 143, 170; matrix 9.

t-distribution 32; non-central—32.
Test for linearity of regression 96.
Test of hypotheses 65ff; —and least squares theory 71.
3^n factorial experiment 217ff.
Total sum of squares 62.
Transformation 18; linear—18.
Transpose of a matrix 9.
Treatment 79; —combination 186; —combination, representation of 208, 217; —contrast 101; —mean 101; —total 101, 129; —trial 79.
Triangular association scheme 173.
Tukey's test for non-additivity 201ff.
2^n-factorial experiment 207ff.

Unadjusted: block sum of squares 135; treatment sum of squares 136.
Unbiased estimate 35.
Unconfounded with blocks 234.
Unit vectors 4.

Variance—covariance matrix 25.
Varietal: component 150, 151, 159; trial 79.
Varieties 79.
Vector 1; —of adjusted yields 257; null—1; zero—1.
Vector space spanned by a set of vectors 3.

Yates' method 211, 219.
Yield 80.

Zero-order interaction 188.
Zero vector 1.

ERRATA

Page	*Line*	*For*	*Read*
13	15	$\sum_{i=1}^{m} \sum_{j=1}^{n} a_{ij}\, x_i y_j a_{12}$	$\sum_{i=1}^{m} \sum_{j=1}^{n} a_{ij}\, x_i y_j$
16	31	$\bar{y}_{ij}. - \bar{y}_{pq}$	$\bar{y}_{ij}. - \bar{y}_{pq}.$
22	21	$(x - \mathbf{Ax})'\, \mathbf{Ax}$	$(\mathbf{x} - \mathbf{Ax})'\, \mathbf{Ax}$
24	11	For given y	For given $\mathbf{y}$
26	17	$\sum_{i<j} 2\lambda_i\, y_j \operatorname{Cov}(y_i, y_j)$	$\sum_{i<j} 2\lambda_i\lambda_j \operatorname{Cov}(y_i, y_j)$
34	22	$\mathbf{y}' = (\mathbf{y}_1, \ldots, \mathbf{y}_n)$	$\mathbf{y}' = (y_1, \ldots, y_n)$
37	4	rank $\mathbf{A}'\mathbf{A} =$	or, rank $\mathbf{A}'\mathbf{A} =$
37	5	Condition (4.)	Condition (4.9)
46	31	$y - \mathbf{A}\boldsymbol{\beta}$	$\mathbf{y} - \mathbf{A}\boldsymbol{\beta}$
49	7	$(\mathbf{y} - \mathbf{A}\boldsymbol{\beta})'\, \boldsymbol{\Sigma}^{-1} (\mathbf{y} - \mathbf{A}\boldsymbol{\beta})$	or, $(\mathbf{y}-\mathbf{A}\boldsymbol{\beta})'\, \boldsymbol{\Sigma}^{-1} (\mathbf{y} - \mathbf{A}\boldsymbol{\beta})$
62	12	$\sum_{i=1}^{n} y_i^2 - {}''\bar{y}^2$	$\sum_{i=1}^{n} y_i^2 - n\bar{y}^2$
66	30	orthogonal	orthonormal
66	31	$\sum_{i=1}^{k} (\mathbf{l}_i'\, \boldsymbol{\eta})^2$	$\frac{1}{k} \sum_{i=1}^{k} (\mathbf{l}_i'\, \boldsymbol{\eta})^2$
72	20	$n\alpha = y.$	$n\alpha = \Sigma\, y_i$
75	9	can,:	can be
80	5	comparisions	comparisons
80	20	comparision	comparison
82	45	comparisions	comparisons
83	7	can: hen	can then
89	23	$\bar{y}_i$	$\bar{y}_i.$

90	20	$\bar{y}.$	$\bar{y}..$
91	1	$\bar{y}_1..$	$\bar{y}..$
91	27	$(\sum_{i=1}^{4} n_i \bar{y}_{i.}^2)$	$(\sum_{i=1}^{4} n_i \bar{y}_{i.})^2$
105	23	5.49	6.49
107	16	$y_{.2}$	$\bar{y}_{.2}$
111	16	parametric contrasts	parametric functions
111	17	orthogonal contrasts	orthogonal functions
118	32	ρ	$\bar{\rho}.$
123	19	79(10)	79(*O*)
124	17	t_M	$\bar{t}_M$
126	22	$\Sigma s_k t_k$	$\Sigma s_k \bar{t}_k$
131	27	$(\sum_i n_{ij} n_{ij'}/k^2)$	$(\sum_i n_{ij} \cdot n_{ij'}/k_i^2)$
132	20	$\sum_i n_{ij} \tau_j$	$\sum_j n_{ij} \tau_j$
156	1	$-(r/v)\tau^2$	$-(r/v)\tau_.^2$
159	1	of $\hat{w}$	of w'
182	23	combinational	combinatorial